The Whole **ART** of

HUSBANDRY:

Or, The Way of

Managing and *Improving*

OF

LAND.

Being a full COLLECTION of what hath been
Writ, either by Ancient or Modern Authors:
With many Additions of New Experiments and
Improvements not treated of by others.

AS ALSO

An ACCOUNT of the particular Sorts of
Husbandry used in several Counties; with Propo-
sals for its farther Improvement.

To which is added,
The Country-Man's Kalendar, what he is to do
every Month in the Year.

The SECOND VOLUME.

By J. MORTIMER, *Esq;* Fellow of the
Royal Society.

The Fourth Edition, with Additions.

LONDON,
Printed by E. H. for R. Robinson at the Golden Lyon, and
G. Mortlock at the Phœnix in St. Paul's Church-Yard,
M.DCC.XVI.

THE
CONTENTS.
BOOK XI.

A 2   XXI. Of

The CONTENTS.

BOOK XII.

BOOK

THE CONTENTS

BOOK XIII.

BOOK XIV.

BOOK XV.

BOOK

The CONTENTS

BOOK XXVI

BOOK

THE CONTENTS

BOOK XVII.

Obser-

THE CONTENTS.

Observations on the Twelve Months, *Viz.*

Advertisement.

BY reason of the Author's distance from the Press,
several Errata have escap'd, which the Reader
is desired to correct.

¶ All Gentlemen who have Collected any Observa-
tions relating to Husbandry, and the improvement
of Lands, are desired to transmit the same to the
Author hereof, directing their Letters to be left with
G. Mortlock, or *R. Robinson,* Booksellers in St. *Paul's*
Church-Yard, and such Observations shall be Printed
in the next Edition of this Book, for the good of the
Publick.

THE.

The ART of
HUSBANDRY:
Or, the Way of improving of
LAND.

BOOK XI.

CHAP. I. *Of the Benefit of Raising, Planting, and Propagating all sorts of Timber, and other Trees, useful either in Building or other mechanick Uses.*

HEN we consider that Trade, Riches, and Strength are inseparable, and that their great Dependance is upon our Navy, we might have hoped so great a Concern to the Nation should have occasioned a greater Care in propagating and preserving of Timber, that is of principal Use to support it. And though we cannot expect to find many in this Age publick-spirited enough to have

such a Regard to the general Good, as to prefer it
before their private Interest ; yet the particular Profit that Timber brings to the Owners of it, as well
as its Advantage to the Publick, might, if it had not
caused more Care in propagating of it, have at least
prevented those that have had Opportunities of ex-
periencing its advantage from making that Destruction and general Spoil, that hath every where of late
been made of Woods, had they kept particular Accounts of the Profit, or been able to make a true
Judgment of their own Advantage, which, I think,
in most Places to exceed that of the Plough, or most
other sorts of Husbandry, where I having my self
transplanted an Elm, that in twenty Years time had
above twenty Foot of Timber in it, and must have
had a great deal more had it not been transplanted ;
where the Soil and other Circumstances are proper for
it : Without a due Consideration of particulars, no great
Advantage can be expected from this, or any other
sort of Husbandry ; for though Art may improve
Nature, yet the forcing of it commonly requires
more Cost and Labour than will turn to the Advantage of the Undertaker. And therefore, as 'tis from
an Application of such Things as are agreeable to
each other, that Profit must proceed, I shall endeavour according to such Method, to give the best In-
formation I can, of such Things as may be most for
the Advantage and Incouragement of the Planter and
Farmer.

And therefore, as Trees are of several Sorts, and
for several Uses, as some for Building, Utensils, and
Fuel ; and others for Fruit, Ornament, and Pleasure ;
and that some are raised of Seeds, as the Oak, Chesnut, Ash, &c. and some spring from the Roots, as
Elm, Alder, and others ; and some are raised of Sets,
as Willows, Oziers, and the like : And considering
that some Trees, even those that are the most useful, have been lately cultivated amongst us, and that
there are more that will deserve the Care of our Propagation,

pagation, I shall not tie my self up only to the common Sort; but add something for the Improvement of such kind of Trees, as may be of Use, though not commonly known amongst us, that so there may be an Improvement of the Species, as well as of the way of ordering of them, which work, though 'tis so well performed by that learned and ingenious Gentleman Mr. *Evelyn*, that it may justly be thought needless for me to meddle with this Subject; yet as the design I first proposed, was a compleat System of Husbandry, and that there are several Things for the Improvement of this part of it, that I could not otherways have had an opportunity of mentioning and getting Intelligence about, that I think none have treated of yet, I was upon this account under a Necessity of making a small Treatise of this Subject, pursuant to something of a different Method, than hath hitherto been done; what I shall meddle with being but a small part of what Mr. *Evelyn* hath done, and I shall take Care in the following Design to avoid as much as I can, what may any Ways be prejudicial to one that deserves the chief Honour of so useful a Subject, he having been the only Promoter of this advantageous Part of Husbandry.

For the Propagation of Trees, I shall not recommend the waiting for a spontaneous Product, except where the Ground is very full of Roots and young Wood, because of the length of Time that such Production requires, and because that neither Planting nor Sowing are any hindrance to it; nor shall I determine in this Place which is the best way of raising of Trees by Seed, or by the Transplantation of such, as we find to have raised themselves from the Seed, or that spring from the Mother-roots of other Trees, because I shall have an Occasion to mention it hereafter, and to give an Account of several particular Experiments relating to each several Way. I begin with the raising of Trees by Seed, as being what must of Necessity be the first Work in most

places,

places, becaufe other Trees are not to be had to make Plantations with; in order to which, it will be neceffary firft to treat,

---

Chap. II. *Of the Soil.*

WHich being the Foundation of this Work, and there being fuch a vaft difference between the growth of fome Woods and others, upon the account of the Ground they grow upon, it may be neceffary (tho' I fhall have occafion hereafter to treat of the particular Soil that each Plant requires) to premife fome general Rules concerning it; for though Trees will many times thrive on courfe Land; yet the beft Sort of Lands for moft Trees is the deepeft and richeft Soil, which always produces the Talleft Trees, all Trees commonly growing fhrubby, unfruitful or fpreading of their tops, where the Soil is either dry cr fhallow: And tho' fome Trees covet to run juft under the Surface, yet I think 'tis generally occafion'd from a want of depth of Earth: and where there is not a fufficient depth to cool the Roots and keep them moift, they are neither lafting nor profperous, though fome Trees, as Beech, Hazel, Holly, &c. affect gravel and fandy Land, and Aquaticks moift and boggy; yet for the moft profitable and ufeful Timber, 'tis neceffary to have a deep Soil, and in fuch places Trees do no hurt to the Land by fucking of the heart away from any thing that fhall be fown upon it. However, I am not for imploying of Land worth twenty or thirty Shillings an Acre for this ufe, farther than by planting of the Hedge-rows, becaufe many forts of Lands not worth above five or Ten Shillings *per* Acre, are near as good; nay, there are fome forts of Lands that are not very good for either Corn or Grafs, that will bear very good Trees, as fome of the hazelly Brick-earths in *Effex*, and fome forts of heathy Lands. If thofe that have opportunities of making any particular Obfervations about the

Nature

Nature of such Soils, and of the growth of the Trees upon them, will send them to the Publishers, I shall take care to add them in my next Volume, and make such use of them, as I believe will be of publick Advantage. Some Earths have a good Soil above, and underneath, Gravel, Sand, Stones, Slate, cold barren Clays, and cold Springs, *&c.* a Planter or Raiser of Trees ought to consider the under Soil, as well as the superficies of the Earth, and to observe that the worse your Land is, the worse it will be for transplanting ; and therefore the raising of Trees by Seed on barren Lands is much to be preferred, because it allies them to it, and makes both the Plant and the Soil the more natural to each other ; but seeing all sorts of Land shew what they are inclined to by their natural Product, and what the profit will be by the growth of the Trees, and the Shoots they make, it will be best to sute your Seeds, and your Soil one to another, and likewise to calculate your profit by the same Rule. Tho' even in this Case, too great Additions and Helps may be afforded to Nature with a little cost and labour, as I shall have occasion to shew hereafter.

Chap. III. *Of Seeds.*

CHuse Such Seeds as are mature, ponderous and sound, and that easily shake from the Boughs being taken from the tops of the youngest and most thriving Trees, and gathered when they are ready to fall, which doth for the most part direct the best time for the sowing of them, (which for most sorts of Seeds is about *November*) except your Land is very cold or moist, there a vernal sowing may be better; Acorns, Mast, and other Seeds, being what may be kept well for the Spring Season, by being barrelled or potted up with moist Sand or Earth, and laid S. S. S. during the Winter, at the end of which, you will find them

B 3 sprouted,

sprouted, which being committed to the Earth by a careful hand, will be as apt to take, as if sown earlier, and by this means better escape the Vermin, who are very greedy of spoiling the Winter Sown Seed, and they are not so easily damaged by the increasing heat, as those sown in the beginning of Winter, especially in loose hot Grounds; and therefore, if you have occasion to preserve much Seed, chuse a fit piece of Ground, and with Boards raise it three foot high, and lay the first lay with fine Earth about a Foot thick, and another lay of Seeds, Acorns, Mast, Keys, Nuts, Haws, &c. promiscuously or separated with a little Mould sprinkled amongst them, and let the third, or upper lay be of Earth or Sand, or you may bury the Seeds in dry Sand or pulveriz'd Earth, either barrelled or laid in heaps in some deep Cellar to preserve them from the rigour of the Winter. If your Seeds be gathered in moist weather, lay them a drying, and so keep them till you sow them, which may be as soon after *Christmas* as you please ; but if they spire out before you sow them, be sure to commit them to the Earth before the Sprouts grow dry.

As for the medicating or steeping of Seeds, or the enforcing or inriching of the Earth by Compost, &c. for Trees of this kind, 'tis a charge that would much discourage the Work, and what is needless, because if one sort of Mould be not proper for one kind of Tree, it may be for another ; but if your Seeds or Kernels prove extraordinary dry, if you lay them twenty four hours in Milk or Water, only impregnated with a little Cow-dung, it may do well to forward their sprouting, especially if you have been hindred from the former Preparation.

What Seed is best. The Shape and weight of the Seeds inform you which are the best, and how they may be set, most of them, when they fall, lying on one side with their small end to the Earth, from which part they put forth the Root first, which when it hath laid hold of the Ground, from the same place sends forth the

Shoot

Shoot for the Tree ; and if they be heavy Seeds, fow them the deeper, as Acorns, Chefnuts, Walnuts Peaches, Apricocks, &c. about two or three Inches deep : But if light Seed, as Elm, Lime, &c. cover them only with a little Mould about half an Inch deep.

Try all forts of Seeds, and by their thriving you'll beft difcern what are the moft proper kind for your Land, and of fuch make the main of your Plantation, which Sorts you may have fome Guefs at by what you find your Land naturally to produce moft.

Being thus provided with Seeds of all kinds, you may raife Woods or Groves immediately from them, *Of raifing* which I think the beft way, where you defign a large *Trees of* Plantation, and refolve to employ the Land for no *Seed.* other ufe, and to keep it well fenced. Firft, becaufe they take fooneft ; fecondly, becaufe they make the ftreighteft and moft uniform Shoot ; thirdly, becaufe they will neither require ftaking nor watering (which are two very confiderable Articles) and laftly, for that all tranfplanting (tho' it much improve Fruit-trees) unlefs they are taken up the firft Year or two, is a confiderable impediment to the growth of Foreft-trees, unlefs where they are removed from a very barren Soil to a rich, and meet with a very moift Summer, efpecially in the tranfplanting of the Chefnut, Walnut, and fome others, that I fhall have occafion particularly to mention afterwards ; but if you defign a Tranfplantation of Trees, it is beft to raife them in your Seminaries and Nurferies firft ; by which means you may tranfplant them as you pleafe, for Coppice-ground, Walks, Hedges, Rows, &c. therefore I fhall refer what I have further to fay about the raifing, tranfplanting and managing of them, to the particular ordering of them in the Seminary and Nurfery.

Chap. IV. *Of the Seminary and Nursery of Forest-trees.*

FOR a Seminary or Nursery of Forest-trees, which is what I only design to treat of here, it is what will be thought by many not worth the taking of Pains about ; but the small cost that attends it, and the small Quantity of Land requisite for such a Use, with the Advantage you will find by it in filling up your Hedge-rows, and other waste uncultivated places, will quickly convince you of its usefulness, and therefore having made choice of your Seeds, chuse out some fit piece of Ground that is well fenced, respecting the South East, rather than the South, and well sheltred from the North and West. Let it be cleared of all trumpery, and if large, it may be plowed up first, because it will make it dig the easier, and after that, I would have it dug up two spits deep, and all the upper part or surface of the Earth cast undermost, and the under Spit laid above, where the Soil is deep enough to bear it ; which tho' it may be a charge at first, it will abundantly answer in the growth of Trees afterwards, because they will every where have loose Earth to root in, and the best of the Soil under them, all Trees shooting of their Roots most downwards, which is the only way I would have Trees advanced in their growth in Nurseries, and not by mending or improving of the Land by Dung, &c. as the Gardiners commonly do for their own profit, and not the buyers, because of the difficulty that you will meet with in making such Trees to grow, if you should have occasion to remove them to a worse Soil: And if your Nursery be dug up the Winter before you sow or plant it, so as to give a Winter and Summer fallowing to make it mellow and fine, it will do well. At one end or side of which make some small Beds of about a yard wide, leaving a small path between them for the Seminary ; cross the Beds, make some small

Trenches

Trenches at about a Foot diſtance, into which throw the Seeds; but not too thick, covering them with a Rake, according to the depth before directed; but if you deſign the raiſing of Oaks, Walnuts, Cheſnuts, &c. the beſt way is to ſet them as they do Beans, and at about a Foot diſtance, which is to be done about the latter end of *October* for the Autumnal ſowing, and in *February* for the Vernal; ſix Buſhels of Acorns will ſow or plant an Acre of Land at a Foot diſtance, which I think enough in the Seminary, becauſe they ſhould be weeded by hand, and the ſpaces between the Beds will give room enough to come at them.

If your Nurſery is of a Gravelly, Stony ſoil you'll do well to pick out the ſtones as often as you dig it; for Stones, lying near the Roots of Trees, do often fret and gall them, and ſpoil not only the Roots, but alſo occaſion Cankers; and likewiſe, if you mix Clay or brick Earth with it, or Marle to make the ſoil deeper, eſpecially your Nurſery being a Shallow ſoil and apt to burn in Summer. What Plants you gather or draw out of Woods plant immediately, for their Roots are very apt to Mortifie or harden and wither by the wind and cold air, becauſe of their coming from a warm Situation.

When your Plants begin to peep, Earth them up, eſpecially after great Froſts, at which time, the ſwelling Earth is apt to ſpue them forth, and when they are about an Inch above Ground, you may in a moiſt Seaſon draw them up where you find them too thick, and ſet them in ſuch places as you have occaſion to beſtow them in.

Your Seedlings having ſtood till *June*, beſtow a weeding or a ſlight howing upon them, and ſcatter a little mungy Straw, Fern, rotten Beans, &c. amongſt them, to prevent the Roots from ſcorching, and to receive the moiſture that falls; and in *March* following, by which time it will be rotten, chop it to pieces, and mix it with the Earth, which continue to do for two or three Years ſucceſſively, for till then, the

the substance of the Kernel will hardly be spent in
the Plant: After which you may transplant them as
you please, only the younger they are removed after
they are three Years old, the better they will grow.
At the removing of them you must consider whether
the Place you design them for be secure to keep Cattle
from them while young; but if not, and you design
to plant them where Cattle come, it will be best first
to remove them into your Nursery, where you may
plant them in Rows three Foot distant, and the Trees
in the Rows to be at least two Foot distant from each
other, because else you will be apt to cut the Roots in taken of them up; in these Rows you may let them stand
till they are big enough to plant out, so as by the help
of some Stakes or Bushes put about them, they may
in a short time be able to defend themselves from
Cattle, the not taking care of which particular has
been a great discouragement to several Planters:
Some at the first transplanting of their Trees out of
their Seminaries cut them off about an Inch from the
Ground, and plant them like Quick; but 'tis not a
good way for any sort of Trees that have a large Pith,
or that you design for Timber-trees, because it lets the
Water into the one, and spoils the But-end of the
other, which is the principal Part of the Tree, by
diverting the Pith, and by consequence the Grain
of the Wood too, and so hinders it from running
clear if you should have occasion to cleave it into La-
thes, Pales, &c. and therefore I am only for pruning
up the side Boughs; but about the Transplantation
of Trees and the pruning of them, &c. I shall have
occasion to take notice afterwards; only you must observe, that having transplanted your Seedlings into a
Nursery, they ought still to be kept clean from Weeds,
and also the Ground to be kept loose, that the Roots
may spread the better, and therefore in the next *Autumn*, before the Leaf is off, your Nursery ought to
be digged a small Spit deep once a Year, only in Spring
or Summer you must pull or howe up the Weeds as
                                                    need

need requires; and I propose the digging amongst
the Trees, while the Leaf is on, because if it be done
while the Sap is up, if a Root should happen to be
cut it will shoot out again, perhaps two for one; but
after the Sap is once down, if a Root be cut it will not
shoot forth that Winter; however be not too early
in the Season, nor yet too careless of the Roots; and
between the Rows you may plant Beans, Pease, or
sow them with Turneps to prevent Weeds from coming up, which if any do, they should be howed up in
June, and laid about the Roots of the Trees to rot
and to keep them moist.

Many Trees may also be propagated by Layers, the *Trees of*
Ever-greens about *Bartholomew* Tide, and other *Layers.*
Trees about *February*; which is done by slitting of the
Branches a little way, and laying of them half a Foot
under Mould, with which, if some of the Earth that
is in hollow Trees be mixed, or some Pigeons Dung,
or if you water the Layers (where the Plants are such
as you are curious of) with the Grounds of Beer or
Ale, it will make them strike Root the better, and if
they comply not well, peg them down with a Hooke
or two, and when you find them compleatly rooted,
the next Winter cut them off from the main Plant, and
plant them forth in your Seminary or Nursery, as is before directed about Seedlings; others twist the Branch
or bare the Rind, and if 'tis out of the reach of the
Ground they fasten a Tub or Basket near the Branch,
which they fill with good mould, and lay the Branch in it.

Some Trees are raised of Cuttings taken from the *Trees of*
young Shoots, the best of which are those Suckers *Cuttings.*
that spring from the Roots (except 'tis a grafted Tree
that you design to have Cuttings from) the Cuttings
must be set a Foot deep in the Earth in a moist shady
place, as near as you can, and stand near a foot out;
but if they are of soft Wood, as Willow, Poplar, Alder, &c. 'tis best to take large Truncheons, so tall as
that they may head above the reach of Cattle; and
if you raise your Trees of such sorts as bear a Knur or
burry

burry fwelling, fet that part into the Ground, and make the hole wide, paring the end of the Cutting fo fmooth that no part of the Bark may be ftripp'd up in fetting; and keeping the Ground moift about it, it will feldom fail of putting out Roots and growing.

Trees rai- Many forts of Trees may be propagated from Suck-
fed by ers coming from the Roots of other Trees, to caufe
Suckers which, dig about their Roots early in the Spring, and finding fuch as with a little cutting may be bent up-wards, raife them above Ground three or four Inches, and in a fhort time they will fend forth Suckers fit for tranfplantation; or you may fplit fome of the Roots with Wedges, or break them, *&c.* and covering of them with frefh Mould they will quickly fprout out; which is one of the beft ways of raifing Elms and fome other forts of Trees; but thefe things I fhall particu-larly mention hereafter.

Chap. V. Of the Oak.

I Shall begin with the *Oak*, as affording the moft ufe-ful and beft fort of Timber, efpecially for the build-ing of Ships, and what was, of all others, the moft efteemed amongft the *Romans*, of which there are fe-veral Species in feveral parts of the World; but Mr. *Evelyn* takes notice only of four forts, two of which he reckons the moft common in *England*, *viz.* the *Quercus Urbana*, which he efteemeth the talleft, being clean, and of a fmooth Bark; and the *Robur* or *Quercus Silveftris*, which hath a kind of a black Grain, and bears a fmaller Acorn, fpreading forth its Roots and Branches more than the o-ther, and keeping of its Leaves all Winter; thefe differences I know variety of Soil will produce, and that the more thriving an Oak is, the more fappy it will be, and the longer the Leaves will hang on it; and therefore whether thefe marks are diftinguifhing of the Species I fhall not determine, but rather advife the gathering of your Acorns from fuch a Tree as you like the kind and fort of beft.   The

The Oak may be propagated by Layers, but not
to that advantage of Bulk and ſtature as from the Seed
nor can it be ſo well tranſplanted as it may be raiſed
from the Acorn: But where you deſign tranſplanting
of them for Walks, Groves, or into Hedge-rows, or
other places where Cattle come, they ſhould be often
tranſplanted ; and the beſt way to make them bear,
it is to raiſe them firſt from Acorns in your Seminaries
and after three Years growth to tranſplant them into
your Nurſery, ordering them as is ſhewed already,
where you may let them ſtand ſeven or eight Years,
or 'till they are about ſeven or eight foot high,
according to their bigneſs and growth, and then
remove them, as you have occaſion, without which
care, moſt Trees are very difficult to remove that are
raiſed of Seed ; the reaſon of which is, becauſe 'tis
the nature of all ſuch Trees to put forth one down-
right Root firſt, and not the ſide Roots till the Tap-
root is got near the bottom of the Soil, eſpecially in
a looſe hollow Ground, and ſo the main Roots going
deep, the ſmall Roots, which are the chief nouriſhers
of the Tree, lie ſo deep that you cannot come at them
to take them up, but, if you take them up young,
while the Tap root is ſmall and not ſhot too far down,
you may, by cutting off the Tap-root about a Foot
long, cauſe it to branch near the Top of the Earth,
which will give you the advantage of taking of it up
with ſmall Roots when 'tis removed again.

But ſome, to prevent this Inconveniency, put under all the Trees they raiſe of Seed about three or four
Inches below the place where they ſow their Seeds, a
ſmall piece of Tile to ſtop the running down of the
Tap-root, which occaſions it to branch when it comes
to the Tile, which is a very good way, and will much
increaſe the number of the ſmall Roots, and is a great
help to its tranſplantation, and many ſay, that it
much helps a Tree where 'tis not removed, but ſuffered to grow from the Seed.

The beſt time for the removing of Oaks and all o-
ther Trees that ſhed their Leaves in Winter, is in
October

October, as foon as the Leaf begins to fall, or in
February juft before the Sap begins to rife, and take
Care in planting of them, not to fet them deeper
than they ftood before, for if the Roots be fufficiently
covered fo as to keep the Body fteady and erect 'tis
enough. The Pofition likewife of the Tree ought to
be carefully obferved, for the fouthern Side of the
Tree being more dilated, and the Pores more expofed to the heat of the Sun by a fudden Tranfpofition
of the Tree in a cold time of the Year, the Tree
will be very much prejudiced.

Soil.   The Oak thrives beft on the richeft Clay, though
it will grow well on moift Gravel, or the coldeft
Clay, which moft other Trees abhor, and even in
fome places ftrike Root between Rocks and Stones,
and grow almoft upon any kind of Land, and will
penetrate ftrangely to come at a marly Bottom, and
often make ftands as they encounter variety of Foot-
ing, and fometimes proceed vigoroufly again; and as
they either penetrate beyond, or out-grow their Ob-
ftructions, or meet with better Earth. But the beft
Timber for Ships is that which grows on the ftiffeft
Land, it being the moft folid, tough, and durable;
whereas what grows on light Land is light and brittle, and not of a folid Grain, which though 'tis beft
for the Joyner's ufe, is not of the value of the other
for Ships and Building; but 'tis in the moft fouthern
warm Parts of *England* that they thrive beft in ftiff
Clay, and not fo well in the northern Parts, becaufe
they have not fo much the heat of the Sun to warm
thofe cold Soils; for Oaks, as to the Soil and Tem-
perature of the Air, as *Vitruvius* well obferves, neither profper in very hot Countries, nor very Cold,
but affect a temperate Climate; which I fuppofe is
the Reafon of our *Englifh* Oaks fo much out-doing
thofe of all other Countries: And where Oaks grow
naturally and in abundance, it is a fign of their being
good, their liking the Soil, and alfo of their Soil being rich.

If you would propagate them for Timber, do not
cut off their Heads, nor be too bufy in lopping of
them, except it be of fear and unthrifty Branches, or
that you are to remove them from a good Soil to a
bad, in which Cafe 'tis neceffary to have as much Root
and as little Top as you can, or that you defire them
for Shade, bearing of Maft, or for Fuel.

'Tis needlefs to mention either the ufefulnefs of *Its Ufes.*
Oak for the building of Houfes and Ships, or to fhew
how much our *Englifh* Oak exceeds that of all other
Countries for that ufe, fome of it being fo tough that
our fharpeft Tools will fcarcely enter it, nor the Fire
it felf confume it but flowly, as having fomething of
a ferruginous metalline fhining Nature, proper for ro-
buft Ufes: It is doubtlefs of all Timber hitherto known
the moft univerfally ufeful and ftrong; for though fome
Trees be harder, as Box, Cornus, Ebony, and divers
Indian Woods, yet we find them more fragil, and not
fo well qualified for the fupport of great Weights,
nor any Timber fo lafting, where 'tis to lie fometimes
wet and fometimes dry. The fine clear grained Oak,
if it be of a tough kind, is beft for the fupport of
Burdens, as for Columns, Summers, &c. except the
Oak that is of a twifted Grain, which may eafily be
difcerned by the Texture of the Bark. And the more
tender fort of a fine clear Bark, as being the beft to
cleave, is the moft ufeful for Pales, Laths, Coopers-
Ware, Shingles, Wainfcot, Wheel-fpoakes, Pins, &c.
as the knottieft and coarfeft is beft for Water-
works and Piles, becaufe it lafts longeft and drives
beft; and the crooked Oak, if firm, is beft for Knee-
timber in Ships, for Mill-wheels, &c. And the knot
of an old Oak juft where a large Arm joins to the
ftem, is often finely veined like Walnut. And if the
planting of Oaks were more in ufe for under Woods,
it would fpoil the Coopers Trade for the making of
Hoops, either of Hafle or Afh, becaufe one Hoop
made of the young Shoots of a Ground-Oak, would
out-laft fix of the beft Afh. The fmaller Truncheons

ons and Spray makes Billet, Bavine, and Coals, and
the Bark is of price with the Tanner and Dyer, to
whom the very Saw-duft is of ufe, and the *Uvæ
Fungus*'s to make Tinder of. Oaks bear also a kind
of a Knur of a cottony Matter, which was ufed of old
for Wicks of Lamps and Candles. *Prævotius* in his
Remedia Selectiora mentions an Oyl extracted chymi-
cally *è quercina Glande,* which continues the longeft of
any whatfoever, fo that an Ounce of it can fcarcely
be confumed in a Month, though kept continually
Burning; and Mr. *Joffelin* in his *New England* Rari-
ties, fays, that they there make an Oyl of the Acorn
growing on the white Oak, by taking of the rotten-
eft Maple-wood and burning of it to Afhes, of which
they make a ftrong Lye, wherein they boil the A-
corns 'till the Oyl fwims on the top in great Quantities, which they put into Bladders to anoint their
Limbs with, which it exceedingly corroborates and
ftrengthens, and ferves them to eat with their Meat,
being exceedingly clear and fweet Oyl; which, perhaps, is what our Acorns will produce too: If not, I
think it would be of ufe to procure fome of thofe
Acorns to plant here, which may be eafily done by
them that have a correfpondency there. *Varro* fays,
they made Salt of Oak Afhes, and fometimes feafoned
Meat with it, but more frequently fprinkled it among
their Seed-corn to make it fruitful; and Acorns are
of great ufe for the fatting of Hogs, Deer, Poultry,
&c. and formerly ferved for the Repaft of Men too.

Chap. VI. *Of the Elm.*

ELMS are of feveral Sorts, and differ much ac-
cording to the Soil and Climate they grow in.
The forts moft worth our Care and Culture are,
Firft, the common Elm, which hath a very rough
Bark and Leaf, fome of which are of a rounder Form
than others; this fort of Tree grows to a very great
Height and Bignefs, efpecially thofe that have the
roundeft

roundeſt ſort of Leaves. Secondly, That which they call the Watch-elm, which kind in Bigneſs and Height is like to the firſt ſort, only it hath a much ſmoother Bark, and in many places is putting forth Burrs and Knobs; the Leaf being alſo ſmooth and long, and varies much for Breadth and Length according to the Soil it grows in. The third ſort of Elm is by ſome called the Witch-haſle or Broad-leaved Elm, the Body and Boughs of which have a ſmooth Bark like the Witch-elm, and the Shoot and Leaf is much like that of the Haſle; upon which account, I ſuppoſe, it hath its Name: The Leaves in ſome ſoils are of a very great Size, and where it thrives it will make very large Shoots. I have one of theſe ſort of Elms that I brought out of *Derbyſhire,* in my Nurſery, that hath Leaves near ſix Inches long, and about five broad. It every Year ſends forth Shoots of fourteen or fifteen Foot long. It will grow to a great bigneſs, being the quickeſt grower of any kind of Elm; but 'tis not ſo apt to ſpire up as the other ſorts, being more inclined to branch into Arms.

Elms may be propagated by Seeds, which are ripe *How rai-* about the latter end of *March,* or beginning of *April, ſed.* which gather and lay in a Room to dry a day or two, and then ſprinkle them in Beds prepared with good freſh Earth, ſifting ſome of the fineſt Mould thinly over them, and water them as need requires. When they are come an Inch above the Ground, which they may do in four or five Months time, ſift ſome more fine Earth about them to eſtabliſh them, keeping of them clean weeded for the two firſt Years, and the ſide Boughs trimmed up 'till they are fit to remove into the Nurſery to be placed at wider Diſtances, and from thence to be removed to ſuch places as you have occaſion to plant them in. But the taking of ſuch up as are of a plantable ſize from Hedge rows and Woods is much more eaſy and expeditious; their not being apt to run down with a Tap-root, like an Oak, makes them more eaſy to tranſplant; only thoſe that you take out of Woods, as they ſtand there very warm, will not thrive ſo well if they are planted directly into an open plᴀce, as if you plant them firſt into your Nurſery for two or three Years, and from thence tranſplant them to Avenues, Hedge rows, &c.

Of Layers or Suckers. Elms may likewiſe be produced of *Layers* from a Mother-plant, as is before directed, or from *Suckers* taken off from the Mother-roots of great Trees after the Earth hath been well looſened from them. Or if ſuch Stubs as have been felled be fenced in ſo far as the Roots extend, they will furniſh plenty of young Plants, which may be tranſplanted from the firſt Year or ſecond ſucceſſively, by ſlipping of them from the old Roots. I ſhall not trouble you about the raiſing of them of Truncheons or Lops, becauſe I could never find them to take ; only ſometimes ſome of the ſmalleſt Suckers, when the Sap is newly ſtirring in them, if they are ſlipt off from the Tree, will grow, though not rooted.

By Trenches. Another way is to ſink *Trenches* at ten or twenty Foot diſtance from Elms that ſtand in Hedge-rows in ſuch order as you deſire they ſhould grow ; and where theſe Gutters are, many young Elms will ſpring from the ſmall Roots of the adjoining Trees, which after one Year being cut off from the Mother-roots with a ſharp Spade, if you tranſplant them they will prove good Trees without doing any Injury to the old ones.

There is a fourth way no leſs expeditious and ſucceſsful, by baring ſome of the chief Roots of a thriving Tree within a Foot of the Trunk, and chopping the ſame with an Ax, or making ſome ſmall Clefts in them with Wedges, into which Clefts put ſome Stones to prevent the Clefts cloſing again, and to give Acceſs to the Wet, which cover three or four Inches thick with good Mould ; and one ſingle Elm thus managed will be a good Nurſery, whoſe Suckers after two or three Years you may ſeparate and plant out as you ſee occaſion.

The

The best time for the Transplantation of Elms is *Of remo-* in *October* or *February* : Of all Trees that grow there *ving of* is none that better suffers Transplantation than the *them.* Elm doth, for you may remove them of what Bigness you will, even of twenty or thirty Years growth; nay, if they are planted in a good Soil, I think those of eight or ten Inches circumference to grow better than smaller ones, provided the Bark be smooth, tender and void of Wens, unless they are removed at the first taking of them off from the Mother-roots, if they are of any great Bigness when you remove them. You must totally disbranch them, leaving only the Summit entire, unless the Soil be very good; it may be necessary to head them too, but then it will be convenient, as soon as you can, to leave a leading Branch near the Top to spire up and cover the Wound; for Elm being a soft Wood, the Wet is apt to sink into it and to spoil the Tree. They must be taken up with as much Earth and Roots as you can, and have a great deal of watering; and the sooner they are removed after the taking of them up the better, except you speedily lay them in a Trench and cover them with Mould 'till you have time to transplant them : They should be planted on the Surface of the Earth, it being a very great Error, in any Soil, to plant Trees deep, and let the Roots be handsomely spread, and the Trees well staked and defended from Cattel : They delight to grow the nearest together of any Trees, which causes them to run up spiring, and protects them from the Winds.

The Elm thrives best in a rich black Mould; espe- *Soil.* cially where they can at some depth meet with some refreshing Springs or Moisture, and will grow almost on any sort of Land; and though they thrive not so well in too dry, sandy, or hot Ground, or those that are too cold or boggy, yet where the Earth is a little elevated above these Annoyances, you will find them to thrive upon the worst Land, as may be observed in several bad Soils, where Mounds and Banks are caſt up, eſpecially ſome ſort of Witch-elms, which in many Places grow on the drieſt Gravel, though not to any great Advantage.

Mr. *Moore* in his Natural Hiſtory of *Northampton-ſhire* ſays, that at a place called *Cranford,* if you in-cloſe any Ground that has had a Witch-elm in it, and newly cut up, in twenty-one Years a Grove will grow from the Roots fit for moſt uſes in that cold Clay, which this Tree ſeems propereſt for, few other Trees liking to grow in it.

The Elm, by reaſon of its aſpiring and tapering Growth (unleſs it be topped to enlarge the Branches and to make them ſpread low) are the leaſt offenſive to Corn, Paſture, or Hedges of any Tree, to which, and the Cattel, they afford a benign Shade, Defence, and agreeable Ornament.

When you would fell them, let it be about *Novem-ber, December,* or *January,* after the Froſt hath well nipp'd them, and that the Sap is fall'n into the Root, which will cauſe all ſort of Timber to be more dura-ble and laſting ; and if you find them any way un-thriving, fell them, and rather truſt to their Suc-ceſſors ; for the leaſt decay in any part of them will quickly ſpoil the whole Tree. In which particular they out-do moſt other Trees, becauſe where they are once planted, you need not plant any young ones, for if an old one be cut down, you'll have young ones enough from their Roots that will thrive better than any you can tranſplant.

But if you have occaſion to uſe them for Firing, ra-ther ſhred up the Boughs than lop them ; the beſt time for the doing of which is in *February,* and by that means you may cauſe them to ſpire up, always taking Care to preſerve the top, becauſe it protects the Body of the Tree from the Wet which common-ly invades that part firſt, and becauſe of its ſpungi-neſs the Rain eaſily penetrates, and will in a ſhort time periſh them to the Heart, by which means they will not only yield more Firing than if lopped, but

Timber

Timber too, efpecially if you take Care to fhred them up once in ten Years, and that you cut the Boughs clofe to the Body, or elfe the remaining Stubs will immediately grow hollow, and ferve as fo many Conduits or Pipes to hold and convey the Rain to the Body of the Tree; but if you lop any let it be the Witch-elm, which is much the hardeft and tougheft Wood, and bears the beft top; or you may form them into Hedge-rows by plafhing of them, and thickning of them to the higheft Twig, which will afford a magnificent Fence, and be a good fhelter againft the Winds and the Sun.

Elm is a Timber of extraordinary Ufe, efpecially in Extreams, where it may be continually Wet or Dry, and therefore proper for Water-works, Mills, Soles of Wheels, Pipes, Aqueducts, Ship-keels and Planks beneath the Water-line; and alfo the Witch-elm, efpecially if knotty, is good for Wheel-naves, becaufe of their being the tougheft Timber, and the ftrait fmooth Elms for Axel-trees, Kerbs for Coppers, Boards, Chopping-blocks, Trunks, Coffins, Dreffers, and for Carvers-work, Shovel-boards of a great length and fine colour; and the Roots of the knotty, curled fort are near as good as Wallnut for Cabinets, and are often fold for it, which they colour by laying in a wet Saw-pit that hath Oak Saw-duft in it, where they let it foak for a Month, which gives the Grain of it a black Colour like that of the Wallnut-tree; and if Elm Timber, as foon as it is fawn, be put into Water and lie three Weeks, and after the taking of it out be kept dry, it will prevent the Worms eating of it, and caufe it to laft as long as Oak, nay fome Elm-trees found buried in Boggs have turned like the moft polifhed and hardeft Ebony, as Mr. *Evelyn* faith, and were only difcernable by the Grain. They make likewife very good Efpaliers, if made to comply well with the Frames, and kept conftantly clipped. Befides, which feveral ufes, it makes the fecond fort of Charcoal, and the

Ufe.

very

very Leaves of it are not to be despised, for being
gathered Green, and suffered to dry in the Sun upon the Branches, and the Spray stripped off in *August*, will prove a great Relief to Cattel in Winter,
or scorching Summers when Hay and Fodder is dear,
which they will eat before Oats, and thrive exceedingly with it, but then you ought to lay the Boughs
in some dry places to prevent their musting : In some
parts of *Herefordshire* they gather them in Sacks for
their Swine and other Cattel; but some say they are
ill for Bees, in that they surfeit of the blooming Seeds,
which makes them obnoxious to the lask, and that
therefore they do not thrive in Elm Countries.

Chap. VII. *Of the Ash.*

OF *Ashes* some reckon that there are two kinds,
the Male and the Female, and that the Male
affects the higher Grounds, and the Female the lower,
which they esteem the whiter Wood and the taller
of Growth : But I could never perceive any greater
difference in them than what the Soil occasioned; but
though there is not much difference as to the form of
the Tree, there is in the Timber, as in that of the
Oak, some being much tougher than others, the
toughest growing on the stiffest Land; but the best
to cleave is that which grows on Gravel, Sand, or
other light Lands.

*How.rai-
sed.*   The Ash is best raised of the Keys gathered when
they begin to fall, which is about the latter end of
October, and during the ensuing Month, which you
must gather to lay by and dry, and then sow them
any time between that and the last of *January*; those
that are gathered from a young thriving Tree, which
produces the fairest Seed, are reckoned the best;
they should be sow'd but shallow, an Inch or an Inch
and an half being deep enough; or you may sow
them upon the top of the Ground and they will come
up, but 'tis best not to sow them in frosty Weather,
they

they will lie 'till the next Spring after before they
appear, except you have a mind not to wait so long
for their Springing, in which case you may prepare
them for spearing by laying the Keys in Earth or
Sand (which is the best) just as you gather them, laying one Layer of Sand, and another Layer of Keys,
and then laying another Layer of Sand, continuing so
to do 'till all your Quantity is disposed of, observing
to keep your Sand moist, and in a covered airy place,
and the next *January* come twelve Month after you
may sow them, and they will come up the next Spring,
but do not let them lie too long uncovered when you
take them out of the Sand left they should spear, and
the Air dry and spoil the Shoot. But if you would
make a Wood of them at once, dig or plow up a
parcel of Land, and prepare it as for Corn; only if
you plow it give it a Summers fallowing to kill and
rot the Turf, plowing it as deep as you can, and with
your Corn, especially Oats, sow your Ash-keys, and
at Harvest taking off your Crop of Corn the next
Spring you will find it covered with young Ashes,
which by reason of their small growth the first Year
should be kept well weeded and well secured from
Cattel, who are very desirous of cropping of them;
the second Year they will strike Root, and quickly
surmount what Impediments they meet with.

The best time to remove Ashes is in *October* and *No-* Of their
vember; they are best transplanted young because of *removal.*
their deep Rooting, which makes them hard to take
up when they grow big; but if you would transplant them large, raise them first in your Seminary,
and then remove them into your Nursery, and from
thence into the Places where you design to plant them
out, which will be a great Advantage in that it will
inable you to take them up with good Roots, as I
said before of the Oak; but as 'tis a Tree that hath
a large Pith, you must avoid heading of them if you
can, and content your self only with cutting off the
side Boughs, which will be better for the Timber, and

be

be likewife fparing of the Roots, except the down-right Tap-root, which you may abate as you fee convenient. Some cut the young Afhes off about an Inch above the Ground, which caufes them to make very large ftreight Shoots, which they call Ground-afh; and 'tis a very good way where you defign them for Under-wood.

Young Afhes are fometimes in Winter-Froft burnt, which makes them look as black as a Coal; you may in fuch Cafes make ufe of the Knife, yet they commonly recover of themfelves, though it is but flowly.

'Tis no way convenient to plant Afh in plow'd Lands, efpecially where the Soil is flete, becaufe the Roots are apt to run upon the Top of the Ground, which makes them troublefome to the Coulter, and the dropping of the Leaves is efteemed hurtful for the Corn, and at *Michaelmas* time makes the Butter bitter when eaten by the Cattel: But in Hedge-rows and Plumps they will thrive very well, where they may be fet at ten or twenty Foot diftance. In planting of a Wood of feveral kinds of Trees for Timber, every third fet, at leaft, fhould be an Afh, it being a Timber that is faleable at any fize.

Soil. Afh delights moft in a light dry Mould, the richer and fatter 'tis the better, yet it will grow on any fort of Land if it be not too ftiff, wet, or boggy, though on Banks or Rifing-grounds, near Rivers and running Streams they will thrive exceedingly, or in wet Lands that are well drained.

Ufe. The Ground-afh (like the Oak) much excels a Bough or Branch of the fame Bulk for ftrength and toughnefs, and is a good lafting Timber where 'tis kept dry, and the Ends of it not laid in Mortar, but Clay or Loam. 'Tis remarkable, that the Afh, like the Cork-tree, grows when the Bark is as it were quite peeled off, as hath been obferved in feveral Forefts and Parks. Some Afh is very finely veined and much prized by the Cabinet-maker, which they often call by the Name of green Ebony, and bring it to it's

luftre

luftre with *China* Varnifh. But they often vein it by
art, efpecially for Gun-ftocks and fuch ufes, by fteep-
ing of filings of Iron in *Aqua Fortis*, with which they
draw Veins or improve thofe that are natural, by
tinging of them with this Liquor, which will fink into the very Grain of the Wood, and give a black
Colour where-ever you touch the Wood with it.

The ufe of Afh (next to Oak it felf) is one of the
moft univerfal forts of Timber we have, it ferves
the Soldier, Seamen, Carpenter, Wheel-wright,
Cooper, Turner, Thatcher, and Husband-man for
Ploughs, Carts, Axle-trees, Harrows, Bulls, Oars,
Blocks for Pullies, Sheffs; and like the Elm, is good
for Mortaffes and Tenons, and likewife for Pallifade-
Hedges, Hop Garden-Poles, Pikes, Spars, Handles,
Stocks for Tools, Spades, Guns, &c. So that in Peace
and War 'tis a Wood of the greateft Ufefulnefs;
the white and rotten part of it compofes a ground
for fweet Powder, and formerly the inner Bark was
made ufe of to write on before Paper was invented;
and the Truncheons make the third fort of the moft
durable Coal, and is of all other the fweeteft of our
Foreft-fuel, and will burn even while it is green.
But the fhade of the Afh is not to be endured, becaufe it produces a noxious Infect; and becaufe of
the late Budding, and early falling of the Leaves,
and therefore 'tis not to be planted for Walks or
Ornaments, efpecially near Gardens, becaufe of their
fpreading Roots and falling Leaves, both which are
prejudicial to them.

The Seafon for felling of this Tree is when the Sap
is at reft, from *November* to *February*; but in lopping of Pollards, it being a foft Wood, it ought not
to be done 'till Spring, that the Bark may quickly
come on to cover the Wound; nor fhould the
Boughs, for the fame Reafon, grow too big, becaufe
they will be fo much the longer before they heal,
and fo give opportunity to the Rain to foak into the
Tree, which will quickly caufe it to decay, fo that
you

you muſt be forced to cut it down, or elſe both Body and Lop will quickly be but of little Value ; and the ſame thing ought to be done when you ſee the Wood-peckers making holes in them, or the top Boughs begin to wither or be unthrifty, which is a ſure Indication of their decay. The Keys of the Aſh are a good Pickle while young and tender; and when near ripe, being gathered about the beginning of *Auguſt*, they are good to preſerve Beer or Ale, I having drank ſmall Beer in Summer-time brewed with them without any Hops in it, that drank well at two Months old ; but if ſome Hops be mixed with them they will do the better.

Chap. VIII. *Of the Beech.*

THE *Beech* is of two Sorts, and numbred amongſt the Glandiferous Trees ; the *Mountain-Beech*, which is the whiteſt and moſt ſought after by the Turner, and the *Campeſtral* or *Wild-Beech*, which is of a blacker Colour and more durable.

How raiſed. Both which ſorts are raiſed from the Seeds, and are to be managed like the Oak, which is the beſt way of raiſing of a Wood, unleſs you will make a Nurſery of them, and manage them as is directed for the management of Aſhes. Sow them in *Autumn* or later, or rather nearer the Spring to preſerve them from Vermin, which are great devourers of them. They may likewiſe be planted of young Seedlings drawn out of ſuch places as the Seed falling from the Trees doth propagate.

How removed. In tranſplanting of them, cut off only the Boughs and bruiſed Parts two Inches from the Stem, within a Yard of the top ; but be very ſparing of the Roots, that is, for ſuch Trees as are of Stature. They make ſpreading Trees and a fine Shade.

Soil. They will grow on Gravel, Sandy, or Heathy Ground, either upon the Declivities, Sides or Tops of high Hills, and on Chalk or Rocks, where they will

ſtrangely

ftrangely infinuate their Roots into the drieft and hardeft Parts, and in Vallies where they grow in Confort they rife to a vaft Height though the Soil be never fo Stony or Barren, if it be but natural to them; but they are very peculiar to what places they affect.

The *Beech* ferves for various ufes of the Houfe-wife, *Ufe.* as for Difhes, Trays, Rims of Buckets, Trenchers, Dreffer-boards, Screws, Chairs, Stools, Bed-fteads, Shovel and Spade-grafts, Floats for Fifhing-nets made of the Bark; it is good alfo for Fuel, Bavin, Felloes of the *London* Cars, and Coal, though of the leaft lafting, not to omit the Shavings of it for the fining of Wine, and the Afhes of it with proper mixtures, is good to make Glafs. If the Timber lie altogether under Water it is little inferiour to Elm. Of the thin *Lamina* or Scale of the Wood they make Scabbards for Swords, Band-boxes, Hat-cafes, and formerly Covers for Books, *&c.* Bees delight to hive in the Cavities of thefe Trees: It is much fubject to the Worm, and therefore it will do well, fo foon as the Timber is cut, to lay it in Water for a fortnight or three Weeks, as is before directed about the Elm. The Beech being pruned heals the Scar immediately, and is not apt to put forth Side-boughs again. The Maft is excellent for the fatting of Swine and Deer, to which Hogs may be drove about the latter end of *Auguft.* Some Families have by the Seed been fupported with Bread in fome parts of *France,* they grinding of them for that ufe, and they afford a fweet Oyl. The Leaves gathered fomewhat before they are too much Froft-bitten make excellent Matreffes to lay under Quilts inftead of Straw, becaufe they continue fweet feven or eight Years without being mufty or hard, as Straw will be; and they are much ufed both in *Dauphine* and *Switzerland* for that ufe.

Chap.

Chap. IX. *Of the Chesnut.*

OF *Chesnut* Trees, *Pliny* reckons several kinds, especially about *Tarentum* and *Naples*, but we have but one sort common in *England* that I can hear of. We have Nuts from several parts, the largest and fairest from *Portugal* and *Bayonne*, our own being very small and not so sweet as the former, but the way to have fair Fruit is to graft them.

How raised. They may be raised by Layers, but the best way to produce them is by the Nut; for which use choose the largest brown and most ponderous for Fruit, and the lesser ones for Timber. The best way of planting them is like Beans, previous to which, let the Nut be first spread on a Floor to sweat, and then cover them with Sand, and after a Month's time take them out and plunge them in Water, and the swimmers reject. Dry them again for thirty Days more, and then Sand them as before; after which try them again in Water, and repeat the same Course with them 'till Spring; at which time, or in *November*, you may set them; but if you set them in Winter or Autumn, it is best for to set them with the Husk on, which is a good Preservative against the Mice; it is also necessary to keep Hogs from them, who will find them if possible, and as they are liable to be spoiled by the Frost as well as by Vermin, I think the latter end of *February* the best time to set them in.

Some sow them confusedly in Furrows like Acorns, and govern them like the Oak, but then the Ground ought to be well plowed, and fallowed the Summer before, and when they spring, be cleansed at two Foot distance after two Years growth; likewise Coppices of Chesnuts may be wonderfully thickned and increased by laying the tender Branches. Such as spring from the Nut are best, and will thrive exceedingly if the Ground is stirred and loosened about the Roots for two or three of the first Years of their growth,

growth, and the superfluous Wood pruned away, which you may esteem most of the side Branches to be : They also shoot into fine Poles from a felled Stem, useful for many purposes.

Being come up they thrive best unremoved, there *How re-* being hardly any Tree that bears Transplantation *moved.* worse, in that they will sometimes upon removal make a stand of four or five Years; and therefore when you design to transplant them, I shall not propose the raising of them in a Seminary, and to give them several Removes, as I directed about other Trees, but to raise them at first in your Nursery, where you design to let them stand 'till you can plant them out into your Grounds ; for they do not much run down with Tap-roots, but may be taken up at any time with good Roots, without that transplanting of them that is upon that Account necessary for the removal of other Trees ; and in the removal of them I am rather for cutting only of the Side-boughs, than heading of them. The best time for removing of them is in *November* and *February*; they may be planted at twenty or thirty Foot distance for Timber, and for bearing Fruit at forty, but for other uses a great deal nearer.

I may here bring in the Horse-chesnut, which be- *Horse* ing easily increased by Layers grows into a goodly *chesnut.* Standard, bearing a fine Flower, and is now all the Mode for Walks and Avenues in *France*, being at first brought from *Constantinople* to *Vienna*, and from thence to *France*, though directly from the *Levant* to our Climate, where it grows very speedily to a large Stature, especially if planted in a rich Soil near Water. They will bear Nuts here, which are ripe about the latter end of *August*, of which they may be raised or by Layers.

The Chesnut likes best a light hazel brick Earth, *Soil* or black Mould, or moist Gravel, and will grow well on Clays, Sand, or any mixed Soil, especially if raised from the Nut without Transplantation, and up-

on

on expofed bleak Places, and the pendent Declivities of Hills to the *North*, in dry airy Places, and fometimes near Marfhes and Waters, where the Water doth not touch the Roots, which when it doth, is very prejudicial both to the Timber and Fruit; but they affect no other Compoft, fave what their own Leaves afford them, the dropping of which makes them injurious to what grows under them. They bear Cold better than Heat.

Ufe. Next unto the Oak, the Chefsnut is the moft fought after by the Carpenter and Joyner for Building; and before the ufe of *Firr*, was much made ufe of, many of the ancient Houfes in *London*, having been built of it. It affords the beft Stakes and Poles for Pallifadoes and Hops, good Props for Vines, and makes extraordinary Hoops, for which ufe only they keep feveral Coppices in *Kent* all of Chefsnut, near the Water-fide, of which they make great advantage by fending of them to *London*. The Timber is likewife good for Mill and Water-works; and the Trees are very good fhelter, being fet on the North-fide of other Plantations, and for Walks and Avenues. The Timber doth alfo well for Columns, Tables, Chefts, Chairs, Stools, Bed-fteads, Tubs, and Wine-casks, which it preferves with the leaft Tincture of any Wood whatfoever: Some fay, that if the Timber be dipp'd in fcalding Oyl, and well pitch'd, that it will become very durable: It is very apt to decay within when it fhews fair to the Eye. The Coals are excellent for the Smith, being foon kindled and foon extinct, but the Afhes are not good for Lee, becaufe they are apt to ftain Linen.

As for the Fruit, it is better beat down from the Trees fome little time before they fall of themfelves, becaufe they will keep the better, or elfe you muft fmoak-dry them, but not if you defign to plant them. And though here in *England* they are Food for Swine, yet in other Countries they are much efteemed, and would certainly be more ufeful and nourifhing to our
Country-

Countrymen than many of the Herbs and Roots that
they eat. In *Italy* they boil them with Bacon; and
they are certainly to be preferred before our Cole
or Beans, for they afford a good robuſt Diet, and
are very nouriſhing, being much commended by *Galen*
above all other ſorts of Nuts; they alſo boil them
in Wine, and roaſt them on Embers, and ſome grind
them to Meal, and make Bread and Fritters of them,
and the *French* Cooks uſe them in ſtew'd Meats, Beatil,
Pies, &c. The beſt way to preſerve them for uſe, is
to keep them in earthern Veſſels in cool places, but
ſome keep them in a Smoak-loft, others in dry Barley-
ſtraw, and others in Sand, &c.

Chap. X. Of the *Wallnut.*

THE *Wallnut* is of ſeveral ſorts, as the large ſoft
Shell, and the ſmall hard Shell; the Wood
likewiſe differs, ſome of which is of a whiter, and
ſome of a blacker Grain than the other; the black
Grain'd bears the ſmalleſt, hardeſt Nut, and the Timber is much to be preferred, the beſt ſort of which
are thoſe that grow in *Virginia,* which bear a kind of
a ſquare Nut, from whence we might eaſily procure
them. Next unto theſe are thoſe of *Grenoble,* which
are much prized by the Cabinet-makers.

The beſt way of raiſing *Wallnuts* is of the Nut, *How rai-*
which you may ſet like Beans: Gather the Nuts from *ſed.*
a young thriving Tree, bearing a full plump Kernel,
which is beſt to be beaten off the Tree (as was pre-
ſcribed of the *Chefnut*) ſome Days before they fall
from the Branches of themſelves, and kept in their
Huſk, or without, 'till Spring; or by Bedding of
them (being dry) in Sand or good Earth. You may
ſet them about the latter end of *February,* or beginning of *March,* or earlier; but if you can, ſet them
with the Huſk on, becauſe the Bitterneſs of it is a
good Preſervative againſt the Worms, and if you chop
ſome Furz and ſtrew about them under Ground, it
will

will be a good Defence for them againſt the Rats and Mice.

How tranſplanted. It is a Tree that bears Tranſplantation but ill, and therefore doth much better to ſtand where it is raiſed of the Nut; but if you have occaſion for the removal of them, let not your Trees be above four Years old when you remove them; and if they be once before removed from your Seminary to your Nurſery it may do well; which firſt Tranſplantation may be performed at two Years old; but ſome ſay, the beſt time to tranſplant them is at one Year old, and to take up the Earth and all with them, that you may not cut the Roots, which ſhould be ſpared as much as you can: And obſerve as near as you can, to remove them to the ſame ſort of Land that they were raiſed on, for it is very difficult to make them either grow or proſper on a different Soil, but in the removal of them by no means cut the Head, only ſhred up the ſide Branches, becauſe of the hollow Pith which is apt to let in the Water, and be as ſparing of the Roots; any thing of Bruiſes is very prejudicial to them. You may likewiſe, as I propoſed for the *Oak*, put a Tile-ſhred under them to occaſion the ſpreading of the Tap roots, and they may be grafted or budded, which will improve the largeneſs of their Fruit. The time to remove them is in *November* and *February*, and the beſt Compoſt for them is Aſhes. They ſhould be planted at forty Foot diſtance at the leaſt, becauſe they love to ſpread both their Roots and Branches, though they will grow amongſt other Wood provided you ſhred up the ſide Boughs.

Soil. The *Walnut* delights moſt in a dry ſound rich Mould, eſpecially if they have Chalk or Marle underneath them, and will grow well on any Land that is dry, where they are rais'd from the Nut, and where they are protected from the Cold (tho' they affect Cold rather than extream Heat) and grow in Pits, Vallies, High-way-ſides; alſo in Lime-ſtone Ground;

If

if Loomy, and on Hills that are Chalky; and they are fo far from prejudicing of Land, that in *France* they plant them in their Corn-fields, efteeming of them a help to their Corn in preferving of it from the Cold ; nor are the Roots any hinderance to the Plough, it being a Tree that will root deep enough where they have a Spil that will allow of it.

The *Walnut-Tree* is the moft beneficial Tree that *Uſe.* can be planted, not only for the value of the Timber, but the conftant Profit they bring by their Fruit; and if the Timber were more plentiful, we fhou'd have far better Utenfils of all forts for our Houfes, as Chairs, Stools, Bed-fteeds, Tables, Wain-fcot, Cabinets, &c. than what is commonly made of *Beech,* and not fo fubject to the Worm. They make fine Avenues, and do excellently near Hedge-rows, of which I am forry Foreign Parts, almoft every where, give more ample Examples than our own Country, tho' many Gentlemen, of late, are much to be commended, for the Improvements they have made this way. Of what Univerfal ufe this Timber is to the *French* may be feen in every Room both of Rich and Poor. It is of particular efteem with the Joyner for the beft colour'd and grain'd Wainfcot, with the Gun-fmiths for Stocks, and excellent for the Bodies and Wheels of Coaches. In *New England* they make Hoops and Bows with it, and the Drum-maker ufes it for Rimbs; the Cabinet-maker for Inlaying, efpecially the part about the Roots when it is curled and knotty, which fells often at very great prizes ; but to render it the better colour'd, the Joyner puts the Boards into Ovens after the Batch is drawn, or lays them in a warm Stable ; and when they work them they polifh them over with its own Oil very hot, which makes it look black and flick ; but as it is fubject to fhrink, it ought to be well feafon'd before it is work'd. It is not good to ufe for Beams or Joyfts, becaufe of the brittlenefs of it : And befides the ufes of the Wood, the Fruit with the Husk, when tender and very young, makes a very good Preferve, and alfo is us'd to pickle like Mangoes, and when ripe they are very good Food, afford a very ufeful Oil for *Painters*, and for Gold Size and Varnifh to polifh Walking-ftaves and other Works with, which is wrought in with burning. One Bufhel of Nuts will yield about feven Pounds of Oil, which in *France* is much ufed to burn in Lamps inftead of Candles ; from whence, in time of Peace, we are fupply'd with great Quantities, when it might with much more advantage be made a home Commodity. The young Timber is held to make the better colour'd Work, but the older being more firm and clofe is the beft for Cabinets. That the Husk may open, lay them up in a dry Room, turning of them with a Broom, but without wafhing, for fear of Mouldinefs, and thofe that come not out eafily of their Husk fhould be laid to mellow in the Heaps, and the reft in the Sun till the Shells dry, elfe they will be apt to deftroy the Kernel ; but the beft way to keep them is to bury them in the Ground in a leaden or earthen Pot clofe cover'd, that the Vermine may not come at them, and they will keep plump all the Year : They are very good for Hogs, but too chargeable a Provifion for them with us. The very Husk and Leaves being macerated in warm Water, and the Liquor pour'd on Walks or Bowling-greens, will kill the Worms without hurting the Grafs ; and the green Husks boil'd makes a good Colour to dye a dark Yellow without any mixture ; and if they, or the Firft peeping red Buds and Leaves be reduced to Powder, it will ferve inftead of Pepper to feafon Meat, or for Sauce, and the Kernel being rubb'd upon the Cracks and Chinks of a leaky Veffel, ftops it better than any Pitch, Wax or Clay. After the Nuts are beaten down, the Leaves fhould be fwept into heaps and carry'd away, becaufe their Bitternefs impairs the Ground, and injures the Trees.

Chap.

Chap. XI. *Of the Mulberry Tree.*

THE *Mulberry Tree* is of two Sorts, that which bears the black *Mulberry*, which I mention first, as most known amongst us, and that which bears the white Berry, having a smoother Leaf, and is much to be preferr'd for its usefulness for Silk-worms, in that it buds near a Fortnight sooner than the black, and the Leaves are finer and tenderer for them when young; which are two very great advantages to them, being much planted in the Silk Countries, and would certainly turn to very great advantage here, if made use of to the same purpose, it being a Tree that will thrive well in our Climate, and may be had of most of the Gard'ners near *London*.

Mr. *Evelyn* says, they raise them in the Countries *How raised.* where they cultivate them for Silk-worms by separating of the Seeds from the Berries, which he says they do, when gather'd thorough ripe, by bruising of them in their Hands, and washing of them in several Waters, and the Seed, which is very small, will sink to the bottom while the Pulp swims on the top; which must be carefully taken off, and the Seeds taken out and laid on Linen cloth, and dry'd, for which an Hour is sufficient; but the sowing of the ripe *Mulberries* he prefers, which should be a little bruis'd or squash'd, and sow'd in rich black Garden Mould, and ought to be well moisten'd at first sowing, tho' rarely water'd afterwards till they peep; they must be kept warm by being thinly cover'd with Straw to protect them from the Heat and the Birds. The season of sowing them is in *April* or *May*, tho' some forbear till *July* or *August*; and some sow them in *September;* which I think the best time; at the first keep them moderately shelter'd and clean weeded; and at two Years growth, about *October* or *February* you may draw them up gently, and dipping of the Roots in Water, transplant them into a warm place in your Nursery, cutting of them within three Inches of the Ground; and

B 3   giving

giving them three dreffings or half-diggings to kill
the Weeds in *April,* *June* and *Auguft.* The fecond
Year purge them of fuperfluous Branches, referving
the Principal Stem; of which, if the Froft injure any
part, cut it off. This way of raifing of them may do
well in hot Countries, but I fhould rather choofe to
raife them of Layers, from another Plant or Suckers
at the Roots of other Trees, which will take very readily: The beft time of doing of which is in *February,*
leaving not above two Buds out of the Ground, which
you muft carefully water and they will be well rooted
in two Years. Dr. *Beal* fays, they may be grafted in
the *Black Mulberry* in Spring or inoculated in *July.*
The Scions are beft to be taken off of an old Tree that
bears broad even Leaves, which being the moft ufeful
part of the Tree is chiefly to be confider'd.

How tranf-planted.

It is a Tree fomething difficult to tranfplant, except it be planted in a rich Soil and while young, and
be kept well water'd; do not cut off the Head in removing of them, but trim up the fide Branches, fo as
to leave but a fmall Head on them. The beft time to
tranfplant the *White Mulberry* is in *February,* it being
a more tender Tree than the *Black.*

Soil.

They affect a light dry rich Mould, which if it is
well manur'd with Afhes or Horfe-dung will be the
better. And if they are expos'd to the Sun and Air,
they will thrive much the more for it.

The beft time to tranfplant them is at about five
Years growth in *September* and *October,* in which Work
fpare cutting of the Roots as much as you can, and
take geat care to fave them in taking them up; and if
you find any of the Branches dry or hurt with the
Froft, cut them off, and where the Branches grow too
thick, a little thinning of them for the firft Year or
two after they are tranfplanted may do well, and likewife digging and ftirring of the Earth about their
Roots greatly improves them.

Of the two forts of *Mulberries* the *White* is much the
fineft, being called fo, becaufe the Fruit is of a paler
Colour,

Colour, more luscious and lesser than the *Black*; the Rind likewise is whiter, and the Leaves of a mealy clear green Colour. It is a beautiful Tree for Walks and Avenues, and gives a fine Ornament to the Silk Country, where they plant them, in Walks or regular Groves about their Fields.

The Timber will last as long in Water as the most solid *Oak*, and the Bark makes good Bast Ropes; it will suffer no Caterpillar or Vermin to breed on it, either as it grows or when cut down, except the Silkworm only. The chief Value of the Tree is the Leaf for the Silk-worm, for which use they are let for great Sums, single Trees yielding to the Proprietors sometimes twenty Shillings *per Annum*, and to the Owners of the Worms six or seven Pounds of Silk, which commonly sells for as many Pounds in five or six Weeks time; besides which, the Leaves are good for Cows, Sheep and other Cattle, especially young Hogs, being boil'd with a little Bran, and the Fruit is very good for Poultry,

The Leaves best for Silk-worms are those gather'd from Trees of about seven or eight Years growth, for the gathering of them from young Trees is apt to burst the Worms; so doth likewise the Leaves of Trees planted in too waterish a Soil, and those that are sick and yellow. The Leaves should be clipp'd off and let fall upon a Sheet or Blanket and not gather'd by Hand, tho' that is better than to slip them off, which galls the Branches, and bruises the Leaves. Gather them as dry as you can, but if you are necessitated to gather them in wet Weather, put the Leaves in a pair Sheets well dry'd by the Fire, and shake them up and down till the moisture be drank up in the Linen, and then spread them on another dry Cloath. The top Leaves and the oldest should be gather'd last, as being most proper to feed the Worms towards their Spinning time.

Chap.

Chap. XII. Of *the Servise-Tree.*

THE *Servise-Tree*, Mr. *Evelyn* says, is of four kinds, but there is little difference of those which we have in *England*, except only that some of them bear a much larger Berry than the others.

How rai-
sed.
They may be rais'd either by Layers or of the Berries, which being ripe, that is, rotten, you may eat or rub off the Pulp, and sow the Seeds in your Seminary, which is best to be done as soon as you separate them from the Pulp; or if you have a mind to keep them longer, you may keep them in dry Sand till after *Christmas*, and from the Seminary remove them to your Nursery, and from thence transplant them as you have occasion; but because they may often be met with in Woods, it being a Tree that admits well of Transplantation, most furnish themselves with them that way. They may likewise be either grafted or budded upon their own kind to a great Improvement of their Fruit.

Soil.
They delight in rich Clay or the Hazel-brick Earth, where it is rather moist than dry. In dry Grounds they never bear well, tho' they will grow almost on any Soil when they are rais'd of the Seeds.

How
trans-
planted.
They may be transplanted of any bigness, it being a Tree that bears Transplantation well: If you head them the Wound will quickly heal up.

Use.
The Timber is useful for the Joyner, Engraver of Wooden Cuts, Bows, Pullies, Screws, Mills, Spindles, Pistols and Gun-stocks, being of a very fine Grain, and is very lasting, being rubb'd over with Linseed-Oil well boil'd, and may be made to counterfeit *Ebony* and most of the *Indian* Woods; also it is used to build with, yielding Beams of a considerable Substance, and the Shade is beautiful for Walks. they give an early Presage of the Spring, and peep out in the severest Winters.

Chap. XIII.   *Of the Maple.*

THE *Maple* Mr. *Evelyn* reckons of several kinds ;
he commends moſt the *German Aier* and the *Maple* of *Virginia* ; as for the ſorts we have in *England* I
can find no great difference in them ; thoſe that we
keep ſhred up to run to Standards, have a fine clear
Grain, and thôſe that are pollarded grow the moſt
knotty and full of Burrs, it being a Tree very Subject
to put out ſide Branches, which fills it with Knots;
but it is prejudicial to let it grow tall where there is
any Wood or Trees under it, becauſe of a clammy
Dew that ſticks to the Leaves, which when waſh'd
off by the Rain, glues up the Buds of what Trees or
Buſhes grow underneath, and ſo kills them ; and
therefore they are not fit to grow in Hedges or a-
mongſt Wood.

It is produc'd, and doth produce it ſelf by the Seed, *How*
Layers, and from the Roots of old Trees, like the *raiſed.*
Elm, and by Suckers, which occaſions their being ſo
plentiful. The Seeds will lie, like the *Aſh,* till the
next Year after they are ſow'd before they come up,
and therefore they may be order'd the ſame way.

They may be tranſplanted almoſt of any ſize, and *How*
may eaſily be remov'd, in that they do not run down *tranſ*
with Tap roots ſo much as many other Trees do ; the *planted.*
beſt time to remove them is in *October* or *February.*

They affect moſt a ſound dry Mould, eſpecially
Banks, in which their Roots delight to run, and deſire
rather to grow on Hills than Bottoms, tho' they
hardly refuſe any ſort of Soil.

The Timber is very uſeful for the Turner for Diſhes, *Uſe.*
Cups, Trays, Trenches, *&c.* and when it is clean and
without Knots it makes excellent Board, and for its
lightneſs is often employ'd by the Muſical Inſtrument-maker under the Name of *Aier*; but that which
is fulleſt of Knots and Burs is of greateſt Value, being much priz'd by the Cabinet-maker.

Chap. XIV. *Of the Sycamore.*

THE *Sycamore*, Mr. *Evelyn* says, is our *Acer majus*, one of the kinds of *Maples*, he prefers the *German Sycamore* much before ours; it's a quick grower, and where an old Tree is near any dug up Ground, it will readily furnish you with Plants enough to set in what places you please.

How rais'd

It is rais'd of the Keys as soon as ripe, which come up the first Spring, and being provided with a large Leaf, the Weeds will not soon choak them; also young ones may be got of Suckers from the Roots and by Layers; they are manag'd like other Nursery Trees, and may, when they are big enough, be planted out for Walks or other occasions.

How transplanted.

They may be transplanted of any size, which they bear very well; and you may in transplanting of them, either head them, or only trim up the side Boughs, which way is to be preferr'd where they do not grow too small and tall.

Soil,

They thrive most in a dry light Soil, in which they will thrive very much, tho' they will grow almost on any sort of Land, and may be planted where other sorts of better Trees will not prosper so well, but they are not esteem'd so good for Walks or near fine Gardens, because of the falling Leaf which is apt to foul them; and it is a Tree whose Leaves hang upon them the shortest time of any Tree, which makes them neither good nor ornamental for Walks; besides which, there is a Honey-dew which hangs upon their Leaves, and breeds Insects; but both their Dew and Flowers are very advantageous for Bees; Mr. *Cook* commends them much for Parks and Underwood, because the Deer are not apt to spoil them, and in that they make large Shoots from old Stubs.

Use.

This Wood is of excellent Use for the Turner for Trenchers, Dishes, &c. and also for Cart and Plough Timber, being light and tough; and not much inferior to Ash it self.

The

The Sap is alſo eſteemed by ſome as good as that of the Birch, and as wholſome. It will run both Winter and Summer, ſo as in a ſhort time to yield a good quantity of Sap. One buſhel of Malt, with this Liquor, will make as ſtrong Ale as four Buſhels, with common Water.

Chap. XV. *Of the Horn Beam.*

THE *Horn Beam* is but of one kind, and is much more valued in Foreign Parts than amongſt us, for its ſhade and the delicate verdure of its Leaves, being the fineſt and the pleaſanteſt green of any whatever, and a Tree whoſe Leaves hang the longeſt upon it, it being one of the forwardeſt in budding, and one of the laſt that falls, the old Leaves ſeldom dropping till the young ones ſhove them off.

They may be rais'd of Seeds, which are ripe in *Auguſt*, and ſhould be ſown in *October*. They will lie in the Ground till the next Spring come twelve Months after, and when they peep, ſhould be carefully ſhaded and weeded: But the beſt way of raiſing them is by Layers from another Plant, or by Sets which may be eaſily procured in Woods, and other places where they grow, they being apt to run up with Suckers from the Roots. *How raiſed.*

They delight moſt to grow on cold Hills, in ſtiff Clays, and in barren moiſt places in the Woods, there being few Soils that they refuſe, except it be thoſe that are very dry and burning; it being a Tree that grows beſt in the ſhade, and under the dropping of other Trees.

The common way of tranſplanting of them is like Quick, when they are about the bigneſs of your Finger, and to cut them off about two or three Inches above the Ground; but they may be planted, when they are about ten or twelve Years growth, for Standards like other Trees.

The

Use. The Timber is useful for Mill-coggs, and other things of that kind (for which purpose it excells all others,) and for Yoak Timber, Heads of Beetles, Handles for Tools, Stocks, Lasts, and for the Turner's use. It's very tough and white, and is good Firewood, for which use it is often cut. When the Boughs are too large, they commonly decay its Body, occasioning in it often rottenness and hollowness. It is the quickest grower of any of the hard Woods, and preserves it self the best from the nipping of Deer ; and upon that account is more common in Forests and Parks than other sort of Wood, where the Soil is natural for it. It bears clipping the best of any Wood, and makes the thickest Hedges and covered Walks. Is grows thickest and fullest of Boughs at the bottom, even to resemble the thickness of Walls it self ; upon which account most of the fine Grottos in *Italy*, and the Walks and shady places of *Verſailles*, are of this sort of Wood, which they keep about fifteen or twenty Foot high, cutting them with a Scythe faſtned to a ſtreight Handle, which diſpatches that Work much ſooner and eaſier than the Shears.

Chap. XVI. *Of the Lime Tree.*

THE *Lime-Tree* or Linden, Mr. *Evelyn* reckons to be of two ſorts, *viz.* The broad leaved Lime which comes from *Flanders*, being a very quick growing Tree, and bears a very fine broad Leaf, only it is apt to ſhed too ſoon; and the wild kind, which, he ſays, grows naturally in many places of *England*, bearing a ſmaller Leaf than the other ; by which, I ſuppoſe, he means the Tree which they call the Pry-tree, which grows the moſt plentifully in *Eſſex* of any part of *England* that I have ſeen ; the Timber of which may be as good, if not better, than that of the right Lime, and the Bark is much us'd there to make Baſte-ropes, which are much cheaper than Hemp Ropes, are reckoned, and not to rot ſo ſoon ; but

for

for the propagating of it for Walks and such uses, I
can no way commend it ; first because of its being of
much flower growth than the *Flanders* Lime ; and secondly, because it's a Wood of that ill favour, that
if you break a piece of it, you cannot well hold it to
your Nose: It yields an ugly stench in burning, which
I mention, because it is a good way of distinguishing
the true Lime from it, that hath none of that scent.

   They may both be rais'd by Seed, which doth not *How rais-*
every Year ripen with us ; but when they are ripe, *ed.*
you may know it by their being plump and full under
the Husk, and the Body white when bit or cut in two ;
whereas if the Year be not kind for them, the Husk
will be full of nothing but a chaffy substance, or have
some small lank Seeds in them. The Seeds are ripe in
October, which gather in a dry day, and after you
have laid them to dry about a Week in an open Room,
put them into a bed of Sand indifferent moist, and so
keep them till about the middle of *February*, sowing
of them under some Wall or other shelter, that may
preserve them from the North and North-west
Winds ; and if the Spring and Summer be dry, keep
them indifferent moist, and stick some Boughs over
them to shade them, keeping of them clean from
Weeds, where let them stand two Summers, and then
transplant them into your Nursery, ordering of them
as is directed for other Trees.

   But the common way of raising of them is by Layers from a Mother-plant, or by laying of such Suckers as come from the Root. It being a Tree very apt
to put forth Suckers, it's best to lay them betimes,
as about *September* or *October*, and the next Year they
will be fit to transplant into your Nursery ; it being
a Tree that strikes Root well, and is very easily
rais'd by the same Methods that you are directed for
ordering the Elm.

   They will grow almost in any Soil, but they de- *Soil.*
light most in a fresh, rich, Mould, tho' they will grow
well on a loamy Ground, that is rather strong than
                        light,

light, or any moist Land : But the first Year of planting, if the Spring is any thing dry, they should be
kept well water'd. It is a Tree will grow any where
in Cities and Towns amongst the Smoak of Houses,
which choaks other Trees, and therefore is a very
proper Tree to plant, as the *Dutch* do for Walks in
the middle of their Streets, and round the tops of
their Walls; so that some of the Cities at a distance,
seem rather Walks of Trees, than to be compos'd of
Houses.

*How trans-
planted.* You may, if you have occasion, transplant them very
large, as about the bigness of your Leg, but when
they are about two Inches Diameter and eight or ten
Foot high is the best size; but if they never were
transplanted before, it is better to remove them
while smaller. They bear removing well, especially
if once or twice remov'd in the Nursery before you
plant them out into Walks or Hedge-rows.

It is a Tree that loves pruning, and heals the wound
the soonest of any, and naturally delights to grow
tall; and therefore if in your Nursery they shoot
much, leave some side-Boughs to check the Sap, lest
by forcing them too much up into the Head, it make
the Head too large for the Body to bear it, which many times in high Winds is apt to break off the top;
and therefore whenever you plant them for Walks or
in Hedge-rows, a discreet thinning of the Boughs is
necessary to let the Wind through; for being a large
Leaf and a light Wood, the Wind is very apt to damage them. If you plant them for Walks, set them
at about twenty or thirty Foot distance.

Use. It is a noble Tree for Walks and Avenues, casting
a large shade, and growing of a fine shape for that use,
and is of a pleasant green Colour. The Timber is fit
for any use that the Willow is good for, and is much
to be preferr'd both for its strength and lightness, being of a fine white Colour, and works easily without
being subject to split, and is much used by Carvers for
the Images they adorn Ships with; for which, if it is
                                                    large,

large, it fells for as good a price as most other forts
of Timber. It is also employ'd by Architects for
Models for defigned Building, and for fmall Statues
and Figures, and the Coal is efteem'd for Gun-powder better than that of Alder it felf. They make
alfo fine white Boards, and the fmooth fide of the
Bark makes Tablets to write on, and the Twigs
Baskets.

Chap. XVII. *Of the Quick Beam.*

THE *Quick Beam* or wild Sorb, by fome called
the *Irifh* Afh, is a fpecies of wild Afh, having
both a Bark and Leaf much refembling the Afh, only
the Leaf is jagged on the Edges, and fomething fmaller and longer in Proportion to its Bignefs; and inftead of Keys bears red Berries, which are very ornamental to them, and are preceeded by Bloffoms
of an agreeable Scent.

They are beft raifed of the Berry, which is ripe *How rai-*
in *October*, when is the beft time to fow them. They *fed.*
muft be cover'd but fhallow, an Inch deep being deep
enough to lay them. Whether they come up the
firft Year, or lie in the Ground like the Afh 'till the
next Spring after, I cannot yet learn.

They delight to grow on Hills, and in Woods, *Soil.*
and in any dry Ground, and grow more commonly
in the Northern than Southern Parts of *England*, as
upon the coldeft, bleak, and moft expofed of the peak
Hills in *Derbyfhire*, where they thrive without fhelter
upon dry ftony Land, which makes me think them
a very hardy Tree, and worth propagating.

I fuppofe in Tranfplanting of them, they may be *How*
ordered as the Afh. Whether they will grow to the *tranf-*
fame Stature, I cannot determine, becaufe I never *planted.*
faw any of them but on the barren peak Hills. Mr.
Evelyn fays, they plant them every where in the
Church-yards in *Wales.*

It

Use.   It is a very tough Wood, and all heart, being good for the Wheel-wright, and all forts of Husbandmen's Tools. If large, 'twill make good Plank-boards and Timber. Mr. *Evelyn* fays, it is commended by our Fletchers for Bows, next unto Yew; that the Berries fermented by themfelves, if well preferved, make an excellent Drink against the Spleen and Scurvy; and that Ale or Beer brewed with them makes an incomparable Drink, which he fays is much ufed in *Wales.*

Chap. XVIII. *Of the Birch.*

THE *Birch* is a very common Tree, and needs no Defcription, being to be found almoft in all Parts of *England.*

How raifed.   It increafeth commonly from the Roots and Suckers, though it bears Seeds which it fheds in the Spring; but whether the Seeds will produce them, I have not yet heard of any that have made the trial.

Soil.   It affects moft a dry barren Soil, where hardly any thing elfe will grow, and will thrive on any fort of Land, let it be wet, dry, fandy, gravelly, rocky, or boggy, and the barren heathy Lands that will hardly bear any Grafs.

How tranf. planted.   The beft way of tranfplanting of them, is to remove the Suckers that have Roots to them, which cut off about three or four Inches long, and plant as you do Qnick, from which will come many Shoots, which you may let grow for Underwood, or reduce them to one Stem, which in a few Years will make it fit for the Turner.

Use.   *Birch,* though it is the worft of Timber, yet it is of ufe for many Occafions, as for Ox-yoaks, Hoops, Screws, Wythes for Faggots, Brooms, &c. and for Difhes, Bowls, Ladles, and other Utenfils; especially the Roots, of which, in *Ruffia,* they make very fine Bowls and Difhes that are very tough, and not fubject to fplit, covering their Houfes with the Bark,

of

of which the *Indians* in the Northern Parts of *America,* as *F. Hennepen* says, make Canoes that are very swift and large; and of several other Parts of the Tree, they make fine Baskets, Boxes, &c. It is good Fuel, and makes very good Charcoal, and some say that the Bark will tan Leather quicker and better than that of Oak, besides the Wine made of the Sap, which I shall refer to another place.

Chap. XIX. *Of the Hazel.*

THE *Hazel* is of several kinds, and differs both as to the Leaf and Nut, even amongst the wild kind, without reckoning the several sorts of Filberts, and the large *Spanish* Nut, the Fruit of which last is much improved by Transplantation and Grafting.

The Hazel is best raised of Nuts, which you may *How rai-* sow like Mast in a pretty deep Furrow; the best *sed.* time for the doing of which is in *February,* because Vermin are very great devourers of them. They are very much prejudiced by Frost, 'till which is over they should be kept moist in their own Leaves or in Sand, and not suffered to Mould. They may likewise be raised by Layers and Suckers from their own Roots, which they are apt to put forth in great plenty.

They will grow on any cold, dry, barren Soil, that *Soil.* is either sandy, gravelly, or chalky, and also on Mountains and Rocks, but they thrive best on moist Bottoms, sides of Hills and Banks; and the Filbert loves a rich black Mould.

They are commonly transplanted small, being sel- *How* dom removed, but to fill up or thicken Woods with, *trans-* or for Hedges, being cut like Quick-set, about six or *planted.* seven Inches long.

The use of the Hazel is for Hoops, Poles, Spars, *Use.* Rake handles, Angle-rods, Fuel-bands, Hurdles; and Mr. *Evelyn* says, that the Chips of it are the best Wood of any to fine Wine with.

Chap.

Chap. XX. *Of the Poplar, Aspen, and Abell.*

THESE three sorts of Trees are much of a
kind; only the *Poplar* is esteemed of three sorts;
the white Poplar which is the most common amongst
us; and the water Poplar, the Leaf of which is of a
pale green Colour, shaped something like the other,
but is not so white underneath; and the black Poplar.

The *Aspen* or Asp-tree hath Leaves much the same
with the Poplar, only much smaller, and not so white.

The *Abell* is a kind of white Poplar, only much
finer, bears a larger Leaf, makes a much stronger
Shoot, and is a much quicker grower; the best sort
of which comes from *Holland* and *Flanders*.

*How rai-
sed.* They may be raised by Layers or Suckers taken
from the Roots, which they are very apt to put
forth; especially if any way lopped or cut down,
often to the prejudice of the Land they grow in,
especially if it is any thing good; so that a small
Place inclosed where they grow, will furnish you with
Sets enough, though I think those raised by Layers
from a Mother-plant make the best Trees.

Soil. They will grow on any sort of Land, wet or dry,
but thrive best on a rich moist Soil, especially the
water Poplar. It is esteemed one of the quickest
growing Trees that is; especially the *Abell*, which is
one of the best Trees to plant where you desire a
speedy Shelter and Walks; they many times making
Shoots of eighteen or twenty Foot long in a Year, and
any sort of Trees or Shrubs will thrive under their
shade.

*Trans-
plantati-
on.* It is a Tree that bears Transplantation well, and
may be planted out small like Quick, cutting of them
two or three Inches above the Ground, which you
must keep clean weeded, and you may prune up the
most thriving Shoots for two or three Years. It
will make a fair Standard; or you may transplant
them when pretty large, cutting off their Heads,

which

which they will quickly recover, and fet them at ten or twenty Foot diftance, but it is a Tree that doth not grow to any very great Age.

The Timber is very good for all forts of white *Ufe.* wooden Veffels, as Trays, Bowls, and other Turners Ware, Bellows-makers, Ship-pumps, Wooden-heels, Lafts, Carts, Hop-poles, *&c.* and makes good Timber for Building where it lies dry, and very good Boards, and is ferviceable for Fuel ; great quantities of it being rais'd in places where Wood is fcarce; for the lopping the Boughs of which the beft time is *January.*

The Leanes of Poplar are good for Cattle, which may be ftripped from the cut boughs before they are faggeted, which fhould be done in the decreafe of the Moon in *October,* and referved in Bundles for winter fodder.

Chap. XXI.   *Of the Alder.*

THE *Alder* is reckon'd chiefly to be of two kinds ; the common fort which only affects moift Ground, and the blacker fort which thrives better on drier Lands.

The Alder is propagated by Truncheons, and, *How rais-* fome fay, may be rais'd by Seeds; but large Roots *fed.* or fmall Suckers, which they put forth very plentifully, are the beft to raife them of.

They Thrive moft in the moifteft boggy places *Soil.* where nothing elfe will grow, and are a great Improvement of fuch Lands; and will grow on the fides of Rivers and Springs.

The beft way of removing them, is to tranfplant *How* the Suckers, which trim up to one Branch ; and after *trranf-* they have ftruck Root, you may cut them within three *planted.* or four Inches of the Ground, which will caufe them to fpring in Clumps and to increafe their Roots and Suckers, or you may tranfplant the Roots which will fpring into Branches ; and if you raife them of Truncheons, fteep one end of them in Water fome time before you plant them, making of holes for them, and not forcing them into the Ground to ftrip up their Bark; the beft time of doing which is in *February* : And as they are to be planted in moift places or near Springs, it will be good to plant them deep, that the Streams may not wafh away the Roots, and to preferve them fteady from the violence of the Winds. If you have occafion to cut or prune them, let it be done in *February*, which is the beft time for all Aquaticks or foft Woods, becaufe they will be the lefs time expos'd to the wet of the Winter.

Ufe.   It is excellent Timber for Water-works where it may conftantly lie cover'd. It will grow as hard as a Stone. The Coal is good for Gun-powder, and the Wood for Piles, Pumps, Hop-poles, Water-pipes, Troughs of Sluices, Wheels, &c. and is in much requeft with the Turner. The Bark is much fought after by the Dyer, and fome Tanners, and mix'd with the Fruit, or macerated in Water with a little rufty Iron, it makes a black Dye, and may be us'd for Ink. Faggots made of it, thrown upon Flints, Brick-bats and other Rubbifh that is laid in Drains, and Earth thrown on them, will laft a great many Ages.

Chap. XXII. *Of the Withy, Sallow, Ozier and Willow.*

EACH of which kinds (there being feveral forts) I fhall not particularize, becaufe they may be all propagated the fame way, and delight in the fame Soil, efpecially the Withy, Sallow and Willow.

How rai-   They may be rais'd of Cuttings ftuck in the Ground,
fed.   or large Truncheons of eight or ten Foot, long, where they are in danger of Cattle coming at them, becaufe they will by their heighth be the better fecur'd from them, by putting of Bufhes or other Fences about them; only you muft obferve when they put

their

their Buds out firſt, to rub off all the Under-buds, leaving only a few near the top to draw up the Sap.

They may likewiſe be rais'd of Layers, by which means, you may thicken your Woods, and have Plants to ſet out, which if rooted, are much better to plant on dry Ground than Sets. They may be rais'd of Seeds; but as they ſeldom come to be ripe in *England*, and the other ways of raiſing them being eaſie, it will be but a needleſs Experiment to endeavour the propagating of them that way.

All the ſeveral ſorts delight in a moiſt Soil, as I *Soil.* ſaid before; but the Withy and Sallow, will grow on the drieſt Land, and the Willow in Banks near Rivers and moiſt Ditches, where they can reach the moiſture with any of their Roots; but the Ozier will grow the beſt on the moiſteſt Lands, eſpecially ſuch as are overflow'd with the high Tides, and left dry at the Ebb.

The beſt time to tranſplant them and to lop them, is in *February*, juſt before the Sap begins to riſe, eſpecially if you deſign the raiſing of them by Cuttings or Truncheons, thoſe Cuttings being eſteem'd the beſt which grow neareſt the Roots. In many Places where Wood is ſcarce, they make great advantage of them by planting them in the Quincunx Order at ten Foot diſtance; tho' I think fifteen or twenty Foot a better diſtance; but a fat rich Soil requires their being planted at greater diſtances than a more barren Soil doth.

It is a good Wood for fire, if kept dry, and is very *Uſe.* uſeful for Stakes in Hedges, to prevent Hedge-breakers pulling of them up, and likewiſe to thicken your Hedges; it is alſo good for Rake and Scythe-handles, Heels, Clogs, Pattons, Hurdles, Seives, Lattices, and for ſeveral Uſes of the Turner, &c. and the Oziers for Baskets, Panniers, and ſeveral other Utenſils, which makes the Ozier Ground of very great Value, even beyond that of Wheat; many Ozier Grounds being Let for ten Pounds *per* Acre, which conſidering

their

their conſtant Crop, the ſmall charge that attends them, and the little pains that is taken to renew any of the old Plants when they decay, which is done by only cutting off a piece of Ozier and ſticking of it in the Ground, makes it one of the greateſt Improvements that is of moiſt Lands.

Chap, XXIII. *Of the Fir, Pine,* &c.

FIRS are of ſeveral ſorts, as the *Norway,* Spruce *Dram,* Scotch; but the beſt ſort both for Beauty and Timber, is that which they call the Silver-Fir, becauſe the under-ſide of the Leaf is of a white Colour, and the Leaf is longer than any of the former; except the *Scotch* Fir, which, whether they may not rather be eſteem'd a Pine than a Fir is diſputable. This Silver-Fir, being, I ſuppoſe, the ſame as Mr. *Evelyn* calls the *Spaniſh* Fir; it grows to the greateſt highth of any of the ſorts of Firs which we have in *England,* and is of great value for Maſts of Ships.

How rai- The common way of raiſing the ſeveral ſorts of Firs,
ſed. is of Seeds; the beſt way to get which out of the Cone or Clogs, is to lay them in the Sun, which will quickly occaſion them to open; or in Water a little warm, you muſt take care to gather them before they open, which they commonly do in *April* or *May,* when the Weather begins to be hot, it being the ſecond Year before they are ripe; they may likewiſe be rais'd by Slips or Layers interr'd about the latter end of *Auguſt,* and kept moiſt. The Silver-Fir, the Gard'uers tell you, is to be rais'd no other way than by Seed, tho' I am of another Opinion in that I could never produce them that way, and I am rather inclin'd to think them produc'd of Layers; but this is a Secret I could never yet diſcover. As to the Seeds, do not take them out of the Cone or Clog till you uſe them, in which they will keep good two or three Years. The beſt time to ſow them is in *April* or *May*; they ſhould be very carefully kept from the
Mice

Mice before they get up, who are very desirous of them ; and when they begin to peep, shelter them with Furz, or such sort of Fence, from the Birds, who are very apt to pull them up by taking hold of the Cap, which they commonly bear upon their tops when they first spring. The Beds wherein you sow them, had need be sheltred from Southern Aspects. Sow them in shallow Rills, not above half an Inch deep, and cover them with light Mould ; and observe that those Seeds which bring up the Shell of the seed on their Heads, will either not grow at all ; or but with Difficulty, if the sharp end of the Seed, be set downward ; because in that position, it must turn it self before it can get out of the Ground ; for they shoot first from the sharp end. They will peep in about five or six Weeks time, and being risen two Inches high, establish their weak Stalks by sifting of some more Earth about them ; for being heavy they are apt to swag so as often to blow out of the Ground. When they are of two or three Years growth, you may transplant them where you please ; and when they have gotten good Root they will make very large Shoots, but not for the first three or four Years. When you Sow any Seeds of Fir, Pines, &c. there is some times a worm which will destroy them at first coming up. Which, when you find, take some Tobacco dust and lay it in a small Circle, round the Plant, and I am told it will prevent their medling with them.

They will grow on any dry Soil, especially the light *Soil.* hazelly Brick-earth, and refuse not moist barren Gravel, or any sort of poor or rocky Grounds or Clays, except they are too moist and spue, tho' the *Scotch* Fir, delights in a moist Soil, and grows, as I am told, very well in the boggy part of *Ireland.*

The best time to transplant them, is from the mid- *How rain* dle to the latter end of *August* and *March*, or as some *sed.* say, at the several times of the fall of their Leaves. Remove them with as much Earth about the Roots as you can, tho' the Fir will bear a naked Transplanta-

E 3   tion

tion better than the Pine. You may tranfplant them
from the place where they are rais'd of Seeds; but
if you defign to plant them where the Cattle come,
you muft firft remove them into a Nurfery, where
you may let them ftand till they are eight or ten Foot
high, and then plant them out, obferving to water
them well, and neither to bruife nor cut the tops;
and when you prune them, leave the Stories about a
Yard afunder; the beft time of doing of which, is in
the beginning of *March*, and be careful to rub a little
dry Earth upon the Wound where you cut them, to
ftop the Turpentine, and to prevent their fpending
of themfelves too much, which thefe Trees are very
fubject to do; and it will caufe them to grow taper,
to fhoot in heighth, and to let the Wind thro' them,
which may prevent their being blown down, and the
breaking of their Boughs, which thefe Trees are very
liable to during the Winter Gufts; and therefore,
where you plant any of them, I fhould advife the
planting of other Trees round them to fhelter them,
but be fure, on whate'er Soil you plant them, to let
no Dung touch either their Body or Root; tho' if laid
at a diftance it will advance their growth, and if you
plant them on Gravel-ground, Mud or Clay mix'd
with it will do well to temper it, and if you mix Sand
with Clay where the Soil is Clay, it will be better
than Dung; but the fheltring of them with fome Litter will do well to preferve them from the parching
heat in Summer, and the cold in Winter,

There are feveral forts of Pines, but amongft us
not above three or four forts; but as they are to be
rais'd of Seeds, and order'd the fame way, and delight
in the fame fort of Soil as the Fir, I fhall refer you to
the Directions of the Fir-tree for the ordering of
them. Only note, that no ever greens will bear Loping
well, tho' they will fpare many of their fide Branches
in *April*. If they are of the tendereft fort, cut them
three or four Inches from the body of the Tree, and
the next fpring, cut them clofe to the ftem, and cover

ver the place with wax, and well tempered clay, and it will heal them.

They do not bear Transplantation so well as the *How* Fir when large, and should be always removed with *transf.* as much Earth as you can, and both Firs and Pines *planted.* should be well water'd at first removal, and planted as shallow as their Roots will allow of.

There is likewise the *Piceaster* (a wilder sort of Pin,) out of which the Pitch is boil'd, which grows both in the cold and hot Countries; the Body of which being cut or burnt down, Mr. *Evelyn* says, will emit Suckers from the Roots, which neither the Fir nor Pine will do.

I need not say any thing of the Use of these Trees *Use.* for Building, Masts, &c. it being so well known to most that have any occasion.

Chap. XXIV. *Of the Larch, Platanus, Lotus and Cornel Trees.*

THE *Larch* or *Larix* Tree, Mr. *Evelyn* says, may be rais'd of Seed, and that it will grow in *England*, which he commends much for the largeness, durableness and usefulness of its Timber, and gives you an account of several Buildings in *Italy* made with it, which he says no Worm will touch nor hardly any fire burn.

The *Platanus* is a very beautiful Tree, and grows very well in *England*. It may be had at most of the Gardeners near *London*. It's rais'd by Seeds or La-yers. It affects a moist Soil, and should be well wa-ter'd when transplanted. It is a fine Tree for Walks, and 'tis pity it is not more propagated.

The *Lotus* is a Tree frequent in *Italy*, that affords a fine shade, and very durable Timber. It affects a moist Soil, and the Roots of it are very fine for Hafts of Knives, and other Tools; and of the Wood are made Pipes and Wind Instruments.

The

The *Cornel* is a very durable Wood ; and, as Mr. *Evelyn* says, is useful for Wheel-work, Pins, Wedges, &c. and will grow to a good stature with us. Its Berries are commonly very much used for Preserving and Pickling.

Chap. XXV. *Of the Cypress-Tree.*

THE *Cypress* is of two sorts ; the Sative or Garden-tree, which grows in the best form, and is the most beautiful ; and that which is call'd the Male, which bears Cones, and is of a more irregular shape. This Tree is commonly much prejudic'd by the staking and binding it up to a pyramidical Form, which heats the inward Branches for want of Air, and occasions their Moulding, hindring of their growth, and is very troublesome and chargeable ; whereas with clipping only, they may be brought to a much finer shape, and not be so liable to prejudice from their bandage, nor from the Frost. They also make fine Hedges when kept clipt, branching from the very Roots ; the best time of doing which is in *April* and *August.* If they run too much without branching from the bottom, a descreet cutting of the principal Stem may be of advantage to make them shew beautiful, but a good Management of them while young may prevent that occasion.

How rais'd They are rais'd of the Seed, Procure them in the Nuts, and when you have occasion to use them, expose them to the Sun, or put them in warm Water, and the Seeds may be easily shaken out. The time to sow them is in *April,* which do after this manner : Prepare a bed of fine Earth, and make it even, upon which strew the Seeds pretty thick, and sift more Mould upon them about half an Inch thick, keeping of them well water'd every Evening, except when the Season waters them, and after a Years growth you may transplant them : When they are come up well, be sparing of your watering of them.

They

They will grow on any dry Ground, even Gravel *Soil.*
and Sand, especially the Male sort, and will never be
injur'd by Frost if they are not planted in a cold moist
Soil.

The Timber is very lasting, and never cleaves but *Use.*
with great violence. The bitterness of its Juice preserves it from Worms and Putrefaction. It is the best
of all Timber for Building, and will last when either
wet or dry.

Chap. XXVI. *Of the Cedar.*

IT is a great pity the *Cedar* is not more propagated among us, being so easily rais'd, and a Tree
that will grow so well with us. They are of several
sorts and kinds, some of which do very much resemble the *Juniper,* which I cannot but think a Species
of it, and therefore it might do well to be better en-
courag'd where the Soil is proper for it, to see what
magnitude it will grow to. Other sorts of it are
more like Cypress, as (according to the *London* Gardeners Opinion,) those kinds growing in *New-England*
and *Virginia* are: But the Cedar of *Lebanon* bears the
severest Weather we have. I have rais'd several of
them of Cones I had from thence, and have now a
Walk planted with them; and wheresoever they are
to be had of any sort, the Seeds may be brought from
the furthest part of the World in the Cones, for I
had some two Years old that grew as well as those
that were brought me directly from *Mount Lebanus;*
and I am apt to believe, if they are kept in the Cones,
and not taken out till just you sow them, they may be
kept three or four Years without prejudice.

They are rais'd of Seeds which seldom fail of grow- *How*
ing if order'd right, and if care be taken to preserve *rais'd*
them from the Mice, who are very greedy of them.

They bear a Cone as the Pines do, but it is rounder and more like Scales; the same Observations are
to be minded in the gathering of them, as that of the
Pine;

Pine; only to open them, let them be steep'd in cold water forty eight Hours, and the Seeds set as soon as taken out. The time of the setting of them is about the latter end of *March*, which sow on a Bed of good rich Mould, and lay it at least two Foot deep, but let no Dung come near them; and if your Bed be made a little sloping, it will do well, that the water may run off from them, for too much wet is apt to burst the Seed. They cannot well stand too dry, if they are but shaded in dry Weather. As they come up, sift Earth about them to establish their Roots, as is before directed about the Pine.

Soil.　　They delight most in a rich dry Soil, but they grow very well with me in *Essex*, both on the hazelly Brick-earths and on Gravel, that hath something of good Mould about a Foot and a half deep on the Surface of it.

How transplanted.　The best time of transplanting of them is at three or four Years old from the place where you rais'd them of the Seed. If the first Year you water them with a Lift, it may do well, and be of great advantage to them. Whether they may be removed at a larger growth, I have not experienc'd, and so can say nothing of it. They grow but slowly the first seven or eight Years, but I am told that after that, they grow with as much speed as most other sorts of Trees do.

Use.　　I need not say much of the Usefulness of the Timber, being so much known. The fragrancy of which, its fine Grain for all sorts of Work, and its durableness being able to recommend it for all Uses, besides the stateliness of it for Walks and Avenues, several of them being reported to be two hundred Foot or more in heighth. A Friend of mine assur'd me that he cut down one in *Barbadoes*, that had above four hundred Foot of Timber in it; but I am told there are some of a far greater magnitude.

Chap.

Chap. XXVII. *Of the Cork, Ilex,* &c.

THE *Cork-tree*, with us, is of two forts (and there are divers other Species in the *Indies*) one of which is of a narrower lefs jagg'd Leaf than the other, being a conftant green, whereas the other is broader and falls in the Winter, it grows near the *Pyrenaan* Hills, and in feveral parts of *Italy* and the North of *New England*, efpecially the latter fort, which is the hardieft and beft for our Climate; and that upon the worft of Soils, as dry Heaths, ftony and rocky Mountains, where there is hardly Earth enough to cover the Roots. They may be had of the Gardeners at *London*, and I am told they grow very well with us, and bear our fevereft Winters.

This Tree hath three Barks, on the outer of which is the Cork, which they ftrip once in two or three Years in a dry Seafon, becaufe the wet is apt to prejudice the Tree; and one of the other Coats being red, when they fell the Tree, bears a good price with the Tanner. The Wood is good for Fire, and ufeful for building Pallifado Work, &c.

The *Ilex*, or great fcarlet Oak, thrives well in *England*. They are a hardy fort of Tree, and eafily rais'd of the Acorn. If we could have them brought to us well put up in Earth or Sand, they might be brought any where from Foreign Parts. The *Spaniards*, Mr. *Evelyn* fays, have a fort they call *Enzina*, which bears Acorns, of which they have profitable Woods and Plantations: The Wood of which, when old, is finely chambletted as if it were painted; which Wood is ufeful for Stocks of Tools, Mallet-heads, Chairs, Axle-trees, Wedges, Beetles, pins and Pallifadoes for Fortifications, being very hard and durable. Of the Berries of the firft fort is extracted the Painters Lac, and the Confection of Alkermes. Their Acorns are good Food, little inferiour to Chefsnuts: But the *Kenne* Tree doth not always produce the *Co-*
cum,

cum, except it grow near the Sea, and where it is very hot, and therefore they frequently cut down the old Trees that they may put forth fresh Branches, upon which they find them.

Thuya, or *Arbor vitæ*, grows of Layers or Slips to a tall ſtraight goodly Tree, hardy in all Seaſons. The Wood makes incomparable Boxes, Bowls, Cups, and other Curioſities ; and the Leaf makes one of the beſt Oyntments for green Wounds, that is, cloſing of them ſuddenly.

Box. *Box* deſerves our Care becauſe of the excellency of its Wood, and in that it will proſper on the declivity of cold, dry, barren Chalky-hills, where nothing elſe will grow ; of which there are two ſorts, the dwarf Box, and a taller ſort that grows to a conſiderable Heighth. The dwarf Box is very good for Borders, and is eaſily kept in order with one clipping in a Year. It will increaſe of Slips ſet in *March*, or about *Bartholomew-tide*, and may be raiſed of Layers or Suckers.

It is of uſe for the Turner, Ingraver, Carver, Mathematical Inſtrument-maker, Comb-maker, &c. for which they give great Prices by Weight as well as Meaſure, eſpecially the Roots (as even of our neglected Thorn) which is of great value for Inlaying and Cups.

Yew. *Yew :* Since the uſe of Bows is laid aſide, the Propagation of this Tree hath been neglected, though it will grow on our coldeſt and barreneſt Hills, eſpecially if chalky, where it may be of uſe to propagate it for the ſame uſes as Box, for moſt of which Purpoſes it is as good, beſides which it makes extraordinary Axle-trees.

It is eaſily produced of the Seeds, which muſt be waſhed and cleanſed of their Mucilage, and then buried in Sand made a little moiſt any time in *December*, and ſo kept in ſome Veſſel in the Houſe all Winter, and in ſome ſhady cool place abroad in Summer. The Spring come twelve Months after you have put
them

them in Sand, fow them on a Bed, the Ground not too ftiff. Some bury them in the Ground like Haws. It is commonly the fecond Winter before they peep, and then they rife with their Caps on their Heads. At three Years old tranfplant them. They may likewife be raifed by Layers or Slips, and fo planted out for Standards, Walks, and Hedges, being to be clipped in what Form and Order you pleafe, and therefore are much valued by our modern Planters, to adorn their Walks and Grafs-plats.

Juniper is of three forts, whereof one is much taller than the other, the Wood whereof is yellow, and if cut in *March* is fweet like Cedar, of which this is accounted a fpurious kind. They fhould neither be fhaded much nor dropt upon. They may be raifed of Seeds, which will peep in two Months after fowing. They fhould neither be watered nor dunged ; and being managed like Cyprefs, will make fine Standards ; efpecially where they are not obnoxious to eddy cold eafterly Winds, which are apt to difcolour them, but they foon recover it again. To make it grow tall, prune it clofe to the Stem, and loofening the Earth about the Roots, haftens its growth much. It may be clipt for Hedges. It loves a gravelly Soil, and is raifed moftly of Seed.

Chap. XXVIII. *The Laurel.*

THE *Laurel* or Cherry-bay is moft commonly ufed only for Hedges, but being planted upright and kept for Standards by cutting away the fide Branches, fo as to maintain only the principal Stem, it will rife to a large Tree, and carries a fine fpreading Head, that is very ornamental and fhady for Walks or Groves, and may this way be of much better ufe than to plant them in Hedges, as moft do, where the flower Branches growing fticky and dry, by reafon of their frequent and unfeafonable Cuttings (the Genius of this Tree being to fpend much in

Wood)

Wood) caufes them never to fucceed after the firft fix or feven Years, but are to be new planted again, or abated to the Roots for a frefh Shoot.

They are raifed of the. Seeds or Berries with extraordinary Facility, or propagated by Layers, Slips and Cuttings fet about the latter end of *Auguft*, or earlier at St. *James's-tide* in a fhady moift place, they delighting moft in a moift cool Soil.

BOOK XII.

Chap. I. *Of Coppices.*

S for Coppices or Underwood, I have already fhewed how each particular fort of Tree is to be raifed both by fowing and planting, and for the raifing of Coppices, great Care ought to be taken that the Wood they are to be compofed of, be fuch as is proper for the Soil you raife them on, and that the fort of Wood is proper for fuch Ufes as you defign to fell your Wood for, which you muft be regulated in by the vent you have, as whether it is for Fire-wood, for which the Oak, Horn-beam, and other hard Wood is beft ; or for Hoops, Hop-poles, *&c.* for which the Afh, Chefs-nut, Oak, Hazel, *&c.* is the moft ufeful, as I have already fhewed ; and according to the Profit of your Under-wood, regulate the thicknefs of your Stand-ards, which as they are thicker or thinner, do more or lefs Injury to your Under-wood. You are like-wife to confider at what Growth you can fell your

Under-

Under-wood, only remember, that the older and taller your Under-wood is, the better it is for Fire-wood, and the better it is for what Standards you leave, because they will be the taller and straighter by being forced up by the Wood that grows about them; though a deep Soil, as I observed before, con-tributes much to their spiring, and according to the time of your felling, it is necessary to lay out your several Falls, that so you may have an annual Suc-cession to yield a yearly Profit, which in many Places is from eight Years to twenty or thirty: But though the seldom felling of Wood yields the more and the better Timber, yet the frequent cutting of Under-wood makes it the thicker, and gives room for the Seedlings to come up. If many Timber Trees grow in the Coppice which are to be cut down, fell both them and the Under-wood together, cutting off the Stubs as near the Ground as may be, and the Stubs of the Under-wood asloap and smooth, and not a-bove half a Foot from the Ground, and stock up the Roots of the Timber Trees, if they send forth no Shoots, (which they are not apt to do if sawn down, which is the best way of felling of Timber Trees) to make way for Seedlings and young Roots to shoot; but where you design to sow Seeds, you must prepare your Ground with good Tillage, as much as you do for the sowing of Barley; and about *February* sow them, and if the Soil be shallow, plow your Ground into great Ridges, and it will make the Soil lie the thicker on the top of each Ridge, by which means the Roots will have the more depth to search for Nourishment, and the Furrows will in a little time be filled up by Leaves, which when rotten, will lead the Roots from one Ridge to another; and if you sow them on the sides of Hills that are dry, plow your Ridges cross the descent of the Hills, that the Water may be kept on the Land without having too sudden a descent; and if your Ground be very wet, observe just the contrary. Some sow their Seeds with a Crop

of

of Corn ; but as the Season for the sowing of Corn is too late for the Seeds, it is better to sow them by themselves, and be sure to keep them well weeded the first and second Years. But if you have a mind to raise Wood on very barren dry Land, sow it with what Fruit or Seeds you design it for, and with them sow Furz or such Trumpery, as will grow on the worst Land, and it will become a shelter to your Trees, which when they have once taken Root, will soon out-grow the Furz, and kill it with their dropping. For the raising of Coppices, the nearest distance for the Plantations ought to be about five Foot for the Under-wood ; but as to what number and scantlings of Timber, you are to leave on each Acre, the Statutes direct, and it is an ordinary Coppice which will not afford three or four Firsts, that is, Bests, fourteen Seconds, twelve Thirds, and eight Wavers, &c. according to which Proportion the sizes of young Trees in Coppices are to succeed one another. By the Statute of the 35th of *Henry* the VIII. in Coppice or Under-wood felled at twenty-four Years growth, there were to be left twelve Standlings or store Oaks upon each Acre ; and in defect of so many Oaks, the same number of Elms, Ash, Asp, or Beech, and they to be such as are likely Trees for Timber which are to be left, and so to continue without felling, 'till they are ten Inches square within a Yard of the Ground. In Coppices above this growth when felled are to be left twelve great Oaks, or in defect of them, other Timber Trees as above, and so to be left for twenty Years longer, and to be enclosed seven Years.

However, I think it better to leave a much greater number of Timber Trees, especially where Underwood is cheap ; and as to the felling, begin at one side, that the Carts may enter without detriment to what you leave standing ; where your Woods are large, it is best to have a Cart way along the middle of them, by which means you may fell on each fide where you will, and have a Cart-way always ready without prejudice to the reſt of the Wood. The Under-wood may be cut from the beginning of *October* to the latter end of *February*, but *February* is the beſt Month to cut Wood in, where you have but a ſmall quantity to fell, that you may do it before the Spring comes on too much : Take great care to prevent the Carters bruſhing againſt the young Standards; and let all your Wood be carried out by Mid-ſummer, and made up by the end of *April* at the lateſt ; for where the Rows and Bruſh lie longer unbound or unmade up, you ſpoil many of the Shoots and Seedlings. If the Winter before you fell, you incloſe it well, ſo as to keep all Cattle out of it, it will recompence your Care and Trouble.

By the Statute Men were bound to incloſe Coppicewood after felling, if under fourteen Years growth, for four Years ; thoſe above fourteen Years growth to be ſix Years incloſed ; and for Woods in Common, a fourth Part to be ſhut up, and at felling the like Proportion of great Trees to be left, and ſeven Years incloſed ; this was inlarged by the 13th of *Elizabeth*. Your elder Under-wood may be grazed about *July*, or in Winter ; but for a general Rule, newly wean'd Calves are the leaſt prejudicial to new cut Wood where there is an abundance of Graſs, and ſome ſay Colts of a Year old, but then they muſt be drove out at *May* at fartheſt ; but if nothing at all be ſuffered to come in, it is better, every Man's Experience being able to direct him.

If your Woods happen to be cropt by Cattle, it is beſt to cut them up, and they will make freſh Shoots ; whereas what is bit by the Cattle, will elſe ſtunt for ſeveral Years before it will take to its growth.

If your Woods are too thin, lay down Layers of the longeſt and ſmalleſt Shoots you can find of ſuch kinds of Wood as you like beſt to have your Coppice of, or that is neareſt to the bare place where you want a ſupply, according to the Method already proposed for the laying of Layers of Trees, and they will send forth abundance of Suckers, and thicken and furnish a Coppice very speedily.

As to the size of Faggots and Wood-stacks, &c. it differs in most Countries, and therefore you must in all those things be guided by the Custom of the Country where you live; the Prices of which, and the stacking up of Wood, Roots, Stumps of Timber Trees, &c. I shall give you an Account of hereafter, when I come to consider the Prices of the Husbandman's Labour and Charges. I shall at present only Note one thing, and that is, that when the Workmen have bound up the Faggots, with their Bills they trim off all the straggling small Branches to make the Faggots more neat and tight, which Trimmings they commonly gather up and put into the middle of the next Faggot, where it is of little Advantage, but would be of much greater Profit to the Land, if it were left to rot in the Wood, for which it is as good as Dung, and would much advance the growth of your Trees, as I have known by Experience; for though the Leaves falling and rotting in Woods do much improve them, yet it is not to be compared with the Advantage that they receive from rotten Wood, which will turn any Soil whatsoever into a rich black Garden Mould, as may be found by Experience by any that will make Observation of it where Wood-stacks have stood; and though those Sticks are but small, and cannot do much the first time, yet a constant Repetition of it every Fall is a much greater Improvement of Wood than can easily be believed.

The best time to fell Timber is in _January_ or _February_, because the Sap is then all down in the Root, but the Oak they commonly fell about _April_ or _May_, when the Bark will run, which they are obliged to do by the Statute, because of the Bark for the Tanner, which is a very great prejudice to the Timber: But the Opinion and Practice of Men have

been

been very different concerning the best time to fell
Timber. *Vitruvius* is for an Autumnal Fall ; *Cato*
was of Opinion, that Trees should not be fell'd till
their Fruit was ripe, and tho' Timber unbark'd be obnoxious to the Worm, yet we find the wild Oak and
many other sorts of Trees fell'd late (when the Sap begins to be proud) to be very subject to the Worm too;
whereas being cut about Mid-winter, it neither casts
Rifts nor Winds, because the cold of the Winter
both drys and consolidates it.

Some Authors advise in felling of Timber to cut it
but into the Pith, and so let it stand till it be dry, because, say they, by drops there will pass away that
moisture which would cause Putrefaction ; others advise, to bore a hole in it with an Auger for the same
Purpose, but I suppose a nipping Frost will effect the
same, by causing the moisture to descend into the
Root, not that I would have them fell'd in frosty
Weather, but not fell'd till a hard Frost hath been
upon them. In *Staffordshire*, they bark their Timber
Trees in *April*, when the sap will run ; and the
next Winter about *December* or *January* ; they fell
them. This they say makes the Timber firm and good,
and not subject to the worms.

When the Stubs of your Under-wood are great,
stock them up, this is a good piece of Husbandry, because it makes way for Seedlings and young Roots that
are thriving, whereas when the Stumps are old and
large, they are apt to let in the Water and be unthrifty ; the time of doing which is in the Winter Season.

When you fell your Woods, leave young Trees
enough, you may take down the worst at the next
Fall, especially if any grow near a great Tree; that
you think may be fit to fell the next Season, to supply
its place, because several may be spoil'd by its fall.

When Trees are at their full growth, there are several signs of their decay, as the withering or dying
of any of the top Branches ; or if they take any Water
in at any Knot ; or are any ways hollow or discolour'd ;

it

if they make but small Shoots; if Wood-peckers make any holes in them; also a very spreading Tree in a Wood, is many times very prejudicial, because of the young Trees it drops upon, according to the Directions of the Poet.

> *To fell thofe Trees can be no lofs at all,*
> *Whofe Age and Sicknefs would your Ax foreftall;*
> *A youthful Succeffor with much better Grace,*
> *And Plenty will fupply the vacant Place.*

Lastly note, That if you sell your Wood by the Acre, you must take great care before-hand to mark out what Standards shall be left, or else the Wood-Buyers will be very apt to deceive you; and observe that all Wood-Lands are measur'd by the eighteen Foot Pole.

Chap. II. *Of Transplanting of Trees.*

THE smallest Trees, and those that have been transplanted in Nurseries, &c. are the best to remove, as I have already observ'd; but as the removing of Trees is commonly upon the account of the making of Walks, Avenues, Groves, or to fill up Hedge-rows where Cattle come, it's neceffary that they should be of a size so big, as with some shelter they may be out of danger of being spoil'd by Cattle; for which purpose, I reckon Trees that are of about five or six Inches in Circumference, and six Foot and a half or seven Foot high, to be the best, and the best size both upon account of the Trees and of the Cattles reaching to crop them. In cases of neceffity, Trees of very great Stature have, according to the account given by Mr. *Evelyn*, been remov'd upon particular Occasions; the way of doing of which, tho' it is too troublesome for ordinary planting, yet as it may be of use upon some Occasions, I shall propose it according to the Method he has laid down, which is,

To

To chufe a Tree as big as your Thigh, remove the Earth from about it, cut thro' all the fide Roots till you can with a competent ftrength enforce him down on one fide, fo as to come with your Ax at the Tap-root, which cut off, and redreffing your Tree, let it ftand cover'd with the Mould you loofen'd from it, till the next Year or longer, if you think fit, take it up at a fit Seafon, and it will have drawn new tender Roots fit for Tranfplantation.

Or elfe a little before the hardeft Froft furprize you, you may make a Trench about the Trees at fuch diftance from the Stem, as you judge fufficient for the Root, which dig of a competent depth fo as almoft to undermine it by placing of Blocks and Quarters of Wood to fuftain the Earth ; this done, caft in as much Water as may fill the Trench, or at leaft fufficiently wet it, unlefs the Ground were very wet before ; let it ftand till fome very hard Froft doth firmly bind the Earth to the Roots, and then convey it to the Pit prepar'd for its new Station, which you may preferve from freezing by laying ftore of warm Litter in it, and fo clofe the Mould the better to the ftraggling Fibres, placing what Earth you take out about your Gueft to preferve it ; But in cafe the Mould about it be fo ponderous as not to be remov'd by any ordinary force, you may then raife it with a Crane or Gin, and by this means you may tranfplant Trees of a large Stature to fupply any defect, or for the removal of a Curiofity : The beft fort of Trees to remove large is the *Elm*, efpecially if to be plac'd in a moift place.

The beft time for the removing of all Trees, ex-cept Winter Greens, of which I have particularly treated, is either in *October* or *February*, as I faid be-fore : But if the Soil be moift, 'tis better to plant in *March*, that fo the Trees may not ftand fobbing all the Winter to chill their Roots.. Though I have fe-veral times for a Curiofity remov'd fome fort of Trees at *Midfummer*, that have profper'd very well, which I did after this manner ; I made a hole large e-

nough

nough to contain the Roots of the Tree I defigned to remove, into which I pour'd Water, and in that Water I put the Earth I took out of the hole, fo as to make it a meer foft Sludge or Mud, and having taken up my Tree with as many Roots as I could, and abated the Head with the fame Caution, I plung'd the Root into the Mud in the hole, where I let the Tree ftand without taking any further care of it, except daily watering for about a Fortnight or three Weeks : And fuch Trees have grown as well as thofe planted in Winter, but then they were fmall Trees. But as for the common Rules of Planting ;

1. Obferve to fet your Trees deeper in lightGround than in ftrong, but fhalloweft in Clay ; fix Inches is fufficient for the drieft, and two or three for the moift, provided you eftablifh them from the Winds, and fhade them from the heat of the Sun ; the beft way of doing which is by Stakes, and round the Stem of the Tree to raife a fmall Hill about two Foot thick, and four or five Foot in Diameter, which cover with Stones, Tiles, or mungy Straw, to keep it moift, and to prevent the Weeds growing, taking care after a competent time to remove them, elfe the Virmin, Snails and Infects which they produce and fhelter, will gnaw and injure the Bark, and be fure not to plant any Trees deeper than they grow, before they were removed. Abate about half a Foot of the heighth of the Hill, every Year, till they become level with the reft of the Earth round the Tree, and carefully pull up what Weeds grow about them, becaufe they draw away the Heart of the Soil which fhould give Nourifhment to the Tree.

2. Where you dig your Holes for Trees, if it is in a Gravel Soil or Sand, mix Clay, or, which is better, Earth, Loam or Mud with the earth you fill into the Holes again ; and if it is a ftiff Clay, trench it with Straw, Thatch, Litter, Wood-Stack-Earth, &c. but let not the Roots touch any of thefe Mixtures, nor yet any Dung or Turf, but lay your Dung rather

round

round upon the Surface of the Earth, and dig it in a little, covering it with Mould to keep the Sun from drying of it. When you dig down your Hills, or dig about the Roots of your Trees, (which you should mind carefully to do once a Year, the Advantage of which I will prove to you afterwards) and wherever you plant Trees, make your Earth as fine as you can.

3. For Trees, or most sorts of Plants, the strong blue, white or red Clay are some of the worst Soils; but if any of these Lands have some Stones naturally in them, or the nearer they are to a Loam, by any mixture of Sand, they are much the better; so likewise is gravelly or sandy Grounds, the nearer they are to a Loam, by a mixture of Clay; for a Loam or light brick Earth, compos'd of a due mixture of Clay and Sand, I reckon to be the best Land for Trees.

4. Plant in a warm moist Season, the Air being tranquil and serene, the Wind westerly, but never when it freezes, rains, or is misty, for it moulds and infects the Roots; and if you Water any Trees you have new planted, it will settle the Earth the better to the Roots, and keep them moist; only observe, not to transplant if you can help it, any Trees after *Michaelmas*, till you have had some Rain to Moisten the Ground; because the Trees will rise with less Labour, and be better Root'd, the Roots being apt to break when the Ground is dry. Large Trees may sooner be removed in *October* than smaller ones.

5. Trees that have not been transplanted, or others that have, if their Roots go deep, you must have them abated, or else you will be necessitated to place them too deep; only the small fibrous Roots must be spar'd as much as you can, for they are what affords the chief Nourishment to the Trees, and take them up with as much Earth as you can, letting the Holes, into which you transplant them, be left open for the Rain, Frost and Sun to mellow the Earth sometime before you plant them.

6. If you take up a Tree, mind how the Roots grow, and difpofe of them in the fame order where you new place them, fpreading of the Roots carefully, obferving to place the Tree to the fame Afpeſt that it grew before.

7. In the Spring rub off the fide Buds to check the exuberancy of the Sap in the Branches, and to caufe it to run up to the Head.

8. Tranfplant no more Trees than you can fence well from Cattle; efpecially from Sheep, the Greafe of whofe Wool is very prejudicial to Trees, where they can only come at them to rub them. All young Trees fhould likewife be defended from the Wind and the Sun ; efpecially thofe of a tender fort from the *North* and *Eaſt*, till the Roots are fix'd, and that you find them begin to fhoot; the not exaſtly obferving of which Point is the caufe of the perifhing of moſt of our Plantations in Summer; and in Winter there is more danger to be fear'd from wet and cold, in Conjunſtion one with another, than the fevereſt Froſt alone.

9 Wood well planted will grow in moorifh, boggy, heathy and the ſtonieſt Ground; only the white and blue Clay is the worſt for Wood, as I faid before ; and what large Timber you find in either of them (the Oak only excepted) is of a very great Age.

10. If the Seafon require it, all new Plantations are to be well water'd, in *April* at their firſt budding, efpecially the firſt Year of their planting, upon which depends much of their future growth, and what Water you pour on them let it be in a Circle, at fome diſtance from the Roots, which fhould continually be bar'd of Grafs and Weeds ; and if the Water be rich, or impregnated with any Manure, the Shoots will foon difcover it, for the Liquor being percolated or ſtrain'd thro' the Earth, will carry the nitrous virtue of the Soil with it : By no means water at the Stem, becaufe it wafhes the Mould from the Roots, and lets the Water come too crude to them, which often en-

dangers

dangers their rotting. If your fear dry Weather, do not defer too long before you water your Trees or Seeds, but water while your Ground is yet moist that it may keep so.

11. Young Trees will be strangled with Corn, Oats, Peas, Hemp or any rank growing Corn or Weeds, if a competent Circle and Distance be not left (as of near a Yard or so) from the Stem.

12. Cut no Trees that have any large Pith in them, especially being young and tender, when either Heat or Cold are in their extreams; nor in wet or snowy Weather, tho' the discharging Trees of unthrifty broken wind-shaken Boughs is a very great advantage to them when done in a good Season. *Ever-greens*, especially such as are tender, prune not just after planting, but when you find by some small fresh Shoot they have taken Root.

13. If you plant Fruit-trees, reduce them to a Head as soon as you can. As for Timber-trees, it is best not to head them at all, but to shred them up to one single Bough, if the Soil be good that you plant them in; but if bad, the Sap will hardly run so high; and therefore in such case it is better to head them; and when they are shot out, reduce the Head to one single Branch; for which purpose, leave one of the most upright and thriving Boughs; and if your Top die, or your Tree meet with any prejudice from Cattle, so as to occasion its breaking out of the Sides, which impedes both its Growth and Spring, prune off some of the Shoots, and quicken a leading Shoot with your Knife at some distance beneath its Infirmity; but if it be in a very unlikely Condition at Spring, cut off all close to the Ground, and hope for a new Shoot, which nurse up by cutting away all superfluous Branches. If you would not have a Tree put forth side-Branches, prune them up in *February*, and whatsoever side-Branches it puts out after, cut off at *Midsummer*, when the Sap is in them, and they will hardly ever sprout again; tho' you must be cautious not to

cause

cauſe your Tree to have too great a Head for the Body; eſpecially if it is of a tender Wood, leſt the Wind break it; the way to prevent which, is lopping or thinning of the Head, or letting of the more ſide-Boughs grow out of it to check the Sap from running all into the top Branches.

14. *Wallnut, Aſh* and Pithy Trees you muſt by no means head when you tranſplant them; eſpecially the *Wallnut*; and if you have occaſion to lop off any of the Boughs, do it where they may be the leaſt expos'd to the wet, which I reckon the ſide-Boughs to be; and late in the Spring, as about the latter end of *February*, or beginning of *March*, that the Bark may the ſooner heal the Wound.

15. Trees will grow well upon almoſt any Soil, that is full of Fern; becauſe it ſhades the Roots and keeps them Cool; even upon the hotteſt burning Gravel.

16. If you plant on a ſhallow or a very moiſt Soil, plant it into large high Ridges, and plant your Trees on the Middle of the Ridges.

To preſerve Trees from Winds and Cattle that are expos'd to them, empale them with three or four quarter Stakes of a competent heighth, ſet triangular or quadrangular, and faſten them by one another with ſhort pieces above and beneath, in which a few Brambles may be ſtuck; except you will be at the charge of Pales, which will ſecure them without that fretting which Trees are otherways liable to that are only ſingle ſtak'd and buſh'd; but where Cattle don't come, a good piece of Rope ty'd about the Neck of the Tree upon a wiſp of Straw to preſerve them from galling, and the other end lightly ſtrain'd to a Hook or Peg in the Ground, will ſufficiently ſtabliſh the Tree againſt the Weſtern Blaſts, the Winds of other Quarters ſeldom doing them much miſchief. If the Cords are well pitch'd they will laſt many Years.

Chap. III. *Of Planting of Avenues, Walks,* &c.

MOST Walks fhould be made to lead to the Front of an Houfe, Garden-gate, High-way-gate, or Wood, or to terminate in a Profpect; in all which cafes moft People are apt to plant their Walks too narrow, fo as not to give a fair Profpect of what they are defign'd for; as fuppofe them planted for an Avenue to an Houfe, whatever the length of the Walk is, it ought to be as wide as the whole breadth of the Front; and if it be long, the wider it is the better: And for Walks to Woods, Profpects, &c. they ought to be at leaft fixty Foot in breadth; and becaufe fuch Walks are a long time before they are fhady, I would propofe to plant a narrower Row on each fide, rather than loofe the ftatelinefs that the main Walk will afford by being broad, efpecially where any thing of a Profpect is to be gain'd; and if the Trees are any thing of a fpreading kind, I would not have them planted nearer together than thirty five or forty Foot in the Row; and the fame diftance is to be obferv'd if they are planted for a regular Grove.

Chap. IV. *Of the Planting of Trees in Hedges.*

THE beft way of raifing Trees in Hedges, is to plant them with the Quick, if you can preferve them well from Cattle; but where Hedges are planted already, and Trees are wanting, I fhould propofe to plant them after this manner, as doing leaft damage to the Hedge, and as affording the beft Shelter, and giving room for the greateft number of Trees to be

planted,

planted, Let *a a* be the Bank that the Hedge ftands on, and *b b* the Ditch, and let all the Trees be planted

not on the Bank where the Hedge ftands, as the common way is, but at the bottom of the Bank, about a Yard from the Hedge; which will prevent their dropping on it, as at *o o ooo:* And over-againft them, on the other fide of the Ditch, about a Yard from it, not in a direct Line, but in the Intervals, let another Row be planted, as at *c c c c*; and if each other of thefe Trees be a fpring Tree, and the odd one between, a Fruit-tree to fpread, they may be planted the nearer together, and will afford the better Shelter.

Chap. V. *Of the Infirmities of Trees.*

THere are feveral Difeafes and Cafualties that fpoil Trees, and affect the feveral parts of them, that are carefully to be look'd after.

1. Weeds, fuch being diligently to be pluck'd up by hand after Rain as can be fo eradicated; efpecially while the Trees are young, and not able to over-drop them; but for the ftronger Weeds, they muft be extirpated with the Howe, Spade, or other Inftrument, being very prejudicial to the Trees in robbing them of their Nourifhment, and in choaking fuch as are young.

2. Suckers fhould be cut off clofe to the place they put out from, opening the Earth that you may come well at them; and if you find them rooted, you may fet them again: But they fay, Trees grafted upon them are more apt to produce Suckers than other Trees.

3. Over-much Wet often prejudices Trees, efpecially fuch kinds as require drier Ground, which is to be

be help'd, by Drains; and if a Drip fret the body of
a Tree by the Head (which will certainly decay it) cut
first the Place smooth, and apply to it, so as to cover
the Wound, some Loam or Clay mixed with Horse-
dung, which keep to it till a new Bark succeed, or
take refined Tallow, which mix with a little Loam
and Horse-dung newly made, and apply; only note,
that wounds made in a Tree in Winter, are much
harder to cure than those made in Summer.

4. If a Tree is Bark-bound, slit through the Bark
from the top to the bottom in *February* or *March*,
which will do most Trees good, but no harm to any;
and if the gaping be much, fill the Rift with Cow-
dung : Also the cutting off of some Branches is profitable, especially such as are any ways blasted or Light-
ning-struck ; and so is digging about the Tree, it being many times occasion'd from the baking of the
Earth about the Stem.

5. The *Teredo*, *Cossi*, and other Worms lying between the Body and the Bark, poison the passage of
the Sap, to the great prejudice of the Trees ; but the
Holes where they lie being found out, open them,
and make a small slit from the bottom of them to let
any moisture that may fall in them, run out, and do
the Place over with Loam.

6. Trees, especially Fruit-bearers, are often infected with the Measles by being burn'd and scorched
with the Sun in great Droughts. To this commonly
succeeds Lowsiness, which is cur'd by boring a hole
into the principal Root, and pouring in a quantity of
Brandy, stopping the Orifice up with a Pin of the same
Wood.

7 Excorticated and Bark-bar'd Trees may be preserv'd by nourishing up a Shoot from the Foot or
below the stripp'd place, cutting the Body of the Tree
sloping off a little above the Shoot, and it will quickly
heal and be cover'd with Bark like a Tree new grafted;
and if you cover the top with Clay and Horse dung in
the same manner as you do a Graft, it will help to
heal the sooner. 8. Deer,

8. Deer, Conies, and Hares, by barking of Trees, often do them very great Mischief, and many times destroy them quite. To preserve them from Deer, fence them with Pales; but to preserve them from Conies and Hares, Mr. *Evelyn* proposes the anointing of them with *Stercus humanum* temper'd with a little Water or Urine, and lightly brush'd on, this to be renewed after every Rain, or to sprinkle Tanners Liquor on them, which they use for dressing their Hides; also tie Thumb-bands of Hay or Straw round them as far as they can reach. I have not experienced any of these ways; but Tar and Lime, which I have known some use, will bind the Bark, and make it so hard that the Tree will not thrive.

9. Moss is to be rubbed and scraped off with some fit Instrument of Wood which may not hurt the Bark of the Tree, or with a piece of Hair-cloth, after a *Vid. Book* soaking Rain. But the most certain way to cure it *14. Ch. 8.* is by taking away the Cause; which is, to drain the Land well from all superfluous Water, and to prevent it in the first planting of your Trees by not setting them too deep. But Moss growing on Trees is of several sorts, cold and moist Ground produces a long shaggy, moist and dry Ground a short thick Moss. If the Moss is much and long, so as to smother the Branches, it may in such Case do well to prune off the greatest part of the Branches, and to Moss the rest, or to take off all the Head, and the Tree will shoot, and as it were become young again, and if your Plantation is too thick, which will in cold Ground occasion Moss, you must mend the fault by thinning of them, but if it proceed from the dryness of the Ground, open it, and lay Mud on it, which will both cool it, and also prevent the falling of the Fruit, and of its being Worm eaten, which is what is Incident to dry Grounds.

10. Ivy, Briony, Honey-suckles, and other Climbers, must be dug up, lest they spoil your Trees by pinching and making them crooked.

11. Wind-

11. Wind-shock is a Bruise and Shiver throughout the Tree, but not always visible, twisting the Warp from smooth-renting, being occasioned by High-winds, and perhaps by subtil Lightnings, those Trees being most in danger of it, whose Boughs grow more out on the one side than the other. The best prevention is Shelter, choice of place for the Plantation, and frequent shredding up while young.

But as the Winds often spoil Trees by twisting them, they many times do them as much Mischief in prostrating of them ; which, though it cannot properly be called an Infirmity of the Tree, yet the Winds are many times a principal Cause of rendring them infirm, for which there is no better Remedy than what is already proposed : But in case any Trees should chance to be blown down which you desire to preserve or redress, be not over-hasty to remove them, but cut off their Heads, and let them lie, and many times the weight of the Roots will bring them up ; but if not, take some of the loose Earth out of the Hole that the Tree hath made, and cut off some of the straggling Roots that hinder it from falling back, and you may easily redress them.

12. Cankers are caused by some stroke or galling, or by hot burning stony Land. They must be cut out to the quick, and the Scars emplastred with Tar mingled with Oil, and over that Loam thin spread ; or else with Clay and Horse-dung, but best with Hogs-dung alone bound to it with a Rag, or by laying Ashes, Nettles, or Fern to the Roots, &c. But if the Canker be in a Bough, cut it off; if a large Bough at some distance from the body of the Tree ; but if a small one, cut it close to it. But for over-hot stony Land, you must cool the Mould about the Roots with Pond-mud and Cow-dung: And for Fruit-trees, the best way to raise them on such Land, is to graft them on Crab stocks raised in the same Mould.

13. Hollowness is contracted by the ignorant or careless Lopping of Trees, so as that the wet is suffered

fered to fall perpendicularly upon any part of it, especially the Head: In this case, if there is sufficient found Wood, cut it to the quick close to the Body, so as to make it as sloaping as you can, that the wet may fall from it, and cap the hollow part with a Tarpaulin, or fill it with good stiff Loam, Horse-dung, and fine Hay mixed together. This is one of the worst Evils belonging to Trees, and what all soft Woods are very liable to if lopped; especially the Elm, which is much better to be shred up, the side Boughs of which will yield a constant Lop, and the Bodies afterwards be good Timber, whereas when lopped they soon decay and perish; though many times a spire Elm will begin to grow hollow at the bottom when any of its Roots happen to perish; but the Unthriftiness of its Branches will quickly discover it.

14. Hornets and Wasps do mischief to Trees by breeding in them, which are destroyed by fuming of their Cells in the Night with Brimstone, or by stopping up their Holes with Tar and Goose-dung.

15. Ear-wigs and Snails do seldom infect Timber Trees, but are very prejudicial to Fruit; and so are likewise Pismires, Caterpillars, Mice, Moles, &c. of which I have already treated.

Mice, Moles, and Pismires, cause the Jaundice in Trees, which is known by the discolour of their Leaves and Buds.

16. Blasted parts of Trees are to be cut away to the quick; and to prevent it in the Blossoms, smoak them in suspicious Weather by burning moist Straw, or the superfluous Cuttings of Aromatick Herbs, as Rosemary, Lavender, Juniper, &c.

17. Rooks do great prejudice to Trees by cropping off the tops of old ones, and by lighting on young ones, whose weight breaks the tender Branches and often spoils their Tops: They also destroy Seedlings where they breed, and their Dung breeds Nettles and Weeds.

18. If

18. If a Tree has grown well for several Years, and begins to abate of its thriving, lop off some of the Branches, and see if that will cause it to shoot with Vigour; if not, dig away the Earth 'till you come at the Roots, and see if they are spoiled with any Rottenness, which may be occasioned by their being planted too deep, or from a cold Soil underneath; but if you find the Roots sound, you may conclude, the disease of the Tree proceeds from the poorness of the Soil; to mend which, put new fresh Earth next the Roots, mend the upper part of the Soil with Cow-dung if the Soil is hot, if cold with Horse-dung; and when the time of pruning comes, cut away most part of the old Wood, and you'll find it shoot again afresh; if not, you may conclude, it decayed with its Roots or Trunk, for which there is no Remedy.

Chap. VI. *Of Pruning Forest-Trees.*

AS to the pruning of Trees, it is a Work that requires a great deal of Skill and Care, and for which general Rules cannot well be given, because of the great Variety which is met with in doing of it; only you may observe, that whatsoever Shape you design your Tree shall have, form it to that Shape as much as you can while it is young, because young ones will best bear pruning, when their Boughs are small, and soonest heal when cut off: But for those Trees which you design for Timber, be cautious of cutting off their Heads, as I told you before, especially those that have great Piths, as the Ash, Walnut, &c. and all soft Woods, as the Elm, Poplar, &c. But if your Trees grow too top heavy, you must abate the Head to lighten them, which in many Trees it is better to do by thinning of some of the Boughs that shoot out of the sides of the main Branches, so as to let the Wind have a passage through them, than by cutting off the main Branches themselves, especi-

ally

up of the side Branches 'till you come above the Crook, where they are young.

If any Boughs are cropt by Goats, or other Cattle, cut them off close to the Body, for Cattle leave a drivel where they Bite, which not only infects the Branches, but sometimes endangers the whole Tree.

The best time to prune Trees is in *February*, which should be repeated where need of pruning is every Year, or every second Year, that so the Tree may easily over-grow the Knot, and the place will not be very subject to put forth Suckers, because the Sap hath had no great Recourse to it ; only observe, that if you are to cut a Bough of any bigness, that you give it a chop or two underneath, lest when it falls it strip part of the Bark away with it ; and likewise, that if you keep any Trees for Pollards, that you head them every ten Years ; for if you let the Boughs grow large, they will be the longer before the Bark covers them, and be apt to let Water into the Body, which will soon spoil their bearing of Lop. Vid. *Pruning of Fruit Trees*, Book 14. Chap. 18.

Chap. VII. *Of the Age and Stature of Trees.*

AS to the Age, Stature, and Growth of Trees, I shall refer you to Mr. *Evelyn*, who is very copious in this particular ; and only observe, that the growth of most Trees, for the Circumference of them (it being easily seen what length the Top shoots grow) is from about one to two Inches in a Year, and that the Increase of small and large Trees is much the same, provided they are alike thrifty. I have an Oak that grows in the middle of a Corn-field that is constantly plowed about, and the Cattle often lie under it and dung it, growing upon a red brick Earth, that is at least forty Foot deep, whose Increase is some Years four Inches in a Year, whereas the common growth of other Trees is but about an Inch and a quarter, or an Inch and an half in Circumference ;

which

which shews the Advantage of what I propofed before, of digging and dunging about Trees, and of
killing the Weeds about them, which I reckon the
greateft Prejudice of any thing to the growth of
Trees, in drawing away the heart of the Ground
from them : And I am fatisfied by feveral Meafures
that I have taken of the growth of Trees, that Bufhes
and Under-wood (though they are by many efteem'd
to be as prejudicial as Weeds) are a very great help
to the growth of Trees ; and the greateft of any,
except digging and dunging about them ; for they
both improve the Land, and keep their Roots moift.
But as I find an Account of very great and quick
growth of Trees in feveral Authors, I would defire, that where any fuch growth of Trees is, that
they that are willing to encourage Husbandry, would
be pleafed to meafure the Circumference that they
grow in a Year ; and likewife to be particular in inquiring into the Nature of the Soil, and likewife into the Depth of that Soil they grow upon, and to
communicate it to the Publifhers, that I may be able
to give an Account what fort of Land it is that is
moft likely to be fo improved by Trees ; which will
be an Advantage for thofe to know that have Land
of the fame kind, and likewife an Encouragement to
the Planting and Raifing of Timber.

Chap. VIII. *Of the Felling of Trees.*

I Have already given an Account of the Signs that
fhew the Unthriftinefs of a Tree : And therefore
when you are refolved to fell any of them, the firft
thing to be taken care of is a skilful disbranching
fuch Limbs as may endanger them in their Fall, wherein much Fore-caft and Skill is required, many Trees
being utterly fpoiled for want of this care ; and
therefore in Arms of Timber that are very great,
chop a Nick under them clofe to the Bole ; and fo

meeting it with downright Strokes, it will be fevered without splitting, as I said before.

2. In felling of Timber, take care to cut them as near the Ground as poffible, unlefs you defign to grub them up, which to do is of Advantage both for the Timber and Wood, becaufe they do not reckon the Timber good that grows out of old Stools. The price of Felling of Trees is 12 *d.* *per* Load, and the fame for hewing or fquaring.

When your Tree is down, ftrip off the Bark, and fet it fo as it may dry well, and be well covered from the wet in cafe of Rain ; and then cleanfe the Bole of the Branches that are left, and faw it into Lengths, if you do not fell it to the Timber-buyers to do it for themfelves.

Note, That Trees that are nine Inches girt about a Yard from the Ground, they commonly reckon Timber Trees, but none under, becaufe fuch will be about fix Inches girt in the girting place when the Bark is off, which will fave the labour of climbing of them to meafure them.

The common way of dealing with whom, is to fell your Timber as it ftands, (which is a very uncertain way) or by the Ton, Load, or Foot, forty Foot being reckoned a Ton, and fifty a Load, and in fome places juft the contrary ; therefore 'tis good in all Contracts to mention particularly how many Feet makes a Load or Ton, or elfe you may have great Contefts about it, which Trees you meafure either by girt or fquare Meafure. They reckon that forty Foot of round Timber, or fifty Foot of hewn or fquare Timber weighs the fame, that is, twenty Hundred, which is commonly accounted a Cart-load ; and as they feldom ftrip the Bark off of Elm or Afh, they commonly allow one Inch for the Bark, which is a great deal more than it comes to : And therefore if you can ftrip off the Bark in the meafuring-place, which fhould be always about the middle of the Tree,

it

it will be better. Some allow four Foot out of every Load for Ash, and five Foot for Oak and Elm:
And as for the Computation of the Feet, if it is square
Measure, the square is taken by a pair of Cannipers,
or two Rulers clapped to the side of the Tree, measuring the distance between them; and if the sides
are unequal, they add them together, and take half
the Sum, which they account the true side of the
square; but if girt Measure, by girting of the middle of the Tree with a Line, and taking a quarter
part of that girt for the square, measuring the length
from the But-end so far forwards 'till the Tree comes
to be six Inches girt, that is, twenty-four Inches in
Circumference; and if the Trees have any great
Boughs which are Timber, that is, which hold six
Inches girt, they measure them by themselves, and
add them to the whole: For the casting up the Contents of which, they make use of *Gunter's* Line, upon
which, if you extend your Compasses from 12 to the
number of Inches contained in the square, and placing one Foot of the Compasses at the length, and
keeping of the same extent with your Compasses, if
your square is under 12 Inches, turn your Compasses
twice towards 12; if above, twice from 12, and it
will shew you the Contents: The way of doing
which, any one that understands it will shew you in a
very little time; which way, though it is a false way
of measuring, being near a fifth part short of its true
measure, yet it being the common Practice, you must
be guided by it. But as many of the Rules are false,
and that upon several Occasions and Disputes it may
be necessary to try your Measure several ways, I shall
first propose the doing of it by common Arithmetick:
As, suppose a Tree 40 Inches girt, and 30 Foot
long; the 4th part of 40 Inches is 10 Inches: Now,

The Rule is as 12 to 10 the square Inches,

So is 30 Foot the length to a fourth Number;
and that fourth Number tells you the Contents in
Feet.

But to work this the common way, take the fourth part of the Circumference to be the fide of the fquare of the Tree (though erroneous) and meafure it as a Cylinder. The fourth part of 40 is 10, which multiply'd by it felf is 100: And 30 Foot the length multiplied by 12 makes 360, which multiplied by 100, the fquare of the Tree, gives 36000, the number of fquare Inches in the Tree: Which Sum divided by 1728 the fquare Inches that are in a folid Foot, gives 20 Feet, and about three fourths, and odd Fractions ,being no ways material in Timber-meafure.

$$
\begin{array}{r}
10 \\
10 \\
\hline
100 \\
30 \\
12 \\
\hline
60 \\
30 \\
\hline
360 \\
360 \\
100 \\
\hline
36000
\end{array}
$$

$$
\begin{array}{r}
14 \\
12640 \\
36000 \ (20\,\dfrac{1440}{1728} \\
17288 \\
172
\end{array}
$$

Now to try this by the following Table, look for 10 Inches in the Left-hand Column, and for 30 Foot at the top, which is the length, and you will find 20 Foot, and (82 parts of a Hundred, which is about) three fourths of a Foot.

The following Table of girt and fquare Meafure, Numb. 1. is what you may fee the Contents of any piece of Timber by, according to the common way of meafuring Timber, from half an Inch fquare to 36 Inches, and in length from 1 Foot to 30.

E X A M P L E.

Suppofe a Tree, the Circumference of which is 136 Inches as girt by the Line, which doubled four times mikes the fquare or quarter part 34 Inches: Which Number look for in the firft Column of the following Table, and fuppofing the length of the Piece to be 9 Foot, in the Column under 9, againft 34 Inches, you will find 72 : 25, which is 72 Feet, and 25 of the hundredth Parts of a Foot, which makes 72 Feet and a quarter; which is the Contents of a piece of Timber of that Dimenfion.   A

A Table of Girt and Square Meaf. *N.* 1.

The Length of the Timber in Feet.

The Square of Timber in Inches and half Inches.

In.	1	2	3	4	5	6
	Ft. Pt	Ft. Pt	Ft. Pt	Ft. Pt	Ft. Pt	Ft. Pt
	0 00	0 00	0 00	0 01	0 01	0 01
1	0 01	0 01	0 02	0 03	0 03	0 04
	0 01	0 03	0 05	0 06	0 08	0 09
2	0 03	0 05	0 08	0 11	0 14	0 17
	0 04	0 08	0 13	0 17	0 21	0 26
3	0 06	0 12	0 18	0 25	0 31	0 37
4	0 08	0 17	0 25	0 34	0 42	0 51
	0 11	0 22	0 33	0 44	0 55	0 66
5	0 14	0 28	0 42	0 56	0 70	0 84
	0 17	0 35	0 52	0 69	0 81	1 04
	0 21	0 42	0 63	0 84	1 05	1 26
6	0 25	0 50	0 75	1 00	1 25	1 50
7	0 29	0 58	0 88	1 17	1 46	1 76
	0 34	0 68	1 02	1 36	1 70	2 04
	0 39	0 78	1 17	1 56	1 95	2 34
8	0 44	0 89	1 33	1 77	2 22	2 66
	0 50	1 00	1 50	2 01	2 51	3 01
9	0 56	1 12	1 68	2 25	2 81	3 37
10	0 63	1 25	1 88	2 51	3 13	3 76
	0 69	1 39	2 08	2 87	3 47	4 16
	0 76	1 53	2 29	3 06	3 82	4 59
11	0 84	1 68	2 52	3 36	4 20	5 04
	0 92	1 84	2 76	3 67	4 39	5 51
12	1 00	2 00	3 00	4 00	5 00	6 00

A Table of Girt and Square Meas. N. 1.

The Length of the Timber in Feet.

The Square of Timber in Inches and half Inches.

In.	7		8		9		10		20		30	
	Ft.	Pt.	Ft.	Pt.	Ft.	Pt.	Ft.	Pt.	Ft.	Pt.	Ft.	Pt.
	0	01	0	01	0	02	0	02	0	04	0	06
1	0	05	0	05	0	06	0	07	0	14	0	21
	0	11	0	13	0	15	0	16	0	32	0	48
2	0	19	0	22	0	25	0	28	0	56	0	84
	0	30	0	34	0	39	0	43	0	86	1	29
3	0	43	0	49	0	56	0	62	1	24	1	86
	0	59	0	68	0	76	0	85	1	70	2	55
4	0	78	0	89	0	99	1	11	2	22	3	33
	0	98	1	12	1	26	1	40	2	80	4	20
5	1	22	1	39	1	56	1	74	3	48	5	22
	1	47	1	68	1	89	2	10	4	20	6	30
6	1	55	2	00	2	25	2	50	5	00	7	50
	2	05	2	34	2	64	2	93	5	86	8	79
7	2	38	2	72	3	06	3	40	6	80	10	20
	2	73	3	12	3	51	3	90	7	80	11	70
8	3	11	3	55	3	99	4	44	8	88	13	32
	3	51	4	01	4	52	5	02	10	04	15	06
9	3	93	4	49	5	06	5	62	11	24	16	86
	4	29	5	01	5	64	6	27	12	54	18	81
10	4	86	5	55	6	24	6	94	13	88	20	82
	5	35	6	12	6	88	7	65	15	30	22	95
11	5	88	6	72	7	56	8	40	16	80	25	20
	6	43	7	35	8	27	9	19	18	38	27	57
12	7	00	8	00	9	00	10	00	20	00	30	00

A Table of Girt and Square Meas. N. 1.

The Length of the Timber in Feet.

In.	1		2		3		4		5		6	
	Ft.	Pt.	Ft.	Pt.	Ft.	Pt.	Ft.	Pt.	Ft.	Pt.	Ft.	Pt.
	1	08	2	17	3	25	4	34	5	42	6	51
13	1	17	2	35	3	51	4	69	5	87	7	04
	1	26	2	53	3	80	5	06	6	33	7	59
14	1	3:	2	72	4	08	5	44	6	80	8	16
	1	46	2	92	4	38	5	80	7	30	8	76
15	1	55	3	12	4	68	6	25	7	81	9	37
	1	67	3	33	5	00	6	67	8	34	10	01
16	1	78	3	55	5	33	7	11	8	89	10	67
	1	89	3	78	5	67	7	56	9	45	11	34
17	2	1	4	01	6	02	8	03	10	03	12	04
	2	13	4	25	6	38	8	51	10	63	12	76
18	2	25	4	50	6	25	9	00	11	25	13	50
	2	38	4	75	7	13	9	51	11	88	14	26
19	2	51	5	01	7	52	10	03	12	53	15	04
	2	64	5	28	7	82	10	56	13	20	15	64
20	2	78	5	55	8	33	11	11	13	89	16	67
	2	92	5	83	8	75	11	67	14	59	17	51
21	3	06	6	12	9	18	12	25	15	3	18	37
	3	11	6	42	9	63	12	84	16	05	19	26
22	3	36	6	72	10	08	13	44	16	8	20	16
	3	51	7	03	10	55	14	06	17	58	21	09
23	3	67	7	34	11	02	14	69	18	36	22	04
	3	83	7	67	11	50	15	34	19	17	23	01
24	4	00	8	00	12	00	16	00	20	00	24	00

The Square of Timber in Inches and half inches.

A Table of Girt and Square Meaſ. *N.* 1

The Length of the Timber in Feet.

The Square of Timber in Inches and half Inches.

In.	7		8		9		10		20		30	
	Ft.	Pt.	Ft.	Pt.	Ft.	Pt.	Ft.	Pt.	Ft.	Pt.	Ft.	Pt.
	7	51	8	68	9	76	10	85	21	70	32	55
13	8	22	9	39	10	56	11	87	23	48	35	22
	8	86	10	13	11	39	12	66	25	32	37	98
14	9	53	10	89	12	25	13	61	27	22	40	83
	10	22	11	68	13	14	14	60	29	20	43	80
15	10	93	12	49	14	06	15	62	31	24	46	86
	11	67	13	34	15	01	16	68	33	36	50	04
16	12	44	14	22	16	00	17	78	35	56	53	34
	13	24	15	13	17	02	18	91	39	82	58	73
17	14	95	16	95	18	06	20	07	40	14	60	21
	14	89	17	01	19	14	21	27	42	54	63	81
18	15	75	18	10	20	25	22	50	45	00	67	50
	16	62	19	01	21	39	23	77	47	54	71	31
19	17	55	20	05	22	56	25	07	50	14	75	21
	18	49	21	13	23	77	26	41	52	82	79	23
20	19	40	22	22	25	00	27	78	55	56	83	34
	20	42	23	34	25	26	29	18	58	36	87	54
21	21	43	24	49	27	56	30	62	61	24	91	86
	22	47	25	68	28	89	32	10	64	20	96	30
22	23	53	26	89	30	25	33	61	67	22	100	83
	24	61	28	13	31	64	35	16	70	32	105	48
23	25	77	29	38	33	06	36	73	73	46	110	19
	26	84	30	68	34	51	38	35	76	70	115	05
24	28	00	32	00	36	00	40	00	80	00	120	00

A Table of Girt and Square Meaſ. *N.* 1.

The Length of the Timber in Feet.

The Square of Timber in Inches and half Inches.

In.	1 Ft.	Pt.	2 Ft.	Pt.	3 Ft.	Pt.	4 Ft.	Pt.	5 Ft.	Pt.	6 Ft.	Pt.
	4	16	8	33	12	50	16	66	20	83	24	99
25	4	34	8	68	13	02	17	36	21	70	26	04
	4	51	9	02	13	54	18	05	22	56	27	08
26	4	69	9	39	14	08	18	77	23	47	28	16
	4	88	9	75	14	63	19	51	24	38	29	26
27	5	06	10	12	15	19	20	25	25	31	30	38
	5	25	10	50	15	75	21	00	26	85	31	50
28	5	44	10	89	16	33	21	78	27	22	32	67
	5	67	11	34	17	01	22	68	28	35	34	02
29	5	84	11	68	17	52	23	36	29	20	35	04
	6	04	12	08	18	13	24	17	30	21	36	26
30	6	25	12	50	18	75	25	00	31	25	37	50
	6	46	12	91	19	38	25	84	32	30	38	76
31	6	67	13	34	20	02	26	69	33	36	40	04
	6	89	13	78	20	67	27	56	34	45	41	34
32	7	11	14	22	21	33	28	44	35	55	42	66
	7	33	14	66	21	99	29	33	36	66	43	99
33	7	56	15	12	22	68	30	24	37	81	45	37
	7	78	15	56	23	34	31	12	38	90	46	68
34	8	03	16	05	24	08	32	11	40	14	48	17
	8	26	16	52	24	79	33	05	41	34	49	58
35	8	54	17	01	25	52	34	03	42	53	51	04
	8	70	17	50	26	25	35	00	43	75	52	50
36	9	00	18	00	27	00	36	00	45	00	54	00

A Table of Girt and Square Meas. N. 1.

The Length of the Timber in Feet.

The Square of Timber in Inches and half Inches.

In.	7		8		9		10		20		30	
	Ft.	Pt.	Ft.	Pt.	Ft.	Pt.	Ft.	Pt.	Ft.	Pt	Ft.	Pt.
	29	16	33	33	37	49	41	66	83	32	124	98
25	30	38	34	72	39	06	43	4c	86	80	130	20
	31	49	36	10	40	62	45	13	90	26	135	39
26	32	86	37	55	42	24	46	94	93	88	140	82
	34	14	39	01	43	89	48	77	97	54	146	31
27	35	44	40	50	45	57	50	63	101	26	151	89
	36	75	42	00	47	25	52	50	105	00	157	50
28	38	11	43	56	49	00	54	45	109	35	164	35
	39	69	45	36	51	03	56	70	113	40	170	10
29	40	88	46	88	52	56	58	40	116	80	175	20
	42	30	48	34	54	31	60	43	120	86	181	29
30	43	75	50	00	56	25	62	50	125	00	187	00
	45	22	51	68	58	14	64	60	129	20	193	80
31	46	71	53	36	60	06	66	50	133	00	199	00
	48	23	55	12	62	01	68	40	136	80	204	20
32	49	78	56	89	63	99	71	11	142	22	213	33
	51	33	58	66	65	95	73	33	146	66	219	99
33	52	93	60	49	68	06	75	62	151	24	226	86
	54	46	62	24	70	02	78	80	157	60	236	40
34	56	19	64	22	72	25	80	28	160	36	240	54
	57	84	66	10	74	37	82	63	165	26	247	89
35	59	55	68	05	76	56	85	07	170	14	255	21
	61	25	70	00	78	75	87	50	175	00	262	5c
36	63	00	72	00	81	00	90	00	180	00	270	00

Numb. 1.

Ft.	In.	Ft.	In.	Pts.
0	6	4	0	0
	7	2	11	2
	8	2	3	0
	9	1	9	3
	10	1	3	3
	11	1	2	3
I.	0	1	0	0
	1		10	2
	2		8	8
	3		7	6
	4		6	7
	5		5	9
	6		5	3
	7		4	8
	8		4	3
	9		3	9
	10		3	5
	11		3	3
II.	0		3	0
	1		2	8
	2		2	6
	3		2	3
	4		2	2
	5		2	1
	6		1	9
	7		1	8
	8		1	7
	9		1	6
	10		1	5
	11		1	4
III.			1	3

This Table, *Numb. 2.* is to shew how much in length will make a solid Foot of any Tree, whose quarter-part of the Circumference is from 6 Inches to 36 Inches.

EXAMPLE.

Suppose a Tree of 60 Inches in Circumference, the fourth part of which is 15 Inches, or 1 Foot 3 Inches; which if you look for in the first Column, opposite to it in the second Column you will find 7 Inches and 6 tenth parts of an Inch (which is somewhat above half an Inch) and so much in length will make one Foot square.

By

By this Table suppose a Plank or Board 9 Inches broad, to find how much in length will make one Foot.

First find out 9 Inches in the first Column, opposite to it in the second Column you will find 1 4 0 which is one Foot four Inches, so much in length of a Plank or Board 9 Inches broad going to make up a Foot: So that every 16 Inches in length is a Foot of Plank; and consequently, every 8 Inches half a Foot, every 4 Inches a quarter, &c. Thus again, If a Board hold 2 Foot and 3 Inches in breadth, 5 Inches and 3 tenth parts of an Inch in length will make a square superficial Foot of Plank, &c.

Numb. 5.

A Table to measure Plank, Boards, &c.

Ft.	In.	Ft.	In.	Pt.
	1	12	0	0
	2	6	0	0
	3	4	0	0
	4	3	0	0
	5	2	4	8
	6	2	0	0
	7	1	8	6
	8	1	6	0
	9	1	4	0
	10	1	2	4
	11	1	1	1
I.	0	1	0	0
	1	0	11	8
	2	0	10	3
	3	0	9	6
	4	0	9	0
	5	0	8	5
	6	0	8	0
	7	0	7	6
	8	0	7	2
	9	0	6	8
	10	0	6	5
	11	0	6	2
II.	0	0	6	0
	1	0	5	8
	2	0	5	5
	3	0	5	3
	4	0	5	1
	5	0	5	0
	6	0	4	8
	7	0	4	7
	8	0	4	5
	9	0	4	4
	10	0	4	2
	11	0	4	1
III.	0	0	4	0

Numb. 4.

A Table of Square Measure.			
Inches.	Ft.	In.	Pt.
1	144	0	0
2	36	0	0
3	16	0	0
4	9	0	0
5	5	9	1
6	4	0	1
7	2	11	2
8	2	3	0
9	1	9	3
10	1	5	2
11	1	2	2
12	1	0	0
13	0	10	2
14	0	8	8
15	0	7	6
16	0	6	7
17	0	5	9
18	0	5	7
19	0	4	7
20	0	4	3
21	0	3	9
22	0	3	5
23	0	3	1
24	0	3	0
25	0	2	7
26	0	2	5
27	0	2	3
28	0	2	2
29	0	2	0
30	0	1	9

This Table of Square Measure shews how much goes to make a solid Foot of any piece of Timber, from 1 Inch to 30 Inches square.

EXAMPLE.

I would know how long a piece of Timber of 10 Inches square ought to be to contain a Foot of Timber? Look for 10 in the Left-hand Column, opposite to which you'll find 1 Foot 5 Inches, and 2 tenths of an Inch, which is the length that makes a solid Foot.

Numb.

This Table of Round Measure shews how much in length makes a solid Foot of Timber if any round piece whose Diameter is from one Inch to 30 Inches over.

EXAMPLE.

I would know how much an exact round piece of Timber, containing but one Inch in Diameter, must be in length to make a solid Foot of Timber? Look in the first Column for one Inch, and opposite to it you will find 113 Foot, 1 Inch and 7 tenth Parts of an Inch; which is the Contents sought for.

A Table of Round Measure			
Inches.	Ft.	In.	Pt.
1	113	1	7
2	28	3	4
3	12	6	8
4	7	0	8
5	4	6	6
6	3	1	7
7	2	3	7
8	1	9	2
9	1	4	1
10	1	1	5
11		11	
12		9	4
13		8	
14		6	9
15		6	0
16		5	3
17		4	
18		4	1
19		3	7
20		3	3
21		3	8
22		2	8
23		2	5
24		2	3
25		2	1
26		2	0
27		1	8
28		1	7
29		1	6
30		1	5

IF you have a mind to know the Value of a Tree ftanding, you may girt it, and allowing for the Bark, and fo much as you think it will meafure lefs in the girting place than at the Butt, and taking the heighth of it, compare it with the fore-going Tables, and you may the better guefs at its Worth, becaufe you have a Rule to go by. Now for the taking of the heighth of a Tree, the beft way is with a Quadrant, which the larger 'tis, the more exact you may be in doing of it; which is done after this manner: Hold your Quadrant fo as that your Plummet may fall on 45 Degrees, and go to fuch a diftance from the Tree as you may, through the Sights of your Quadrant, fee the top of it; and meafure from the place of your ftanding to the Foot of the Tree, adding to it the heighth of your Eye from the Ground, and it will give you the heighth defired: Or if you ftand where the Plummet may fall on 22 Degrees and 30 Minutes, it will be half the heighth; or 67 Degrees and 30 Minutes will be the heighth and half the heighth.

Chap. IX. *Of Grubbing up of Woods and Trees.*

THE grubbing up of Woods and Trees may be needful upon the Account of their Unthriftinefs, or to plant better Lands for that purpofe, and to grub up Roots that are decayed to make room for them that are more thriving, &c. which, though a chargeable Work, yet it may much be leffened by a particular Engine, which I thought it might be of Advantage to make more Publick. It is a very cheap Inftrument, only made ufe of in fome particular Places, and will eafe about a third part of the charge of this fort of Labour; it is an Iron-hook of about

the

two Foot four Inches long, with a large Iron-ring to it, the shape of which you have in the Figure, and may be made for about 3 *s.* 6 *d.* charge, which they use after this manner. Where a Stub of Under-wood grows, they clear the Earth round it where they think any side Roots come from it, and cut them; which when they have done, in any Hole on the sides of the Root they enter the Point of the Hook, and putting a long Leaver into the Ring, two Men at the end of it go round till they wring the Root out, twisting the tap Roots asunder, the difficulty of coming at which occasions the greatest Labour of this Work. Stubs also of Trees may be taken up with it; in which Work it saves a great deal of Labour, though not so much as in the other, because the Stubs must be first cleft with Wedges before you can enter the Hook in the sides of it to wrench it out by pieces.

Chap. X. *Of Seasoning of Timber.*

Timber being felled and sawn, is next to be seasoned: For doing of which, some advise that it be laid up very dry in an airy place, yet out of the Wind or Sun; others say, it ought to be free from the Extremities of the Sun, Wind, and Rain; And that it may not cleave, but dry equally, you may daub it over with Cow-dung. Let it not stand upright, but lay it along, one piece upon another, interposing some short Blocks between them to preserve them from a certain Mouldiness which they usually contract when they sweat, and which frequently produces a kind of *Fungus,* especially if there be any sappy Parts remaining.

Others

Others advise to lay Boards, Planks, &c. in some Pool or running Stream; or, which is better, in Salt-water for a few Days, to extract the Sap from them, and afterwards to dry them in the Sun or Air; for by so doing, (say they) they will neither chap, cast, nor cleave: Mr. *Evelyn*, particularly, commends this way of Seasoning of Fir, but against Shrinking there is no Remedy.

Some again commend Buryings in the Earth, others in Wheat; and there be Seasonings of the Fire, as for the scorching and hardening of Piles which are to stand either in the Water or the Earth.

Sir *Hugh Plat* informs us, that the *Venetians* use to burn and scorch their Timber in the flaming Fire, continually turning it round with an Engine 'till they have gotten upon it a hard black coally Crust; and the Secret carries with it great Probability, for that the Wood is brought by it to such a hardness and driness, that neither Earth nor Water can penetrate it. " I my self (says Mr. *Evelyn*) remember to have " seen Chareoal dug out of the Ground, amongst the " Ruins of ancient Buildings, which have in all pro- " bability lain covered with Earth above 1500 Years.

For Posts and the like that stand in the Ground, the burning the Out-sides of those Ends that are to stand in the Ground, to a Coal, is a great Preserva-tive of them. Sir *Hugh Plat* adds, " That a *Kentish* " Knight of his Acquaintance did use to burn (in this " manner) the Ends of the Posts for Railing or Pail- ' ing ". And this was likewise practised with good Success by a *Sussex* Gentleman, *Walter Burrell* of *Cuck-field*, Esq;

This burning of the ends of Posts is practised in *Germany*, as appears by the Abstract of a Letter writ-ten by *David Vonderbeck*, a *German* Philosopher and Physician at *Minden*, to Dr. *Largelott*, registred in the *Philosophical Transactions*, Numb. 92. Pag. 1585. in these Words; " Hence also they slightly burn the

" ends

" ends of Timber to be set in the Ground, that so
" by the Fusion made by Fire the Volatile Salts
" (which by accession of the Moisture of the Earth
" would easily be consumed to the Corruption of the
" Timber) may catch and fix one another.

Chap. XI.   *Of Preserving of Timber.*

WHEN Timber or Boards are well seasoned or
dried in the Sun or Air, and fixed in their
places, and what Labour you intend is bestowed upon
them ; the use of Lin-seed Oil, Tar, or such Oleagi-
nous Matter, tends much to their Preservation and
Duration. *Hesiod* prescribes to hang your Instruments
in Smoak to make them strong and lasting : Surely,
then the Oil of Smoak (or the Vegetable Oil, by some
other means obtained) must needs be effectual in the
Preservation of Timber also. *Virgil* advises the same.

The Practice of the *Hollanders* is worth our Notice;
who, for the Preservation of their Gates, Portcullis,
Draw-bridges, Sluices, and other Timbers exposed
to the perpetual Injuries of the Weather, coat them
over with a mixture of Pitch and Tar, upon which
they strew Cockle and other Shells, beaten almost to
Powder, and mingled with Sea-sand, which incrusts
and arms it after an incredible manner against all the
Assaults of Wind and Weather.

When Timber is felled before the Sap is perfectly at rest (says Mr. *Evelyn*) it is very subject to the
Worm : And to prevent and cure this in Timber, I
recommend the following Secret as most approved.

Let common yellow Sulphur be put into a Cucur-
bit-glass, upon which pour so much of the strongest
Aqua fortis as may cover it three Fingers deep. Distill
this to Driness, which is done by two or three
Rectifications : Let the Sulphur remaining at the bottom (being of a blackish or sad-red Colour) be laid

on

on a Marble, or put into a Glass, where it will easily
dissolve into Oil. With this anoint what Timber is
either infected with Worms, or to be preserved from
them. It is a great and excellent *Arcanum* for tinging the Wood, of no unpleasant Colour, by no Art
to be washed out, and such a Preservative of all manner of Woods, nay, of many other things also, as
Ropes, Cables, Fishing-nets, Masts, or Ships, &c. that
it defends them from Putrefaction either in Waters
or above the Earth, in Snow, Ice, Air, Winter, or
Summer, &c. I am told, that Oil of Spike will kill
the Worm in any Wood.

It were superfluous to describe the Process of making the *Aqua-fortis*; it shall suffice to let you know
that our common *Coperas* makes this *Aqua-fortis* well
enough for our purpose, being drawn over by a Re-
tort: And for Sulphur, the Island of St. *Christopher's*
yields enough (which hardly needs any refining) to
furnish the whole World. This Secret (for the curious) I thought not fit to omit, though a more com-
pendious way may serve the turn. Three or four
Anointings with Linseed Oil, has proved very effectual. It was experimented in a Walnut-tree Table,
where it destroyed Millions of Worms immediately,
and is to be practised for Tables, Tubes, mathematical Instruments, Boxes, Bed-steads, Chairs, &c. Oil
of Walnuts will doubtless do the same, and is a
sweeter and better Varnish. But Oil of Cedar, or
that of Juniper, is commended above all.

H 4   Chap.

Chap. XII. *Of closing Chops and Clefts in green Timber.*

GREEN Timber is very apt to split and cleave after it is wrought into Form, which in fine Buildings is a great Eye-sore : But to close the Chops and Clefts I find this Expedient to do well ; which is to anoint and supple it with the Fat of Powder'd-beef Broth, with which it must be well soaked, and the Chasm filled with Sponge dipped in it : . This to be done twice over. Some Carpenters make use of Grease and Saw-dust mingled ; but the first is so good a way (says my Author) that I have seen Wind-shock Timber so exquisitely closed, as not to be discerned where the Defects were. This must be used when the Timber is green.

I shall conclude this Treatise of Forest-trees with considering, and in some measure proposing of Remedies for two of the greatest Discouragements that belong to the Planting and Raising of them : The one is, the long time that the Owners are forced to wait for the growth of their Timber before they can make any Profit of it ; and the other is, the Timber's being liable to so many Abuses and Cheats from Tenants, Bailiffs, Executors, and others, in case of the Owner's Negligence or Death ; especially if they are forced to leave their Wood to a young Heir. Now as to the first Objection, if the Timber is thriving there is no Stock you can have Money in that will turn to better Account, though you stay long for it ; nor any thing that it can be better secured in, which I think will make amends for the Stay. And as for the second Objection, which I think the most material, and the greatest Inconveniency and Discouragement to Planting and raising of Timber of any ; if I can propose a Method for the taking an exact Account of the Timber-trees, both in Hedges and Woods,

Woods, I shall wholly answer that Objection and
Inconveniency too: For the doing of which, first,
in Hedge-rows you may observe this Method. Suppose the four Fields underneath to be what you have
a mind to take an Account of, which are called by the
Names of, *The Ten Acre Field, The New Mead, The
Road Pleen,* and *Park Pleen,* and lying as in the following Map ;

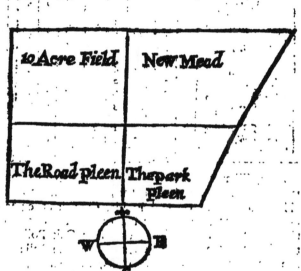

Make a Scheme after the following manner.

Names

Names of the Fields.	Side of the Fields.	Timber Trees			Pollards			Saplings, or young Trees		
		Oak	Ash	Elm	Oak	Ash	Elm	Oak	Ash	Elm
Ten Acre Field.	N. side	10	3	4	6	10	4	10	5	4
	E. end	5	4	6	3	6	3	1	2	3
	S. side		5	4		2		1	3	4
	W. end	4		3	8			6	5	4
New Mead.	N. side	5	3		2	6	5	4		
	E. end		3	4		0		5	4	2
	S. side	3	5	2	2	2	4	6	7	8
Park Pleen.	N. side				5	6		7	5	2
	E. end	4	3	2		0				
	S. side	3	4	2	3	2	2	5	5	6
	W. end		5	4	2	3	2	4	6	5
Road Pleen.	S. side	4	3	2		1	2	8	4	2
	W. end				5	6	7			

Note, That in the above Scheme, the first Column is the Names of the Fields, the second Column is the Sides and Ends of the said Fields, the third the number of Timber, Oaks, the fourth the Timber Ash, the fifth the Timber Elm, the sixth Oak Pollards, the seventh Ash Pollards, the eighth Elm Pollards, the ninth Sapling or young Oaks, the tenth young Ash, the eleventh young Elms ; and the twelfth Column is to add other sorts of Trees in, or to set down when they are felled.

Only 'tis to be observed, that to the *New Mead* Field is reckoned but two Sides and only the East end, because the Ditch being on the *New Mead* side, the Hedge-row between that and the *Ten Acre* Field, is reckoned to the ten Acres ; and though the Hedge-row between the *New Mead* Field and the *Park Pleen* is reckoned to both, 'tis because there are Trees on the Dools belonging to the *Park Pleen,* and not what Trees are in the Hedge-rows.

For

For taking an Account of the Number of Trees in
Woods, where they are long and narrow 'tis easily
done when they are felled; but where Woods are
large, 'tis more difficult: I shall therefore propose to
you the Method I took for the doing of it in a Wood
I have that contains about 40 Acres. A Draught of
which you have as follows.

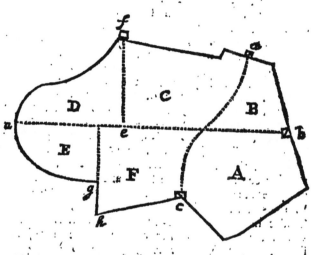

Which Wood being too large to fell at once (I not
having Woods enough to answer such a Fall every
Year) and finding my Wood cut in Patches, and o-
ther parts of it scambled and cut before it was at its
Growth; that they might come at what was fit to cut,
I resolved to cut a Cart-way through the middle of
it, by which means I proposed three Advantages;
First, to cut what part of my Wood I would, and to
have a clear Cart-way to carry off both my Wood
and Timber, which saved my Standers and Wood too
very much. Secondly, To divide my Wood into
two Parts, in order to the counting of my Standards;
and

and Thirdly, it being near my House, to have a fine Walk of it; which Cart-path in the Map is marked with a pricked Line from *a* to *b*. But, I suppose, I shall be asked how, in a standing Wood, I could carry the Path so straight, and keep the middle of the Wood from one end to the other? For the doing of which, the Method I took was this: I drew on the Map the Line *a b*, and taking of the Angle on the Map at *a*, I set my plain Table to that Angle, and by my Sight I directed a Workman to cut a narrow Path of about two Foot wide, and about seven or eight Yards into the standing Wood, and then I stuck up two Sticks of equal Height, on the top of which I made a small Slit, and stuck a small piece of white Paper in them, and then ordered the Workman to go into the standing Wood as far as he could through the Boughs, see the white Papers, and then to cut his way out to them; and this he repeated doing 'till he was so far off the first Stakes as not to discern them well, and then I set up another Stake in the same manner to range with the first, and continued adding of one Stake after another 'till He got through to the Wood's end: This is the way of their cutting their Glades in *Hertfordshire*, which hit as near the middle as you see it laid down in the Draught: This, as I said before, divided my Wood into two Parts; and from the Gate at *b* to the Stile at *d* was a Foot-path, which I marked likewise with a pricked Line, which made in the Wood the two Divisions of *A* and *B*. And having some Woods adjoining to this Wood as *f*, I was obliged to make another Cart-way from *e* to *f*, which made the Divisions of *C* and *D*: And from the Foot-path *c d*, being too large a Division, to lessen it I made several small Hills, in which I stuck Stakes to run parallel with the corner of the Wood *g h*, which made the Divisions *E* and *F*. By which means I divided my Wood into six Parts, which afforded me a part to fell every Year, and gave me an

oppor-

opportunity of counting the Trees in each Division
as I felled them, which I did after this manner.

My Wood confisting only of Oak and Ash, I divided my Trees into three forts, *viz.* first Storers,
which I reckoned all to be that were under 12 Inches
Circumference; secondly, Saplings, which I called all
under 24 Inches Circumference; and what was two
Foot Circumference, or above a Yard from the
Ground, I reckoned Timber-trees: And taking of a
Slate, I drew six
Lines after this
manner; and ta-
king of one with
me with a Paper
ruled after the
fame manner, and
a piece of Chalk,
a black Lead Pen-
cil, and a Line
with two Knots
in it, one of 12
Inches long, and
the other of 24
Inches: Those
Trees that we
were not certain
of being under the Mesfure mentioned he measured,
and as he counted the Trees he drew a Chalk-line a-
bout them, which shewed us which we told, that we
might not tell them again: And as he told them I
scored on the first Line the Oak Storers, on the second
Line the Saplings, and on the third Line the Timber
Oaks, observing the fame Method with the Ash; and
when I had scored twenty on any of the Lines, I rubbed out my Scores on the Slate, and with my Pencil
I scored one on the Paper for one Score. And the
Divisions of *A* and *B* being all the Wood that was
felled at present, when I came home I found my Account of those two Divisions as follows, which I set
down

down in the following Method ; that I think will be
Direction enough for this or any other Woods that
you shall have occasion to take an Account of.

An Account of the Number of Trees in the several Divi-
sions of my Wood, called The Great Wood, *taken in*
the Year 1705.

N.	OAKS.			ASH.		
	Storers	Saplin	Timber	Storers	Saplin	Timber
A.	150	110	50	50	40	60
B.	50	60	55	150	45	50

And though this Exactness may seem more nice
than is necessary, yet no one knowing whose Hand
he may fall into, it may be of Advantage, the Pains
being very little, especially since I have my self,
with only one to help me, taken an Account of three
Acres of Wood-Land in an Hours time, where the
Standards have been very thick, which I think no very tedious Business ; but I shall leave every one to do
as they see most convenient, and proceed to give
some Account of the *Kitchen-Garden.*

BOOK XIII.

Chap. I. *Of such Herbs, Roots, and Fruits, as are usually planted in the* Kitchen-Garden, *or* Olitory.

I Have already shewed the Husband-man how to order his Pasture, Arable, and Wood Land; and what I have now to treat of, is the *Kitchen-Garden* and the *Orchard*, which are parts of Husbandry of no small Advantage to the industrious Farmer; and for the quantity of Ground that the *Kitchen-Garden* takes up, there is no part of his Land that will turn to better Advantage than what is improved this way, it being a great deal of Meat and Corn, that is saved by Beans, Peas, and Roots; and likewise a great deal of Barley, where is plenty of Cyder, besides the Advantage that Fruit brings by what may be sold to neighbouring Towns, and the Cyder that might come to be exported to Foreign Parts, if we could attain to a true Perfection in the Art of making this profitable Liquor; but of these things hereafter. I shall at present only give Directions for the *Kitchen-Garden*; in first shewing what is the best Situation and Figure for a *Kitchen-Garden*: Secondly, How to order the Ground-Beds, &c. for

this

this ufe, that being the Foundation of the Work
And thirdly, Give a Catalogue of fuch ufeful Herbs,
Roots and Fruits (as are therein to be cultivated) in
an Alphabetical manner; which I think will be the
readieft way to find any particular forts of Garden
Commodities that fhall be defired, and will be much
better than the ranking of them under their feveral
Kinds or Species. And therefore,

Situation. Firft, As to the Situation of the *Kitchen-Gardens*,
fmall Valleys or low Grounds are beft, becaufe commonly fuch Places have a good depth of Earth, and
are fatned by the neighbouring Hills, efpecially if they
are not expofed to Inundations, and afford good
Water; but as for Fruit-Trees, a Ground moderately dry (provided the Soil be rich and deep) is the beft;
and for the Pofition, if the Earth be ftrong, and con-
fequently cold, the South Afpect is the beft; but if
it be light and hot, the Eaft is to be prefer'd. The
Southern Afpects are often expofed to Weft Winds
from the middle of *Auguft* to the middle of *Ober*;
the Eaftern is fubject to the North-Eaft Winds, which
withers the Leaves and new Shoots, efpecially of
Peach Trees; and the Weftern to the North-Weft
Winds, which brings Blites in Spring, and ftrong
Gales in Autumn, which commonly fhakes off the
Fruit before it is ripe: But as all Pofitions have their
Perfections and Imperfections, care muft be taken to
make what Advantage we can of the firft, and to ufe
our beft Skill to defend our felves from the laft.

Figure. The beft Figure for a *Kitchen-Garden*, and moft
convenient for Culture, is a Square of ftraight Angles,
being once and a half or twice as long as broad; for
in Squares the moft uniform Beds may be made; the
Walls ought alfo to be well furnifhed with Fruit-Trees,
to be of a good heighth, and placed fo as to afford
good fhelter on all fides; the Beds, Plats and Borders
to be fet with all forts of things neceffary for all Seafons of the Year, and to have the Walks clean, of a
proportionable largenefs, and to afford as much variety as the place will admit of.   Secondly,

Secondly, For the way of managing and ordering *Land to* of Land for this Purpose; as in all other forts of Hus- *order.* bandry, fo in this you are to confider the nature of the Soil, and what is moft proper and beft to make an addition to its Goodnefs; of which Particulars I have treated already, only it is to be noted, that as your *Kitchen Garden* is to afford great variety of Plants, Roots, Herbs, *&c.* your Soil muft be made more rich than for Corn, moft Garden things requiring a richer Soil, if you defign to have them profper well; if you meet with Ground that is naturally good and moift (which is the beft for Garden-ground) it is a great Advantage, and much leffens the Expence; but Land is very rarely found that doth not require a great deal of Labour, tho' it may not want Manure; for many times the Surface of the Earth fhall be good, which (being opened a Spit deep) will be found to afford only a cold Clay bottom; which is a much more pernicious Soil for Trees and Garden-ware than Gravel, it felf, becaufe in Gravel, efpecially if any thing inclin'd to Springs, the Roots of Trees, *&c.* often meet with fome fmall Veins, whence they draw Nourifhment; whereas the Clay, in dry Years efpecially, is fo hard, that the Roots cannot penetrate it, and binds fo firm, that it hinders the moifture that falls upon it from finking through, and fo caufes the Roots to ftand too moift in Winter, and fcorches them with too much heat in the Summer.

To redrefs this defect, the only expedient is, to break up this fort of Land as deep and no deeper than the Earth is good, beginning with a Trench four or five Foot bread, the whole length or breadth of the place, cafting the feveral Moulds all upon one fide; and into your bottom, when your Trench is empty, caft in long Dung, or, which is much better, Fern, Leaves of Trees, rotten Sticks, Weed, or any fort of Trafh you can procure with the moft Eafe and leaft Charge, to rot and keep the Ground from binding.

This Trafh fhould be laid at the bottom about half a Foot thick, and after that fling upon it your top

Spit, casting the Mould which lies upper-most (and which is ever best) upon the Dung, and so making the second Trench as the former, you should fill your first Trench so as that the Mould which you found undermost should now lie on the top; and thus you must continue to do till you have finish'd the whole Work.

· But it may be here objected, that the Earth which you take from beneath will be barren and unfruitful, which indeed it will for the first Year, but being exposed to the Air, Rains and Frost of one Winter, it will be so mellow as to make it fruitful, especially if any thing of Dung be added to it; but if it prove any thing churlish, an addition of drift Sand will quickly loose the binding quality of it; and if it is cold Land, Smiths Ashes, or other Ashes, if you live near Towns where they are to be had, will be the best Manure for it; Pidgeons, Fowls and Sheeps-dung are also very good for all sorts of cold Land. But 'tis to be observ'd, that rich Soils produce large, fair Fruit, and Clay and Chalky Lands the best tasted, and most lasting. Garden-ware eating better that comes off of a natural Soil than what is forced with Dung, or any other Manure, except Chalk.

Time of Trenching. The best time for the trenching up of Land is the beginning of Winter, when the Ground is moist and easie to dig; but as there are two Seasons of the Year to sow and plant Herbs, so there are two times to bring Gardens into order, which is the Autumn and the Spring; the first Labour being to be bestow'd about the beginning of *November*, upon such Ground as you design to sow in the Spring, and to dig in the Month of *May* such Ground as you design for an Autumal Sowing, that the cold in Winter, and the heat in Summer, may have an opportunity to make the Clods short and brittle, to turn them into Dust, and to kill the unprofitable Weeds; which if you do not find your first digging up to destroy, you may give your Land a second and a third Digging in the hottest driest time you can; which is the only time to destroy all such sort of trumpery. But

But where you meet with a gravelly bottom, you should husband it as is already prescrib'd, and the Stones which are mingled with the Earth should be carry'd out; but in case the Gravel lie not very thick, and that when it is broke up, you meet with Sand or small loose Gravel, it may do without flinging it out of the Trenches, because the Trees, by help of the Dung, will strike Root in it.

And tho' these are chargeable ways of ordering of Gardens, yet it is done once for all, and the Charge will be abundantly answered in the Growth of what Plants you set in it, especially for Fruit or other Trees that root downwards; but some are more curious, in that they Skreen all the Earth with which they fill their Trenches; but that is needless for the whole Garden, particular Beds on which you design to raise particular Seeds or Plants being sufficient to be so ordered.

Besides which, you must have an equal Composition *Of Ma-* or mixture of Dung and Earth always ready in some *nure.* corner of your Garden to be laid by, that it may be throughly rotten and turned to good Mould against Spring, to renew the Earth with about your Artichoaks, and for the planting and sowing of Colli-flowers, Cabbages, Onions, &c. *vid.* the Culture of *Orchards.*

And for each sort of Seed and Plants it will be ne- *Laying* cessary to lay out particular Beds, which may be the *out of* more or less in Number, according to the variety of *Beds.* Seeds you have to sow, and Herbs to plant, and which ought not to be wider than you can reach cross; that you may, by Paths left between them, come to weed them, and rake them fine; which is the best way of sowing Seeds, &c. In the warmest sunny part of your Garden or Orchard place your tenderest Plants, or such as you would have forward, observing to keep them, as warm as their nature requires, either with Soil or covering, according to the Season.

Seeds to sow.

The fairest way to have the most Advantage of your Dung or Soil, and to have Seeds prosper, that they may come up most even, and be all buried at one certain depth, is thus; First rake your Bed even, then throw on a part of your mixture of Earth and Dung; which also rake very even and Level, on which sow your Seeds, whether Onions, Leeks, Lettice, or such like; then with a wide Sieve sift on the Earth mixed with Dung about a quarter of an Inch thick, or a little more, and you shall not fail of a fruitful Crop.

Shelter.

In Gardens, as well as in Fields, care must be taken not to sow one sort of Crop too often in one place, or on the same piece of Land, especially Parsnips or Carrots; which being often sown without change of place, are apt to canker, or be worm-eaten.

If your Garden be exposed to the cold Winds, which are very injurious to most sorts of Plants: Next unto Trees, Pales, Walls, Hedges, &c. the best shelter is to lay your Ground after this following manner, that is, Let it be laid up in ridges a Foot or two in heighth, somewhat upright on the back or North-side thereof, and more sloping to the South-side, for about three or four Foot broad, on which side you may sow any of your Garden Tillage; these Banks lying one behind the other will much break the Winds, and these shelving sides will much expedite the ripening of Peas and other Fruits, by receiving more directly the Beams of the Sun; in case the Ground be over-moist you may plant the higher, and if over-dry, then the lower; so that it seems to remedy all extreams except Heat, which rarely injures. Plants newly brought from a hotter Climate than where you plant them, ripen later than those used to it, as those brought from a colder Climate ripen sooner.

'Tis usual to defend several tender Plants from the Cold in Winter to produce them, and to expose them to the Sun in such winter-days as prove clear; which exposure injureth the Plants more than the cold, for

the

the Sun-beams in frofty weather efpecially, if there is Snow upon the Ground, makes a Plant faint and fick, as is obferved in Laurel; which if it grow againft a North-wall, or in the fhade, altho' open to the fevereft winds, yet it will retain its green colour all the Winter; but if it ftand in the Sun, it changeth yellow: the fame is obferved of feveral other tender Plants that are ufually fheltred from the wind, and expofed to the Sun; yet 'tis not improper to give Plants Sun and Air in any time of winter that is mild, but not in Frofty or bleak weather.

To make a hot Bed, in *February* or earlier, for the **Hot Beds.** raifing of Melons, Cucumbers, Radifhes, Colliflowers, or any other tender Plants or Flowers, yo muft provide a warm place, defended from all Winds by being inclofed with a Pale or Hedge made of Reed or Straw about fix or feven Foot high, of fuch diftance or capacity as your Occafion requires, within which you muft raife a Bed of about two or three Foot high, and three Foot over of new Horfe-dung of about fix or eight Days old, treading it down very hard on the top, to make it level; and if you will you may edge it round with Boards or Bricks, laying of fine, rich Mould about three or four Inches thick on it; and when the extream heat of the Bed is over, which you may perceive by thrufting in your Finger, then plant your Seeds as you think fit, and erect fome Forks four or five Inches above the Bed, to fupport a Frame made of Sticks and covered with Straw or Bafs-mat to defend the Seeds and Plants from cold and wet; only you may open your covering in a warm day for an hour before Noon, and an hour after. Remember to earth up your Plants as they fhoot in heighth, and when they are able to bear the cold you may tranfplant them, and the Dung of your hot Beds, when done with, will be of great ufe to mend your Garden.

Many curious and neceffary Plants would fuffer were **Water-** they not carefully watered at their firft removal, or **ing.** in extream dry Seafons, therefore this is not to be

I 3   neglected

neglected early in the Spring; but whilst the Weather is cold be cautious of watering the Leaves of the young and tender Plants, and only wet the Earth about them.

When your Plants or Seeds are more hardy, and the Nights yet cold, water in the Forenoons; but when the Nights are warm, and the Days hot, then the Evening is the best time.

If you draw any Water out of Wells or deep Pits, it ought to stand a Day in the Sun in some Tub, or such like; for your tender Plants in the Spring, before you use it.

But Pond, River, or Rain-water needs it not, and is to be preferred before Well or Spring-water.

If you infuse Pidgeons-dung, Sheeps-dung, Hensdung, Ashes, Lime, or any fat Soil, or other matter in your Water, either in Pits, Cisterns, or other Vessels for that purpose, and therewith cautiously water your Plants, it will much add to their encrease.

For *Colliflowers*, *Artichoaks*, and such like, let the Ground sink a little round the Plant in form of a shallow Dish, and the Water will the better and more evenly go to the Roots.

Water not any Plants overmuch, left the Water carry away with it the vegetative or fertil Salt, and so impoverish the Ground, and chill the Plant.

But it is better to water a Plant seldom and thoroughly, than often and slenderly; for shallow watering is but a Delusion to the Plant, and provokes it to Root shallower than otherwise it would, and so makes it more obnoxious to the Extremity of the Weather.

Slips or cuttings of Herbs or Plants should be planted in moist Ground from *August* to the end of *April*, and be frequently watered, and separated at a Knot, Joint, or Bur, or two or three Inches beneath it, and be stripped of most of their Leaves before you plant them, leaving no side Branches on them; some slit the end where 'tis cut off, and some twist it.

All

All such Plants must stand two Years at the least before they be fit to remove.

If you are willing to have the Ground always moist about a Plant, place near it a Vessel of Water, putting therein a piece of Wollen Cloth or List, and let the one end thereof hang out of the Vessel to the Ground, and the other end be in the Water in the manner of a Crane, causing the List or Cloth to be first wet ; and by this means will the Water continually drop 'till all be dropped out of the Vessel; which may then be renewed, only you must observe to let the end that hangs without the Vessel be always lower than the Water in the Vessel, else it will not succeed ; and if it drop not fast enough, encrease your List or Cloth ; if too fast, diminish it.

If the Weather be never so dry when you sow any sorts of Seeds, water them not 'till they have been in the Ground forty-eight Hours, and the Ground settled about them, that they may be a little glutted with the natural Juice of the Earth, first, lest they burst by too much Water coming on them at once. If the Ground is very dry when you sow Seed, sow them somewhat the deeper.

Such Herbs as are for Physical Uses or to Still, are esteemed to have the greatest Virtue in *May*, or when in Flower, when the full Sap is in them, but the Roots are best in Winter, but then they should be sown between the middle of *July* and the latter end of *August*, or very early in the Spring. Those which are very tender should be sown in Hot-beds, and afterwards being removed into thinner Order be set on good Ground, and constantly kept watered if the Weather be dry : But you must remember, that what is raised on Hot-beds are to be covered, and defended in case of cold Nights or Mornings until they have got some Strength, or are out of danger by the Temperateness of the Season, and that those sorts of Herbs and Flowers are to be sown early, that either Seed, or die the same Year they are sown, that

I 4 they

they may have the benefit of the whole Summer's growth.

The cutting off of the Buds and Branches of Flowers, and other Plants leaving only one Stem or Two more or less, according to the strength of the Root, causes both the Flower and Fruit to be larger, and Herbs also often gather the more strength, yield the fairer Leaf, take the better Root, and endure the Winter the better for their being cut; for the Sap, by this means, has the less to nourish, and the less is expended above, the more the Roots are strengthened, as may be observed, in most Trees that are lopped; nay, some Plants perish in Winter only for want of being cut in the preceeding Summer. Besides, the often cutting of Plants prevents their growing sticky and running to Seed, which often hazards the killing of them; and 'tis observable, that all Herbs wax sweeter both in Smell and Taste for often cutting, especially the latter Sprouts, as may be found in all Esculents.

Flowers and Garden-Fruits are commonly most esteemed for their coming early or late. How to make them early, I have already described, but to retard them, sow or plant them in as late a Season as you can. Remove them often, and prevent as much as you can the usual Excitements of Sun and Air, for the disturbing of the Roots in the removal is a great Hindrance to their Attraction of Nourishment, and new Fibres will not shoot forth to attract new Nourishment 'till several Days after their removal.

Most sorts of Pot and Sweet Herbs may be sown from the beginning or middle of *March* until *August*; in Grounds that are of good temper, but then in extream hot and dry Weather they require a more than ordinary care to be taken of them, but Grounds that are of a moist cold Nature should neither be sown too early nor too late; because, if sowed early, they chill the Seeds and rot them, and the Frosts are apt to spew them out of the Ground, and sometimes the

Worms

Worms are apt to deftroy them ; And if the Seed is fowed late, fuch Grounds are apt to chap or bind, except they are well watered.

Keep or fet all your Herbs that are durable in one quarter in Beds by themfelves, by which means that part of your Garden will be always in order, and do not fet too many in a Heap or Clufter together.

The Sun fhining on Trees or Plants does greatly refrefh and enliven them, efpecially in Winter when 'tis not a Froft, and therefore let the Situation of your Orchard or Garden have as much of the Sun as you can.

The removing of Flowers, Plants, &c. in their proper Seafon, preferves both their Colours and Kinds ; for the long ftanding of them in the fame Soil caufes any Plant to degenerate, becaufe the Plant will exhauft the proper Nourifhment by its long ftanding, and alfo the Soil it felf is apt to change the Nature of the Plant exotick to it, as may be obferved in moft Grains and Plants, which by often removing grow fairer and larger.

The beft time to remove Bulbofe Roots is as foon as the Flower is faded, and the Leaves of the plant withered, which may be done the fooner if when the beauty of the Flower is paft, you cut off the Stalk to prevent its Seeding, fuch Plants being better removed in Summer than Winter. It doing thefe forts of Roots no harm to keep them five or fix Months out of the Ground, but they muft be kept dry leaft they fhould mould or grow ; but Roots of a hollow fpungy Nature, as Frittilarias, Hyacinthis Roots of Ranunculas's, &c. If you keep them long, you fhould mix with them fome fine dry Sand. Thefe general Rules being obferved, I fhall thirdly proceed to the Alphabetical Order promifed ; and begin with

A.
Alefander.

Alefander is propagated only by Seed that is oval and pretty big, and a little more fwelling on one fide

than

than the other, which bends a little Inwards, ſtreaked
all along and croſs-ways on the Edges between the
Sides. It's one of the Furnitures of our Winter Salads, which muſt be whitened in the ſame manner
as *Wild Endive* or *Succory* at the end of Autumn, its
Leaves being cut down, and the Bed wherein it grows
covered over with long dry Dung or Straw, skreen-
ed ſo cloſe that the Froſt may be excluded from it,
whereby the new Leaves that ſpring there-from will
grow white, yellowiſh, and tender. It's ſown pretty thin in the Spring, and the Seed gathered the latter end of Summer, and the Plant being hardy, requires not much watering.

Angelica.

Angelica is of ſeveral ſorts, but that growing in
Gardens is of moſt uſe ; 'tis raiſed by Slips or Seeds,
which it bears in plenty the ſecond Year, and then
fades ; neither the Slips nor Seeds ſhould be planted
or ſown in too dry Ground. It flowereth in *July*
and *Auguſt*, the Roots may be removed the firſt Year.
If you let the Seed ripen the Roots commonly die,
but by careful cutting of it you may prevent its
Seeding, by which means both Root and Plant may
be preſerved many Years.

Arrach.

Arrach, Orrach or Oraga, is propagated only by
Seed, being one of the quickeſt Plants both in coming
up and running to Seed, which laſt it doth the beginning of *June*. They ſow it pretty thin, and ſome
of it, which is good Seed, ſhould be tranſplanted to
a ſeparate place. Its Leaves are very good in Pottage and Stuffings. It ſhould be uſed as ſoon as it
peeps out of the Ground, becauſe it decays quickly ;
and to have ſome the more early, they ſow it in hot
Beds. It thrives very well in all ſorts of Ground,
but grows faireſt in the beſt.

Aſparagus.

Aſparagus is a fine Plant for the Kitchen, and was
much eſteemed, even in *Pliny's* time : They are raiſed of Seeds ſown the latter end of *March*, or about
Michaelmas,

Michaelmas, because they lie long in the Ground before they grow, which some sow in the Shell as they grow, that is, four or six Seeds together; but the best way is to break the Shell, and to beat out the Seed, which put into Water, and the Husk will swim, and the Seeds sink, which dry before you sow : They must be sown indifferent thin, and the Ground kept clear of Weeds, let the Soil be neither too dry, too wet, nor cold; you may sprinkle a few Onions, Radishes, or Lettuce the first Year; but you must draw them as soon as the *Asparagus* begins to spread, and about a Year after, if they are big enough, as they will be if the Ground is good and well ordered; if not, at two Years end at least you may transplant them; which is to be done at the end of *March* and all *April,* planting them in Beds about three or four Foot broad, and raised somewhat higher than the Path-ways that go between them. But as *Asparagus* are most expeditiously raised by Plants bought of the Gardiners of two or three Years old, who raise them on purpose for Sale, you'll find buying of them the most profitable way. In planting of them, mind to spread their Roots as much as you can. These Beds must be well prepared by digging first about two Foot deep, and four Foot wide, and made level at the bottom; mix very good rotten Dung with the Mould, and fill them up. They are planted at two Foot distance in three or four Rows. You must forbear to cut them for three Years, that the Plants may be strong and not stubbed, for otherwise they will be small; but if they be spared four or five Years, they will grow as big as Leeks; the small ones may be left, that the Roots may grow bigger, suffering those that spring up about the end of the Season to run up into Seed, and by this means it will exceedingly repair the hurt that you have done to the Plants in reaping their Shoots. When you have, upon the Winter's approach, cut away the Stalks, the Beds, about the beginning of *November,* must be covered

vered with new Horse-dung four or five Fingers
thick; but some use Earth four Fingers thick, and two
Fingers thick of old Dung, which will keep them
from the Frost : The Beds are to be uncovered about
the middle of *March*, and good fat Mould about two
or three Fingers thick spread over them, and the
Dung laid in the Alleys or elsewhere to rot, and be
fit to renew the Mould with the following Spring,
but Butcher's Dung is the best Soil for them where it
can be had.

If the old Roots of these Plants be taken up about
the beginning of *January*, and planted in a hot Bed,
well defended from the Frost, *Asparagus* may be had
at *Candlemas*; when you cut them, remove some of
the Earth from about them, lest the others that are
ready to peep be wounded, and let them be cut as
low as conveniently may be.

The Beds for this Plant must be covered every
Year with a little Earth taken off from the Path-way,
because they, instead of sinking, are always rising by
little and little; and every two Years they are to be
moderately dunged : About *Michaelmas* the Stems
must be cut down, and the fairest taken for Seed;
and to make them come to bear, an Iron-Fork (the
Spade being dangerous) is to be used to draw them
out of the Nursery Bed. And you must not fail
every Year at the latter end of *March*, or beginning
of *April*, to bestow a small dressing or stirring of the
Ground about three or four Inches deep on every
Bed (taking care not to let the Spade go too deep,
so as to hurt the Plants and to render the Superficies
of the Earth loose) the better to dispose it to drink
up the Rain and *May-dew*, which nourishes the Stocks,
and facilitates the Passage of the *Asparagus*, and kills
the Weeds. The worst Enemies to this Plant are a
sort of Flea that fastens upon its Shoots, and makes
it miscarry; against which mischief there has been
yet no remedy found out; if they are planted in good
Ground, they may stand ten or twelve Years.

Artichoak.

Artichoak

Is by some esteemed one of the most excellent Fruits of the *Kitchen-Garden*, and recommended, as upon other Accounts, so for that its Fruit continues in Season a long time. They delight in a rich deep Soil, and not very dry. The Ground for them must be very well prepared and mixed several times with good Dung, and that very deep. The slips that grow by the side of the old Stumps taken from them at the time of their dressing in the Spring serve for Plants, which are to be set in *March*, *April*, or *May*, according as the Spring is or their Husbandry requires; some plant them in *September*, which by careful covering in Winter may succeed, but the surest Season is the Spring; they must be kept watered 'till they are firmly rooted. You may sow Onions, Radishes or a sprinkling of Carrots or Lettuce between them the first Year, but they must be disposed of before the Artichoaks spread too far. And these, if they be strong, will bear Heads the *Autumn* following; which Off-sets, to be good, should be white about the Heel, and have some little Roots to them. Sometimes *Artichoaks* are multiplied by the Seed which grows in the *Artichoak* bottom, when they are suffered to grow old to Flower, and to grow dry about *Midsummer*.

For the Planting of them they commonly make little Trenches or Pits about half a Foot deep, which they fill with Mould, placing the Roots of them by a Line chequer-ways. They root very deep, therefore plant your sets pretty deep, and if you lay a little Litter thinly about them to keep the Heat of the Sun from them, it will very much improve them. If the Soil be rich, the distance must be three or four Foot; but if not, then nearer. All their other Culture 'till Winter is only weeding, and a little watering if the Spring be dry, and if with the Water you mix Sheep's Dung or Ashes, or if you lay Ashes to the Roots it mightily helps them: But upon the approach of the Winter, for their Security against

Frost,

Frost, be sure to cut the Leaf within a Foot of the Ground, and raise Earth about them in the form of a Mole-hill within two or three Inches of the top, and then cover it with long Dung, which secures them also against the Rain: But others put long Dung about the plant, leaving a little Hole in the middle, and this does very well. An Earthen Pan with a Hole at the top is used by some, a Bee-hive is better; but the most usual way is to cut their Leaves about *November*, and to cover them all over with Earth, and to let them lie in that manner 'till the Spring; but if this be done too soon, it may rot them when they come to be uncovered, and therefore it must be done regularly, at three several times, at about four Days Interval, lest being yet tender the cold Air spoil them. Take off all the old Slips, and leave not above three of the oldest to each Foot for the Bearers, and a supply of good fat Mould must be given to the Roots as deep as conveniently may be. Or you may do this, dig your Artichoak plat all o-ver, and cut off the straggling Leaves both on the top and sides, and lay a coat of Dung all over them, especially about each Stock, and so let it rest 'till Spring; and in *March* or *April* dig your Plat well over, keeping a good open Trench before you, and when you come to a Stock open the Ground pretty deep about it, so low, that you may with your Thumb force the slips from the Stock, excepting two or three of the strongest, unless you find them forward for Fruit; and if any of your Stocks are dead or not thrifty, you may plant young ones in their places: But if your Artichoaks are weak by reason of a sharp Winter that you cannot slip them, you may let them alone 'till they begin to thrive, and then with a slice, without digging, you may force off all the under Slips, maintaining only two or three of the strongest for Fruit; when your Fruit begins to knit, see if any Buds for Fruit appear; if they do, force all off but the principal Head, except such as you spare for latter

ter Fruit. The whole Plantation of them should be remov'd in five Years, tho' they will last much longer in a good deep Mould.

In order to have Fruit in *Autumn*, it is necessary the Stem of such as have born Fruit in the Spring should be cut off to prevent a second Shoot, and these lusty Stocks will not fail of bearing very fair Heads, provided they be dressed well and watered in their necessity, and the Slips that grow on the sides of the Plants (which drain all their Substance) taken away, or you may expect Fruit, from your new set Plants.

As soon as your *Artichoaks* are come to perfection, and fit for use, cut them down close to the Ground Leaves, and all, and by so doing your *Artichoaks* will gather Strength before Winter, and your Plants will be the Stronger, and forwarder in Spring.

The Stalk is blanched in *Autumn*, and the Pith eaten raw or boiled: The way of preserving them fresh all Winter, is by separating the Bottoms from the Leaves, and after parboiling, allow to every Bottom a small earthen glaz'd Pot, burying it all over in fresh melted Butter, as they do Wild Fowl, &c. They are also preserved by stringing them on Pack-thread, a clean Paper being put between every Bottom to hinder them from touching one another, and so hung up in a dry place: They are likewise pickled.

Chards of *Artichoaks*, otherwise called *Custons*, are the Leaves of fair *Artichoaks*, ty'd and wrapp'd up in Straw in Autumn and Winter, being covered all over but at the top; which Straw makes them wax white, and thereby lose a little of their bitterness, so that when boiled they are served up like true *Spanish Cardons*, but yet not so good; besides, the Plants of them rot and perish during the time of whit'ning them.

For *Artichoaks* you have not only the Hard weather and excess of wet to fear, but they have the Field-Mice for their Enemies, which by gnawing of their Roots spoil them.

There

There are three forts of *Artichoaks*, the *White ones*, which are the moſt early; the *Violet* ones, whoſe Fruit is almoſt of a Pyramidical Figure, being the hardieſt ſort; and the red ones, which are round and flat like the white ones; the two laſt are eſteemed the beſt.

Aſarabacca.

Aſarabacca thrives beſt in a moiſt Soil, and is only increaſed by the parting of the Roots.

B

Balm.

Balm is an odoriferous Herb, being multiply'd both by Seed and rooted Branches like *Lavender*, *Hyſſop*, *Thyme*, &c. The tender Leaves are uſed with other Herbs for Salads; the Sprigs freſh gathered put into Wine, or other Drink, during the heat of Summer, give it quickneſs; and beſides, this Plant yields an incomparable Wine made in the ſame manner as Cowſlip, &c.

Barberries.

Barberries are raiſed by Suckers, of which you have plenty about the Roots of old Trees, tho' 'tis not good to ſuffer too many Suckers to grow about them; neither let their tops be cut like cloſe round Buſhes, as many do, which makes them grow thick, that they can neither bear nor ripen Fruit ſo well as if they grew higher and thinner. It is a Plant that bears a Fruit very uſeful in Houſe-wifery, whereof there are ſeveral ſorts, altho' but only one common one; that which beareth its Fruit without Stones is counted beſt; there is moreover another ſort which chiefly differs from the common kind, in that the Berries are twice as big, and more excellent to preſerve.

Baſil.

Baſil is of ſeveral ſorts, as that which bears the biggeſt Leaves, eſpecially if they are of a Violet Colour: but that which bears the leaſt Leaves is moſt Curious, and that which bears the middling ones is the moſt common ſort; all which are propagated by Seed of a black Cinnamon Colour, very ſmall, and a

little

little oval, and by Slips. It is annual and very tender, being seldom sown but in hot Beds, beginning therewith at the beginning of *February*, and continuing so to do all the whole Year. Its tender Leaves are used in a small quantity with the furniture of Salads, among which they make an agreeable Perfume, the same being likewise used in Ragou's, especially when dry. To make it run to Seed (which is gathered in *August*) it's usually transplanted in *May*, either in Pots or Beds. This Plant imparts a grateful Savour, if not too strong: It is somewhat offensive to the Eyes, and therefore the tender tops are to be very sparingly used.

Beans.

Beans are of great Use and Benefit, of which there are several sorts, *viz.* the great Garden Bean, the middle sort of Bean, and the small Bean, or Horsebean, *&c.* the last sort of which I have treated of already. As for Garden-beans, they are usually set betwixt *November* and *February*, at the Wain of the Moon. But if it happen to freeze hard after they are spired, it will go near to kill them, therefore the surest way is to stay till after *Candlemas.* They thrive best in a rich stiff Soil, they should be set five or six Inches deep, and carefully covered from the Mice; and the Ground kept well howed from Weeds. It is a general Error to set them promiscuously, for being planted in Rows by a Line, at three Foot distance, it is evident they bear much better, and may be easier weeded, topped or gathered, and you may sow some early Salletting between the Rows. If they be sow'd or planted in the Spring, they must be steep'd two or three Days in Water, and it's best to set them with Sticks.

In gathering green Beans for the Table, it is the best way to cut them off with a Knife, and not to strip them; and after gathering, the Stalks may be cut off near the Ground, and so probably a second Crop may rise before the approach of the Winter.

VOL. II. K *Some*

Some cut off the Tops while they are about half a Foot high to make them branch, but then they must not be set too thick ; others do not top them till they are about two Foot high or more, to the end they may ripen the better and the sooner ; and that they may sow Turneps as soon as the Beans are gone. The Tops some eat boiled.

Bears-foot.

Bears foot is only raised by Seed.

Beet Raves, or Beet Radishes.

Red Beet produces Roots for Salads, being multiply'd only by Seeds of about the bigness of a midling Peas, and round, but rough in their roundness : They are sow'd in *March,* either in Beds or Borders, very thin, in good well prepar'd Ground, or else they will not grow so fair and large as they should be ; they are best that have the reddest Substance, and the reddest Tops, and are not good to spend but in Winter ; their Seed is gathered in *August* and *September,* for the procuring whereof, some of the last Years Roots that have been preserv'd from the Frost are transplanted in *March,* the Roots being cut into thin Slices and boiled, when cold, make a grateful Winter Salad.

Beet White.

White Beett is also propagated for Chards by Seeds only, like unto that of the Red Beets, but of a duller Colour ; the Rib of it being boiled, melts and eats like Marrow.

Chards of Beet

Are Plants of *White Beet* transplanted in a well prepar'd Bed at a full Foot's distance, producing great Tops, which in the midst thereof have a large, white, thick, downy and cotton-like main Shoot, which is the true Chard used in Pottages and Entremesses. When White Beets have been sown in hot Beds, or in naked Earth in *March,* that which is yellowest is transplanted into Beds purposely prepared, and being well watered in the Summer they grow big
and

and ſtrong enough to reſiſt the hard Winter's cold, if ſo be they be covered with long dry Dung, as we do *Artichoaks.* In *April* they are uncovered, and the Earth dreſt carefully about them, and ſo produced, their Seed is gathered in *July* or *Auguſt.*

Bloodworth.

Bloodworth is raiſed of Seed which is ripe in *June* and *July.*

Bona viſta.

Bona viſta is a kind of *French* bean, and will grow with us, eſpecially if the Summer is kindly; they are raiſed as French Beans are.

Borage

Is propagated only by Seed that is black, and of a long oval Figure, commonly with a little white end towards the Baſe or Bottom that is quite ſeparated from the reſt, being ſtreak'd black all along from one end to the other. It grows and is to be ordered in the ſame manner as *Arach*, but it does not come up ſo vigorouſly. It is ſown ſeveral times in one Summer; the Seed falls as ſoon as ever they begin to ripen, and therefore muſt carefully be watch'd, and the Stalks cut and laid adrying in the Sun, whereby few will be loſt: Its Flowers ſerve to adorn Salads, but they are not eaſily digeſted, tho' their Leaves are, if their String is firſt taken away.

Bugloſs.

Bugloſs is order'd after the ſame manner.

Bucks-born Salad

Is only multiply'd by Seed, which is very ſmall, and is ordered after the ſame manner as Borage. When the Leaves of this Plant are cut, there ſpring up new ones in the room of them.

Burnet.

Burnet is propagated only by Seed that is pretty big, a little oval with four ſides, and as it were all over engraven in the ſpaces between the four Sides. It is a very common Salad, ſeldom ſown, but in the Spring it often ſprings afreſh, after Cutting; the

Shoots are for Salads; the same requires watering in Summer, at the end whereof the Seeds are gathered.

C.

Cabbage and Coleworts.

Cabbage and *Coleworts*, whereof there are divers sorts, such as the *Dutch Cabbage*, which is very sweet and soon ripe; the large sided *Cabbage*, that is a tender Plant not sown till *May*, planted out in *July*, and eaten in Autumn; but the best *Cabbage* is the *White Cabbage*, which is the biggest of all; and the *Red Cabbage*, which is small and low; the *perfumed Cabbage*, so named from its scent; the *Savoy Cabbage*, which is one of the best sort and very-hardy; and the *Russia Cabbage*, which is the least and most humble of them, but very pleasant Food, and quick of growth.

They are raised of the Seed sown between *Midsummer* and *Michaelmas*, that they may gain strength to defend themselves against the violence of the Winter, which yet they can hardly do in some Years; or else they may be raised in a hot Bed in Spring, and transplanted in *April*, or about that time, and that into a very rich and well stirred Mould, if large *Cabbages* are expected. They delight most in a warm and light Soil, and require daily watering till they are rooted. But yet great quantities of ordinary *Cabbage* may be raised in any ordinary Ground if well digged and wrought.

As for the Seed, if you intend to preserve it, it must be of the best *Cabbages*, placed low in the Ground during the Winter, to keep them from cold Winds and great Frosts, they must have Earth Pots, or a warm Soil over them for their covering, and be planted forth at Spring; or you may about *October* or *November*, when the Frost begins to come, take up such as you desire to have the seed of before the Frost surprize you, and hang them up by the Roots about a Fortnight to drain the water from them; they should be hard well grown Cabbages, such being forwarder to seed when the season comes than others. These
                                                                    you

you may forward by cutting off the Cabbage on the top with a cross cut, and you may likewise wrap a piece of old Cloth, Bass-mat, *&c.* about the Roots of them, and lay them in some cold Cellar, or By-room, or hang them up until the end of *February*, or the beginning of *March*, and then plant them in some temperate place, that is neither too hot nor too dry. The Stems of a good Cabbage, if you can preserve them from rotting and frost till Spring, will bear as good seed as a whole Cabbage. You must likewise keep your Cabbages from breaking with the wind, by tying them to Sticks. Besides, this variety of Cabbages, Cauls and Sprouts, springing from old decapitated Stumps, there is a perennial Caul that will continually yield you a green Mess when ever you have occasion.

If your *Cabbages* or *Colliflowers* are troubled with Caterpillars, mix Salt with Water, and water them therewith, and it will kill them.

Calamint.

Calamint is raised by Slipping, or parting of the Roots, and sometimes by Seed.

Camomile.

Camomile double is like the common sort, only the Leaves greener and larger, and the Flowers, are very double, being white, and somewhat yellow in the middle. It is more tender than the common one, and must yearly be renewed by setting of Slips thereof in the Spring, or parting of the Roots.

Carduus Thistle.

Carduus, tho' it is a noisome Weed, yet some of them are received into Gardens, whereof are first the greater *Globe Thistle* with Leaves cut in, and gashed in the middle full of sharp Prickles, its branched Stalk above a Yard high, bearing great round hard Heads, with a sharp bearded Husk of a bluish green colour, from whence come pale blue Flowers spreading over the whole Head, and are succeeded by the Seeds contained in the Husks, which must be preserved, for the Plant dies in the Winter. Secondly, the lesser *Globe*

Thistle,

Thistle, whose Leaves are smaller and whiter, as are the Stalk and Head of the Flowers, the Roots more durable, lasting four Years bearing Flowers.

Their flowering time is usually in *August*, and being sowed of Seeds, they will come to bear Flowers in the second Year ; they prove a great annoyance to some Lands by killing the Grass, Corn, *&c.* tho' they be a sure token of the strength of the Ground. The way to destroy them, is to cut them up by the Roots before Seeding-time.   Our *Ladies milky dapled Thistle* is worth esteem, for the young Stalk, about *May*, being peeled and soaked in Water, to extract the Bitterness, either boil'd or raw, is a very wholesome Salad eaten with Oil, Salt and Pepper ; some eat them sodden in proper Broth, or baked in Pies like the *Artichoak*, but the tender Stalk boil'd or fry'd some prefer ; both are nourishing and restorative.

Carrots.

Carrots are the most universal and necessary Root this Country affords, and hereof there are two sorts, the Yellow, and the Orange, or more Red ; the last of which is by much the better.   They are raised of Seed, and principally delight in a warm, light or sandy Soil ; and if the Ground be so, tho' but indifferently fertile, yet they will thrive therein ; for if the Ground in which you plant them be heavy, you must take the more care in digging of it to lay it as light as you can ; and if you dung the Land the same Year you sow Carrots, you ought to bring it so low, that the Roots may not reach it ; for as soon as they touch the Dung, they will grow forked.   It's a usual thing to sow them with Beans in the Intervals between them, and in digged, not ploughed Lands, because of their rooting downwards ; for after the Beans are gone they become a second Crop ; but Carrots sown among Beans are not so fair nor early as those sown in a Bed by themselves ; and some of the fairest of them being laid up reasonably dry in Sand, will keep throughout the Winter : The same may be reserved till Spring,

and

and planted for Seed, or elfe Seed may be gathered from the biggeft afpiring Branches; obferving to preferve the largeft and faireft Seed for fowing. They may be fown in Autumn or Spring; If you will have them in fpring, fow them in *Auguft*, and preferve them from the froft in Winter by covering of them with Peas haum, or fuch like; or fow them in *February* or *March*; and to make them grow large, do with them as with Turneps, only they will admit of a greater number on the fame quantity of Land than Turneps will.

Cardon Spanish.

Cardon Spanifh are only propagated by Seed that is of a longifh oval Form, and about the bignefs of a Wheat Corn, of a greenifh, olive Colour, ftreaked from the one end to the other. They are fown at two feveral times; the firft from the middle of *April* to the end, and the other time about the latter end of *May*, in a good and well prepared Ground, in fmall Trenches or Pits a Foot wide, and fix Inches deep filled with Mould, and then make for them Beds four or five Foot wide, in order to place in them two Ranks of thofe Pits chequerways, putting five or fix Seeds into every Hole, with Intention to let but two or three grow, and take away the reft if they come up: But if in fifteen or twenty days the Seed doth not come up, they fhould be uncovered, to fee whether they be rotten, or begin to fprout, that their Places may be fupply'd with new ones if need require; they muft be carefully watered; and towards the end of *October*, if you have a mind to whiten them, take the advantage of a dry Day, firft to tie up all the Leaves with two or three Bands, and fome Days after to cover them quite with Straw or dry Litter, well twifted about them, except at the top, which is to be left open; thus ordered they whiten in about three Weeks, and are fit to eat.

K 4   They

They may be transplanted upon the approach of Winter into the green House, removing some Earth with them, some of which may be planted next Spring to run to Seed in *June* or *July.*

Chervil.

Chervil is only multiply'd by Seed that is black, very small, pretty longish, striped longways, and grows upon Plants sown the Autumn before, which knits and ripens in *June* ; the musked sort thereof is one of our Salad Furnitures; and at the beginning of Spring while the Leaves are tender is very agreeable. It remains many Years without being spoiled by the Frost. As for the ordinary ones for Salads it is annual, and a little thereof should be sown Monthly, as there is occasion for it. It runs very easily to Seed, and if you would have some of it betimes, it must be sown by the end of Autumn ; the Stalks are cut down as soon as they begin to grow yellow, and the Seed beaten out, as is done by that of other Plants.

Ciboules.

Ciboules or *Scallions* are a kind of degenerate Onion, and are propagated only of Seed. They are sown in all Seasons, but herein they are different from Onions in that they produce but a small Root, and several Stems or upright Shoots, and those which produce most of them are most esteemed, of which you should be careful to save the Seed which is ripe in *August*, if planted in *March.* They must be thinned as well as Onions, and some that are transplanted will prosper well in dry Summers, if their Beds are well watered, and they are planted in good Earth. The red, hard and sweet are the best. They are reckoned very good to excite appetite.

Citruls, Pumpions or Pumpkins,

Are propagated only by Seeds of a flat and oval Figure, partly large and whitish, and as it were neatly edged about the sides; there are two sorts of them, the green and the whitish ; they are usually sown in hot Beds about the middle of *March*, and beginning

of

of *April*, and being taken up with the Earth about them, are transplanted into Holes two Foot Diameter, and one deep, and at two Fathoms distance, which are filled with Mould; in *June*, when their Vines begin to grow five or six Foot long, some Shovelfuls of Earth are thrown upon them to prevent their being broken with the Wind, and to make them take Root at the place so covered, whereby the Fruit that grows beyond that part will be better nourished, and so grow bigger: If the Weather is dry, they should be well watered, and about the blossoming time, take away all the by Shoots, leaving only one or two main vines, and beware not to hurt the heads of them. Whilst they are about the bigness of a Melon, they eat well pickled.

Cives.

English Cives are multiplied only by Off-sets that grow round about their Tufts; from them a part is taken to replant, being slipt out and separated into many little ones, and transplanted nine or ten Inches asunder, either in Borders or Beds in pretty good Ground; They will last three or four Years without removing, or any other Culture than weeding and watering sometimes during the heat : It is their Leaves only that are used for one of the Sallet Furniture.

Clary.

Clary, when tender, is an Herb not to be rejected in Sallets. It's raised of Seed, but the Root perisheth after the Seed is ripe; which is the second Year after 'tis sown. It flowereth in *June*, *July* and *August*.

Coastmary.

Coastmary is raised by Slips or parting of the Roots, and sometimes by Seed.

Cole-Flower, or *Cauly-Flower*.

Cole, or *Cauly-Flower* is an excellent Plant, whose Seed may be sown at any time between *Midsummer* and *Michaelmas*; the Seeds should not be sown too thick and be covered about an Inch thick at last with

fine

fine Mould ; they ought carefully to be preserved over
the Winter, by Matts or other close shelter ; or else
they may be raised in hot Beds in the Spring, by sowing of the Seeds in *February*, but upon all opportunities,
when the air is temperate uncover them, that you may
harden them by degrees ; and when your plants are
about two or three Inches high, make another bed of
less substance than the first ; and being of fit temper,
that is, near as warm as the bed from which you removed them, set them about three Inches asunder,
not forgetting to water them as often as need requires,
and keep them shaded while they are new planted, in
case of dry weather ; some arch the Beds over with
Pots or Hoops, that they may the better cover them
in cold weather, or when much rain comes, and when
they have indifferent large Leaves they may be removed into good Lands prepared for that end ; tho'
the best way is to dig small Pits, and fill them with
rich light Mould, wherein the *Cole-Flower* must be
planted, and afterwards carefully watered : Thus you
may be furnished with Winter Plants for Seed. Those
that are of one Years growth usually Flower about a
time ; to prevent which, some of the Plants may be
removed once a Fortnight, for two, three, or four
times as a Man pleases, and so they may be had successively one after another ; or else the Flower may
be cut off before it is fully ripe with a long Stalk,
and set in the Ground, as far as may be, and it will
retard its ripening ; but it must be shaded, and have
a little watering left it wither.

Conval-Lily.

Conval-Lily, *May-Lily*, or *Lily of the Valley*, has a
strong Root that runs into the Ground, and comes
up in divers Places with three or four long and broad
Leaves, and from them rises a naked Stalk, with
white Flowers at the top like Bottles with open
Mouths, of a comfortable, sweet Scent. There is another sort differing from these only in Flowers, which
are

are of a fine pale Red, both of them flowering in
May, and bearing beft in a fhady mean Soil, being eafily propagated from Plants.

Corn Sallet.

Corn Sallet is an Herb whofe top Leaves are a Sallet of themfelves, feafonably eaten all the whole Winter, and early in the Spring with other Sallets, it is
raifed of Seed at firft, but afterwards will fow it felf.
Vid. Mulches.

Creffes.

Creffes Garden, *Indian* or *yellow Lark-Spurs*. They
are fown in many Gardens for culinary Ufes; and the
latter from a Flower, are now become excellent Sallets as well the Leaf as the Bloffom; for early Sallets they are raifed in hot Beds: But if fown in *A-
pril*, they will grow very well on ordinary Garden-
ground, and their Leaves and Bloffoms plentifully en-
creafe. *Water-Creffes* are eaten boiled or raw, and
like the other fort of *Creffes* are raifed of Seed.

Cucumbers.

Cucumbers are of two forts, the large green *Cucumber*, vulgarly called the *Horfe-Cucumber*, and the fmall
White which is the more prickly *Cucumber*. The firft
are beft for the Table green out of the Garden; but
the other to preferve. They are planted and propagated after the fame manner as *Melons*, only they
require more watering, and are withal much more
hardy, if planted late, elfe they are as tender; but
though the watering makes them more fruitful, yet
they are more pleafant and wholfome if they have
but little Water; and though they fhould be watered in dry Weather, they are to be defended by
fome covering from the Rain in cold wet Weather,
and if the top Shoot of *Cucumbers* be nipped off when
fhot out three or four Joints, it will caufe them to
knit the fooner for Fruit; if you fow them any time
in *March* it will be foon enough, and if you have
Glaffes you need not make up a Bed for them on purpofe, but only make Holes about the bignefs of a

Bufhel,

Bushel, which fill with warm stable Dung, in the midst of which plant three or four *Cucumber* Plants with mould about them, and Earth them up round like a Dish to hold Water. If you raise them tenderly under Glasses, you must use them tenderly, otherways any cold Rains will be apt to spoil them ; but if you raise them without Glasses, you must not plant them out 'till warm dry Weather, and at first observe to shade them well from the Sun, and to give them Air as often as you can when the Weather is good, only lightly covering of them with Mats or Straw every Night, if 'tis likely to be cold, and remember at first planting of them to Water them ; and if you will not be at the trouble of raising of them on a hot Bed, you may at any time from the middle of *April* to the beginning of *May*, make Holes as is before described ; and in the midst of each put five or six *Cucumber* Seeds, and the Weather being warm, water them now and then as you see Occasion. If your Plants thrive, three in a Hole will be enough to leave, the rest you may pluck up or plant elsewhere. If you desire to have any for Seed, save of those which are ripe forwardest, for the riper and better grown your Seed is, the longer it will last, even to three or four Years old, and the riper it is, the less labour it will require to wash from the Pulp.

Currants.

Currants or *Corinths* first took their Names from their likeness to the small *Grapes* or *Raisins* which come from *Corinth*. They are raised by Suckers or Cuttings stuck in moist places, of which you may have plenty about the Roots of old Trees, which when they have grown for some Years, suffer not many Suckers to grow about them. Do not cut the tops to a round close Bush, as many Gardeners do ; whereby they grow so thick, that they neither bear nor ripen their Fruit so well as if they grew taller and thinner. The *English* red *Currant* (formerly transplanted to *England*) is not now valued, nor yet the Black : The white

Currant

Currant 'till of late was moſt in eſteem, but the red *Dutch Currant* becoming Native of our Soil, has been ſo much improved in moiſt rich Grounds, that it hath obtained the higher Name: Beſides which, there is again another ſort (propagated among us) to be eſteemed only for Curioſity, not for Fruit. Their Culture conſiſts in cutting away the old Wood, and preſerving only that of one or two Years growth; for a confuſed Mixture is not only diſagreeable and pernicious, but the old Branches will bear nothing but very ſmall Fruit, 'till at laſt they quite degenerate; therefore when the Stocks grow old, you ſhould raiſe a Plantation of new ones in ſome other freſh choice piece of Ground, after they have ſtood about ſeven or eight Years.

Currants will thrive mightily, and grow very large, if ſpread upon a Wall even againſt a North-Wall, eſpecially white *Currants.*

Currants and *Gooſe-Berries* may be inoculated on their own kind.

D.
Dandelion.

Dandelion is an Herb which is macerated in ſeveral Waters to extract the Bitterneſs. It is little Inferior to *Succory, Endive,* &c. The *French* Country People eat the Roots of it.

Dill.

Dill is raiſed of Seed, which is ripe in *Auguſt.*

Dittander.

Dittander is raiſed by Sprouts growing from the old Roots.

Dragon.

Dragon is increaſed by Off-ſets or young Roots, and ſometimes by Seed.

Elecampane.

Elecampane delights in a moiſt Soil, is raiſed by Seed, and by parting of the Roots; it Flowers in *June* and *July*, the Roots are beſt which are gathered in *Autumn.*

Endive,

E.

Endive white, or Succory.

Endive, or *Succory,* is of several forts, as the white, the green and the curled, which are only propagated by Seed that is longish, of a white grey Colour, flat at one end, and roundish at the other : It grows upon the Stocks or Stems of the preceeding Years growth, and you would take it but only for little bits of Herbs cut small.   The wild is also propagated in the same manner, from longish black Seed ; and is a fort of a very good annual Plant used in Sallet and Pottage in Autumn and Winter Seasons, if so be it is well whitened, and so made tender. All forts of them agree pretty well with any kind of Ground, and are seldom begun to be sown till the middle of *May,* and then very thin, or they must be thinned afterwards, in order to be whitened in the place where they first grew, without transplanting ; there is also but a little quantity of them to be sown at once, because they are apt to run into Seed ; but for a greater quantity let them be sown the latter end of *June,* and all *July,* in order to have some good to spend in *September* ; after this a great quantity is sown in *August,* for a sufficient supply to serve the Autumn and fore-part of the Winter.   When they are transplanted in Summer-time, they must be set at a large Foot distance, and great Beds of five or six Foot broad are usually made for them, to plant them afterwards in Lines marked out with a Cord ; this Plant requires great and frequent waterings, and when big enough to be whitened, it is tied up with two or three Bands according as its heighth requires, and it is whitened in fifteen or twenty Days : but to preserve it upon the approach of cold, it must be covered with long dry Dung, whether it be tied up or not ; some whiten it, and other fort of Sallets of the same kind by laying of them in sand or Earth, either within or without Doors ; at the end of *September* the Stocks are planted pretty near one another, because it nei-

ther

ther grows so high or spreads so much as in Summer: And in case any Plants can be saved in Winter, they must be transplanted again in the Spring, in order to produce Seed, that they may have a sufficient time to ripen. For the wild Endive it is sown in *March* pretty thick in a well prepared Ground, and fortified by watering and croping, that it may be fit to whiten in Winter; the best way to whiten which, is to interpose some Props from side to side to keep the Dung, where-with it must be well covered, from touching of it. It will shoot under a close cover, and therefore care must be taken to stop up well the Passages on all sides, that no Light or Air at all can get in; for hereby the Roots are much cleanlier, and relish not so much of the Dung. It may be transplanted into Conservatories in Winter. When it is green it endures the Frost well enough, and runs into Seed the latter end of *May*. Many People eat its Shoots in Sallets when they are young and tender: It is eaten with *Mint, Rocket, Tarragon,* and other hot Herbs.

Eschalots.

Eschalots are now from *France* become an *English* Plant, being increased and managed after the same manner as *Garlick,* which may be seen for that purpose; only they are to be set earlier, because they spring sooner, and taken up as soon as the Leaves begin to wither; long after which, they must not lie in the Ground, for either they rot there, or the Winter kills them: They give a fine relish to most Sawces, and the Breath of those that eat them is not offensive to others; but being planted two or three Years in the same Ground they are apt to degenerate.

F.

Featherfew.

Featherfew is both single and double, and is encreased by seeds or slips; and also by dividing of the Roots; it flowereth most part of the Summer, and should have a good Soil.

Fenil.

Fenil.

Fenil is only propagated by Seed that is small, longish, oval and streaked with greenish grey streaks; it is one of our Sallet-furniture, that is seldom transplanted, and resists the cold of the Winter if it is saved in Beds or Borders; it springs again when it is cut, and its youngest and tenderest shoots are the best: The Seed is gathered in *August*, and agrees well enough with any sort of Ground.

French-Beans, or Kidney-Beans.

French or *Kidney-Beans*, are a sort of Cod-ware, that are very pleasant wholesome Food, being but lately brought in use amongst us, and are not yet sufficiently known: There are four sorts thereof. First, The Scarlet-Bean which has a red Husk, and is not the best to eat in the Shell, as Kidney-Beans are usually eaten, but is reputed the best to be eaten in Winter when dry and boiled. Secondly, The painted or streaked Beans which are the hardiest, tho' meanest of all. Thirdly, The large white Bean which yields a fair delicate Pod. Fourthly, The small white Bean which saving in size is like the latter, but esteemed the sweeter. They delight in a warm, light and fertile Ground, which being about the beginning of *May*, or very soon after, planted with them at a Foot distance, and two Fingers deep, will yield an extraordinary Crop. But as they are very uncertain in taking, plant but a small quantity in a day, and two or three days after, a small quantity more, and so on, that you may not plant all your quantity together, for they will often take one Day and miss the next; but where they come up too thick, they may be transplanted, only they must be well watered at first Planting: You may either set up tall Sticks near for them to twine about, or let them lie on the Ground; but if you are straitned in room, those on Sticks will yield the greatest increase.

Garlick.

G.

Garlick.

Garlick is increased by parting of the Cloves or Off-sets in *February* or *March*; and planting of them in a rich good Soil, in which they will encrease wonderfully. Their Leaves about the end of *June*, may be tied in knots; which will prevent their spindling, and keeping down of the Leaves will make the Roots large: Much more of this Root would be spent for its wholesomeness, where it not for the offensive smell it gives to the By-standers, which is taken away by eating of a Beet-root roasted in the Embers; but yet by *Spaniards* and *Italians*, and the more Southern People, it is familiarly eaten with almost every thing, and esteemed of singular Vertue to help Concoction. As soon as it has done growing, it should be taken up, and kept dry for use.

Gentian.

Gentian is of several sorts, but the most valu'd are, 1. The *Great Gentian*, bearing a yellow Flower. 2. The *Gentian* of the Spring; that on the top of its stalks bears a large, hollow, bell-fashion'd Flower, of an excellent blue, with some white spots in the bottom of the inside. Its Roots are small, pale, yellow Strings that put forth Leaves, whereby it yields a great increase. This last flowers from *April* till *May*, as the first does from *June* to *July*, which increases slowly by the Root, and is hardly raised from Seeds: And if there be any got from them, it will be many Years before they come to bear Flowers. The Root must be planted in *September*, in rich Ground, under a South Wall, and carefully defended from Frosts in the Winter; the other will prosper in almost any Soil, so it be in an open Air.

Germander.

Germander is raised by setting of the Tops or Slips; it flowereth the greatest part of Summer.

Goats-rue.

Goats-rue is increased by Seeds or Slips near the Roots ; it flowereth in *July* and *August*.

Goose Berries.

Goose-Berries must be ordered the same way as before is prescribed for *Currants*. They are of six sorts, as white, green, yellow, red, black, and strip'd. *Goose-Berries* may be raised of Seed, they should be removed once in eight or ten Years. They delight in a rich dry Soil, and *Currants* in a moist Soil.

H.

Harts-horn : See *Bucks-horn*.

Harts-tongue.

Harts-tongue is propagated by parting of the Roots, and also by Seed.

Horse-radish.

Horse-radish is increased by Sprouts spreading from the old Roots, or by pieces of Roots left in the Ground, that are cut or broken off ; and if you take up any Roots for use, you may cut off all the Roots except a small part next the Leaves, which you may plant again : If you abate the Leaves to about an Inch long, and water it well ; the best time of increasing of the Roots, is in the Spring. It delights in a rich Soil.

Hyssop.

Hysop, or *Hyssope*, is raised by Seeds, Slips, or the Tops planted out.

I.

Jack by the Hedge.

Jack by the Hedge (*Alliaria*, or *Sauce alone*) is an Herb that grows wild under Banks and Hedges, and has many Medicinal Properties, being eaten as other Sallets are, especially by Country-People, and is much used in Broth.

Jerusalem Artichoaks.

Jerusalem Artichoaks are increased by small Off-sets, and by quartering of the Roots, by which means they

will

will make a very great increase in a small spot of Ground. See *Potatoes.*

Jerusalem Sage.

Jerusalem Sage is raised of Slips, like other Sage.

K.

Kidney-Beans : See *French Beans.*

L.

Lavender.

Lavender is multiplied by Seed, and old Stocks or Plants transplanted, but chiefly by Slips, it serves to garnish Borders of the Kitchen-Gardens, and yields a Flower which is used for several Phyſical Uſes, and to put among Linen to perfume it.

Lavender Cotton.

Lavender Cotton is increaſed by Slips, and makes fine Borders.

Leeks.

Leeks are raiſed of Seed, as *Onions* are, and ſown about the ſame time. They ſeed the ſecond Year, and unleſs they are removed they die : About the Month of *Auguſt* plant them in very fat rich Ground, for which deep Holes are made with a ſetting Stick, but fill not the Holes with Earth ; water them once in two Days with Water, enriched with fat Dung, and they will be very large and white : The beſt for Seed are planted in the ſame manner as *Onions* ; and the Seed bearing Stalks of both muſt be ſupported by Threads or Sticks, otherwiſe they will lean to the Ground.

Lettice.

Lettice of all ſorts ſeed the firſt Year and Die, if they are not tranſplanted for Winter *Lettice,* which prevents their running to Seed, and are multiplied only by Seed, which being ſown in the Spring, ſeed in *July* ; and ſo do the Winter, or ſhell *Lettices,* after having paſſed the Winter in the places where they were replanted in *October* ; they are the moſt common and uſeful Plant in the Kitchen-Garden, e-

ſpecially

specially for Sallets. There are many kinds of them, as the Cabbage *Lettice*, which with the ordinary Culture comes to Perfection: The Shell *Lettice*, so called from the roundness of its Leaf, almost like a Shell, is the first that will Cabbage at the going out of the Winter; otherwise called Winter *Lettice*, because they can pretty well endure ordinary Frosts. They are sown in *September*, and in *October*, and *November*, transplanted into some Wall-border towards the South and East; or else they are sown in hot Beds under Bells in *February* or *March*, and are good in *April* and *May*. Another sort of *Lettices*, called *Passion Lettice*, prosper well in light Ground, and are succeeded by the bright curled *Lettices*, which usually Cabbage in the Spring, and do also well upon hot Beds: Of this sort there are two others, *viz.* *George Lettices* that are thicker and less curled; and the *Minion* which is the least sort, and requires good black or sandy Ground. Near about the same Season comes in curled green *Lettices*, besides the red and short *Lettices* that have small Heads, and require the same Ground. In *June* and *July* come on the Royal Bell-gards, or Fair Looks, Bright *Genoas*, *Capuchins*, &c. to whom frequent Rains are pernicious: Others are called Imperial *Lettices* from their size, delicate in taste, but apt to run into Seed. But to have no more Diversities, the great Inconveniencies that befal Cabbage *Lettices* are, that they often degenerate so far as to Cabbage no more, and therefore no Seed should be gathered but from such as do Cabbage well, and as soon as they are Cabbaged they must be spent, unless you would have them run unto Seed without doing any Service: For if the rot that begins at the end of the Leaves seizes them, which it will often do when the Ground or Season is not favourable unto them, there is hardly any Remedy; only the Ground that is faulty may be mended with small Dung, whether it be sandy or cold gross Earth.

Those

Thofe *Lettices* which grow biggeſt ſhould be placed at ten or twelve Inches diſtance, and for thoſe that bear Heads of a middling ſize, ſeven or eight will do; and ſuch as would be good Husbands may ſow *Radiſhes* in their *Lettice* Beds, for they will be all drawn out and ſpent before the *Lettice* Cabbage; and for the ſame Reaſon, *Endives* being much longer before they come to Perfection than the *Lettices*, ſome of theſe laſt may be planted among the *Endives.* You may alſo blanch the largeſt Roman *Lettices*, when they are at their full growth, by binding of them up with Straw, or raw Hemp, or by covering of them with Earthen Pots, that have Dung put about them. They ſeed the firſt Year and Die, if they are not tranſplanted for Winter *Lettice*, which prevents their running to Seed.

Lily.

Of this Plant there are divers kinds. Firſt, The fiery red Lily that bears many fair Flowers on an high Stalk of a fiery red at the top, but towards the bottom declining to an Orange Colour with ſmall black Specks. Secondly, The double red Lily having Orange coloured ſingle Flowers with little brown Specks on the ſides, and ſometimes but one fair double Flower. Thirdly, The yellow Lily which is the moſt eſteemed of any, being of a fine Gold colour. Fourthly, The common White one like the common Red. Fifthly, The white Lily of *Conſtantinople*, ſmaller every way than the laſt, but bears a great many more Flowers. Sixthly, The double white Lily in all things like the common, except the Flowers which are conſtantly double, ſeldom opening at all but in a fair Seaſon. Seventhly, The *Perſian* Lily rooted like the Crown Imperial, beſet with whitiſh green Leaves to the middle, and thence to the top, with many ſmall Flowers hanging their Heads of a dead purple Colour, with a Pointil, or Chives, in the middle, tipt with yellow Pendants. Theſe (ſave the

laſt,

laſt, which Flowers in *May*,) put forth their Flowers in *June*. All of them are increaſed by the Roots, which hold their Fibres, and therefore like not often removing but when there is Occaſion. The beſt time is when the Stalks are dried down, for then the Roots have the feweſt Fibres, and ought to be ſet five Inches deep in the Earth, and uncovered to the bottom every Year, that without ſtirring the Fibres of the old Roots, the young ones may be parted from them, and they only remain with new rich Earth put to them and covered ; which will much advantage the fairneſs and number of their Flowers. See *Conval Lily*.

Liverwort or *Hepatica*.

Liverwort or *Hepatica*, are of two ſorts, the ſingle and the double : The Seeds of the ſingle one are of uſe, which may be ſown in *Auguſt*, in caſes or well ſecured Beds ; they muſt be planted in rich well dunged Soil, and are increaſed by parting of the Roots when grown into ſeveral Heads. Care muſt be taken when the Flowers have near loſt their Beauty, to tie them up to a Stick, to prevent a rotting of the little Pods, before the Seed ripens, and to prevent the Seeds dropping out of them.

Lupines.

Lupines are an excellent Pulſe, and require little care. They are very advantageous to any Ground they are ſown on, and are a good Manure for barren Land. With us they are annually ſown in Gardens for the ſake of their Flowers ; but in *Italy* they are ſown in the Fields for Food for their Cattle ; for, being ſodden in Water, they are excellent Food for Oxen, and for other Cattle. There are four ſorts of *Garden Lupines* ; the firſt and moſt common is that with yellow Flowers, unto which that with white Flowers is very like ; the other two ſorts are Blue, one ſmall, and the other large, of which the larger are eſteemed the beſt, and afford not only good Nouriſhment for Cattle ; but for Men alſo, they being eſteemed

steemed very easy of Digestion, and very good for
the Stomach, good Bread being made of them, mixed
with Bean or Wheat Flower, the *Lupines* being first
dried in an Oven. They should be kept well and
dry to prevent their growing Mouldy.

M.

Maches.

Maches or *Maskets* are multiplied only by Seed,
which is very small, and of an Orange Colour, they
being a sort of little Sallet, which is termed wild or
rustical. Beds are made for them, which are sowed
about the end of *August*, they are hardy enough to
resist the rigour of Frosts; and forasmuch as they
produce a great many little Seeds that will easily fall,
they will sufficiently propagate themselves without
any other Culture than only weeding.

Marjoram.

Of *Marjoram* there are several sorts, which are easily raised of Seed sown in *May*; the vulgar sort and
pot Marjoram is raised by Slips, whose uses are commonly known: There is also a Distinction of Winter Marjoram, which is the best, and Summer Marjoram that lasts only that Season: It is also propagated
by Slips or Suckers in *April*.

Marygolds.

Marygolds are increased by Seed, they Flower most
part of Summer, and even in Winter if 'tis warm,
they may be transplanted at any time in moist Weather.

Maſtick Tyme.

Maſtick Tyme or *Marum* is increased by Slips, but
'tis apt to be destroyed by cold.

Aſſyrian Maſtick is of the same Nature, only care
must be taken to preserve it from Cats by Thorns
or Fers. These Masticks are best preserved by placing of them within the Earth, and covering of them.

Maſterwort.

Maſterwort is raised of Seeds, or Runners from the
Roots.

L 4   *Maudlin.*

Maudlin.

Maudlin or *Coaſtmary* is raiſed by Slips or Seeds, and Flouriſheth moſt of the Summer Months.

Melons.

Melons or *Musk melons,* as they are uſually called from their pleaſant Scent, are a Fruit raiſed for pleaſure in the Summer time, and diſtinguiſhed by ſeveral Names: Thoſe the moſt uſually known are the large ribbed *Melon,* and the ſmall round *Melon,* the Seeds being firſt ſteeped in Milk for twenty-four Hours, ſome propoſe to ſteep them in Wine and Sugar. They are ſown in *February* at the Full of the Moon, ſetting two or three in a Hole about an Inch deep in a hot Bed, as is directed before.

Towards the end of *April* the *Melon* Plants are to be removed out of the hot Bed, into the Beds, where they are to grow all Summer, and planted at two Foot diſtance; which Beds, or at leaſt ſome large Holes in them, are to be filled with a rich light Mould, only you muſt be careful to prevent both the Roots and Plants touching of the Dung, and to water them moderately, and that only when the Earth is very dry and hot; which repeat the doing of in ſuch Weather about two or three times in a Week; and when you water them, take care as much as poſſible not to wet the Leaves. If too much Rain fall, they ſhould be covered; becauſe either too much wet, or too much drought is prejudicial to them; the beſt time for removing of them is in an Evening after a fair Day, when they muſt be watered and defended from the Sun and Cold, for three or four Days together. If you are obliged to plant them on wet Ground, or ſuch as is apt to hold Moiſture, 'tis a good way to lay bruſh Faggots at the bottom of the Trench, to cauſe the Moiſture to ſink away from the Dung. If you perceive no Fruit, cut off the tops of ſome of the Branches to make them bear; and when the Fruit begins to appear, cut off the Bloſſoms that are likely to bear no Fruit: Alſo the ſmall tendrings,

the

the barren Branches, and every thing you judge may
rob it of its Nourishment, and on one Plant leave
not above three or four *Melons*, they should also be
pegged down with Hooks, Sticks, to keep the Wind
from blowing of them about, and when they are in
Blossoms, heighten them up with some good warm
Mould, which will prevent their falling off. They
may be covered when grown large with glass Bells,
or square Cases of Glass made on purpose, which
must be kept close at Night with some admission of
Air under the Glass, or at the top in the Day time;
the Leaves must not be wet in watering, and a Tile
may be placed under each Melon that it may lie the
warmer upon it, and the small Shoots that do extract
the Sap of the most leading Branches must be nipt
off, taking care to leave not above three or four of
the most vigorous Branches, whose knots grow nearest to one another. When your Fruit is grown as
big as a Tennis-Ball, nip off the Shoot at some distance beyond them, and they will grow large, provided you suffer not above two upon each Foot, chu-
sing such as are nearest the principal Stem, the rest
being of little value.

They are known to be ripe when the Stalk seems
as if it would part from the Fruit, when they begin to
gild and grow yellow underneath, and by the fragrant Odour they yield, which encreases more as they
ripen; but if they be to be carried far, it is necessary
they be gathered when they begin to ripen; and when
the Fruit is ripe, turn it three or four days before
you cut it, that the Sun may ripen it the better on all
sides; and do not hasten their ripening too fast with
glasses. Before they be eaten, they must be put in a
Bucket of cold Water, which will make them eat cool
and pleasant.

The Seeds of the most early ripe ought to be preserv'd, and those Seeds that lodg'd on the Sunny side
of the Melon, are to be preferr'd before the rest.

Mint.

Mint.

Mint is multiply'd by Runners, that are as so many Arms that spring out of its Tuft, and take Root, but chiefly by Slips. There are divers sorts, whereof the Garden Mint is the best.

It must be remov'd every three Years, and placed always in good Earth, at about a Foot distance: Some thick Tufts of it are likewise planted in hot Beds in Winter.

Motherwort.

Motherwort is raised by Seed or parting of the Roots, it Flowereth most in the Summer months, and tho' the Stalks and Leaves perish in Winter, the Root endureth.

Mustard.

Mustard is of a hot and dry Nature, is rais'd of Seed, and will grow in any sort of dry Soil.

N.

Nasturces.

Nasturces, commonly called Capuchin Capers, are multiply'd only by the Seed. The Leaf of it is pretty large, and the Flower of an Orange colour; the figure of the Seed is a little pyramidical, divided by Ribs, having all its Superficies engraven and wrought all over, being of a grey Colour, inclining to a light Cinnamon. They are sown in hot Beds about the end of *March,* or the beginning of *April,* and afterwards are replanted by some Wall. The Seed easily falls as soon as ripe, as does that of Borage, and therefore they must be carefully gather'd.

O.

Onions.

Onions are sown the latter end of *February,* or beginning of *March,* and are of two sorts, the Red and the White, being rais'd of Seeds: the White is esteem'd the best, whose Roots are much in request for the several Uses they are put to in the Kitchen; they delight in a fine fat and warm Mould, and are to be sown in *March* or soon after; but if sooner, they must

be

be at firſt covered. They do not extend their Fibres far downwards, and therefore at the time of ſowing, the Bed is to be trod and beat flat, and the Seed as eqoally diſperſed as may be. When they ſpring you are to ſift ſome fine Earth a Finger thick almoſt over them, and if when they begin to appear they are trod down, the Roots will grow the larger; They have proſper'd exceeding well when ſown with Bay-ſalt, and are uſually ripe in *Auguſt*, when they are to be taken up and dried in the Sun, and reſerv'd in a dry place for Uſe. But they may be ſown all the Year for the uſe of young Onions, or Scallions; ſuch as are ſown in Autumn, muſt be cover'd with Straw, or Peas-hawm, and being preſerv'd all Winter, they will be early Cibouls, or Scallions in the Spring. The beſt Onions are ſuch as are brought out of *Spain*, whence they of St. *Omers* had them, ſome of which have weigh'd eight Pounds, chuſe therefore the large, round, white, and thin ſkinned ones.

Orach.

Orach is raiſed of Seed.

P.

Parſley.

Parſley, of all Garden Herbs, is the moſt univer-ſally us'd in the Kitchen, it being an excellent Ingre-dient in moſt Pottages, Sawces and Sallets. There is the common and curl'd ſort multiply'd only by Seed, that is ſmall and of a greeniſh grey colour, and a lit-tle bending inward on one ſide, and all over ſtreak'd from one end to the other. It muſt be ſown in the Spring pretty thick when the Froſts are over, and in good well dunged Ground that is moiſt, or elſe it ſhould be well watered. Its Leaves when cut ſhoot out new ones like Sorrel; it can bear any moderate, but no violent Cold, and therefore it is beſt to beſtow ſome covering on it to defend it: In order to its pro-ducing ſmall Roots, it muſt be thinned in Beds, or Borders, where it is ſown, and in hot Weather it re-quires pretty much watering. Its Seeds are gathered in *Auguſt* and *September*, the ſecond year after ſown.

Stone

Stone Parsley is order'd the same way as Alesander,

Patience.

. There is an Herb called *Patience*, that is planted by sets in some Gardens; it makes a very good boiled Salad.

Parsnips.

This is an excellent sweet Root, and must be sown in the Spring, in a rich mellow and well ordered Soil, whose tops, when they are grown to any bigness, should be trod down, whereby the Roots will be made to grow the bigger; when you have raised them towards the Winter, they may be disposed of in Sand, to be preserv'd in the same manner as Carrots, Turneps, &c. and the fairest may be kept for Seed, or else the fairest and oldest of the tops of those Seeds may be taken in Summer and sown, whereby the fairest Roots may be attained unto.

Peas.

Peas is the chiefest of Pulse, whereof there is almost a different kind for every sort of Land and every Season; in a stiff fertile Ground they yield a considerable Crop, without such frequent Fallowings as other Grain requires, in that they destroy the Weeds, and fit the Land for after Crops, being an Improver, and not an Improverisher of Land. Of such as are planted or sown in Gardens the Hot-spur is the speediest of any in growth, for being sown about the middle of *May*, it will in about six Weeks return dry into your Hands again; or, if sown in *February* or *March*, they will spring earlier than any sort sown before Winter. But if you sow them in *September*, and can by Fences of Reed, or otherwise defend them from extream Frosts, you may have ripe Peascods in *May* following; but the best way is to sow them so as to have them successively one after another. The next is the Sugar Peas, which being planted in *April* is ripe about *Midsummer*, its Cods are very crooked and ill shaped, being boiled with the unripe Peas in them, is extraordinary sweet; the great Inconveniency that attends them

them is, that their extraordinary fweetnefs makes them liable to be devoured by Birds. The large white and green Haftings are tender, and not to be fet till the cold is over, and then not very thick, for they fpread much and mount high, and therefore require the help of tall Sticks, the Rouncival Reading Sandwich, white and grey Tufted, or Rofe Peafe of two forts grey, Windfor great Maple, great Bowling, great Blue Peafe, marrow Peafe, &c. befides which, there is another very large, grey and extraordinary fweet Peas that is but lately propagated, and deferves a large Bed in your Kitchen-Garden.

They delight in a warm and light Soil. if it be rich, the Peas are the fairer; but if lean, they are the more early, and fpend better, efpecially when dry; fome fow them at random, as they do Corn, but that is not a good way; others fet them in Ranges with a Dibble, or Setting-ftick at a convenient diftance, which is a very excellent way both for the faving of the Peas, and to give liberty to pafs between them for the hoeing, gathering, &c. But that which is moft ufed and beft approved of, is the hoeing of them in, which makes a quick riddance of the Work, and covers all at a certain depth, and does not harden nor fadden the Ground as fetting doth, but care muft be taken to cover them well, when you fow them without fcattering, becaufe it will occafion the Mice to fearch after them; and when they are to undergo the Hazard of the Winter, they fhould be fowed fomething thicker than in Spring; if you fhould fow them too thick, you may, when the danger of the Frofts are over, take up thofe which grow too thick, and tranfplant them, only they muft be watered a little at the firft removal. If the Ground between them be kept bare, and when the Peas are about three or four Inches high, if you hoe the Earth up againft them on each fide; and hoe up all the weeds, and if you lay up the land in deep furrows from Eaft to Weft, and fet your Peas, on the South fide of each

furrow

furrow they will ripen the fooner by the reflection of the Sun ; and fecure themfelves much better in Winter ; and if you can furnifh them with Sticks to climb on they will yield a greater increafe, but on the Ground they will ripen fooner.

If Peafe are fown on binding Lands, their produce is very uncertain.

As for Salleting, the Pod of the Sugar Peas, when firft it begins to appear with the Husks and Tendrels, affords a pretty acid Compofition of Sallet, as do thofe of the Hops and Vine.

Peas Everlafting.

Peas Everlafting are Plants eafily propagated, and in good Land thrive exceedingly. Their Roots yield yearly a great burden of excellent Provender for Horfes. They muft be fown early in the Spring on digged Ground in rows, and fo hoed in the intervals between the Seed ; for the Seed is long in coming up, and affords no profit the firft Year. They require Care and Pains to preferve them from Weeds, but the fucceeding Years will recompence you abundantly: Some fow them firft on a fmall Bed, and next Year remove them into Ground new d refs'd with Plough or Spade, and plant them about twelve or eighteen Inches afunder, whereby they may be eafily weeded or hoed.

Pellitory.

Pellitory of the wall is raifed of Slips or Seeds which ripen in *July* and *Auguft* ; and tho' the Leaf wither in Winter, the Roots remain.

Penny-Royal.

Penny-Royal is of three or four forts. 'Tis a common Plant in every Kitchen-Garden propagated from Slips or Branches fet in *April*.

Peony.

Peony is a plant of two Sexes, Male and Female ; the firft being fingle, and known by its Leaves coming conftantly whole, without any divifion, its Root being long and round, and the Flower of a purplifh red ;

the

the Females many times bearing single, others double, the Leaves of all being divided on the edges, the Roots more tuberous, growing in Clods, with many round pieces fastned to them with smaller Strings. Of the best double ones there are several sorts: As First, The double purple Peony, smaller in all its parts than the common red ones, the Leaves of a whiter green, and those of the Flower of a bright shining Colour. Secondly, The double Carnation Peony, of a bright shining Carnation Colour at the first opening, but daily waxing paler, till almost white; the Leaves never fall off, but wither on the Stalk. Thirdly, The double Blush, or white Peony, large Flowered, and at first opening tinctured with a light Blush, but in a few Days turns perfect white, and continues so long before it decays, and then withers on the Stalk, which is the best yet come to our Knowledge. Fourthly, The double strip'd Peony, that is smaller than the last in all its parts, the Flower of a fine red, strip'd with white, lasts long, and falls no Leaf.

All these Flower in *May*, are hardy Plants, and endure long in the Ground without stirring. *October* is the only time to remove them; and of those Roots none will grow but such as have Sprouts, or Buds at the end, or rather top of them, of which sort each piece thereof will grow, the double ones some Years, bring Seeds to perfection, which being sown very thin in *September*, where they may stand unremov'd in the Ground for two Years, may produce new Varieties.

Pimpernel : See *Burnet.*
Pumpions : See *Citruls.*

Potatoes.

Potatoes are planted in several parts of our Country to a very good advantage, being easily encreased by cutting the Roots into several pieces, each piece growing as well as the whole Root. A good fat rich Mould is best for them; but they will grow indifferently in any, provided 'tis well dung'd: The Root is very near the Nature of the *Jerusalem* Artichoak, but not so good or wholesome. These are planted either of Roots or Seeds, and may probably be propagated in great Quantities, and prove good Food for Swine.

Purslain.

Purslain is of two sorts, the green and the red or the Golden, and is raised only by Seed; to have a good Crop of which, the Plants should be replanted by the end of *May*, and set a Foot distance one from another: It is a Sallet Herb propagated with some difficulty, being tender in the Spring, and the Frosts usually nipping of it: But to have it early, it may be sown on a hot Bed, or in *April* on any rich Soil finely dreſt; when the Seeds are sown, clap over the Bed with the back of the Spade, and water it, for it delights in moisture. If it be sown thin, or transplanted apart, it will yield fair Plants either for Seed to pickle, or to boil: as soon as the Seeds look very black, the Stalks must be gathered and laid abroad in the Sun, (which will the better maturate the Seed) lay them on a Board, or Cloth to preserve them from scattering or spilling. House them in the Night, and expose them again in the Day time till they are ripe. Some have affirmed, that the Seed of three or four Years old is better than new.

R

Rasberries.

Rasberries are of three sorts, the common wild One, the large red Garden Rasberry, which is one of the pleasanteſt of Fruits, and the white, which is little inferiour to the red. They are propagated only by Slips that sprout out of their Stocks every Year in the Spring time, and are fit to replant the next Spring after. All of them begin to ripen about the beginning of *July.* They are planted in *March* either in Beds or Borders, observing the distance of two Foot between Plant and Plant. They shoot out during the Summer many well rooted Suckers, some of which you may take away to make new Plantations, by which means the old ones are likewise renewed, for they are dry as soon as their Fruit is gathered, therefore let not the tops be cut to a round Bush, whereby they grow so thick that they will neither bear nor ripen their Fruits so well as if they grew taller and thinner; the only Culture used to them is first in the Month of *March*, to shorten all their new Shoots which grow round about the Stock, and which ought only to be thickest and handsomest; and secondly, to pluck away all the small ones as likewise the old ones that are dead. Thirdly, just above the bearing part, a fortnight or three Weeks before they are ripe, cut them to let the Sun into them, and it will make them bear the better. They are also much infected with green Lice that spoil them; to get rid of which, sprinkle them with water, in which Lice have been dissolved. They should also be removed once in eight or ten Years.

Radishes.

Radishes are multiplyed by Seed, that is round, somewhat thick, and of a Cinnamon Colour, growing in little kind of Cods. It is a very good Garden Root, of which there are three sorts; the small eating one which is raised of Seeds on a hot Bed (to have 'em early) with a sufficient thickness of good rich light Mould, that they may have depth enough to Root in before they reach the Dung, and in order to have large and clean ones, make holes as deep as your Finger about three Inches distance, into each of which a sound Seed or two is to be dropped, and a little covered, leaving the rest of the Hole open, whereby they will grow to the height of the Hole before they dilate their Leaves, and yield a long transparent Root. But such of them as are sown after *Midsummer*, will not run to Seed that Year. The second is the Horse-Radish, which is encreased by Plants, or pieces of the Roots planted out, and by many made use of as an excellent wholesome Sauce. And note, that if you dig up any of the Roots for use, that you leave to the upper part that joins to the Leaves about an Inch in

length of the Root to plant again, which will grow
and increase; only if 'tis dry Weather it will do well
to water them, and to abate some of the Leaves in
Proportion to your having lessened the Root.   The
last is the black Radish, which is so mean a Root as
to find no place in a good Garden. Those Radishes
are best that grow on brackish Lands, and are watered with brackish water.

The best Seed for Radishes, is that which produces
few Leaves, and a long red Root. The time of its
ripening and gathering is the end of *July*, when all
the Stems are cut down, and when they have been
dried some Days in the Sun; the Seed is beat out and
winnow'd. The Stocks that run to Seed shoot their
Branches so high, that it is good to pluck them off to
a reasonable heighth, that the first Stocks may be better nourish'd ; you may sow them all the Year, only
those that are sown in Winter must be sown on hot
Beds, which are the first Radishes that are eaten, and
by that means some of them may be had during the
Months of *February, March,* and *April.* And in order
to be supplied all the other Months, some must be
sown among all manner of Seeds, they coming up so
very quickly, that there is time to gather them before they can do harm to other Plants. The
bigger Roots (so much desired) should be such as being transparent, eat short and quick without stringi-
ness, and not too biting.

Rampion.

Rampion is a Plant, whose tender Roots are eaten in
the Spring, like those of Radishes, but much more
nourishing.

Reponces.

Reponces, or wild Radishes, are propagated only
by Seed, being a sort of little wild ones that are eaten in Sallets, and grow without any pains in the
Fields.

Rhubarb.

Rhubarb is of several sorts, which are raised all by
Seed, or by parting of the Tops.   *Rocamboles*

Rocamboles : See *Shallots.*

Rocamboles are a fort of wild Garlick, otherwife called *Spanish* Garlick, which is multiplied both by Cloves and Seed, which latter is about the bignefs of ordinary Peas.

Rocket.

Rocket, being one of the Sallet furniture, is multiplied by Seed which is extream fmall, and of a Cinamon or dark tanned Colour; it's fown in the Spring, the Leaf being pretty like that of Radifhes.

Rofemary.

Rofemary is fmall, but a very odoriferous Shrub, that is propagated by Seed, or Branches that have fome fhare of Root, or by Slips. The principal ufe whereof is to perfume Chambers, and in Decoctions for Wafhing, being multiply'd much like Rue, and other border Plants, it lafts feveral Years. And being planted upon dry Ground, hardly any Froft injures it. There are feveral forts of it, as the broadleaved, which is bigger than the common, and the gilded, and varioufly ftrip'd with yellow, as if gilt, the Silver denominated from its Silver colour'd Leaves, and the double flower'd Rofemary that has ftiffer Stalks, bigger Leaves, and many pale blue double Flowers. *Rofemary, Sage,* &c. if planted in a dry Soil, feldom receive any hurt from Frofts; but if planted in a moift are ufually deftroyed; and 'tis the fame with young tender Fruits, which a Frofty night, after a wet day, fpoils more of, than ten dry Frofts.

Rofe-Tree.

Rofe-Tree is of divers kinds, and one of the chiefeft Ornaments of our *English* Garden, but it's more particularly diftinguifh'd into four kinds. Firft, The red, whereof there are feveral forts, as the *English* red Rofe, only obferve that the Flowers of fome forts are of a far deeper red than others.

The Rofe of the World, which differs not from the former, but in the colour of its Leaves, which are of a pale Blufh colour, directly fpotted thro' every

M 2 Leaf

Leaf of the double Flower, of the same red colour which is in the Rose, and is the most beautiful of any. The *Hungarian* Rose, whose Shoots are green, and Flowers of a paler red Colour, as are those of the red *Provence* Rose, whose Branches and Leaves are, bigger and greener than those of the common red Rose; the red *Belgick* Rose that is much Taller than the common dwarf Red, or Gilliflower Rose, which grows lower than the ordinary Rose, whose Flowers are of a pleasant Carnation colour. The double Velvet Rose that hath young Shoots of a sad reddish green Colour, with few or no Thorns thereon, it seldom bears any store of Roses. The Marbled Rose, much like the last in growth, but its Leaves are larger, of a light red Colour marbled and vein'd. The Rose without Thorns, that has green and smoother Shoots and Leaves than the Marble one, without any Thorns at all, and the Flowers of a pale red, spreading their Leaves. The *Frankfort* Rose, that hath strong reddish Shoots full of Thorns, thick Flowers, and the Button under the Rose bigger than ordinary. Secondly, The Damask, or pale colour'd Rose, whereof the common Damask Rose is the ancient Inhabitant of *England*, and well known without describing. The Party colour'd Damask Rose, *York* and *Lancaster*, only differing from the other in its parted and marked Flowers. The Chryftal Rose, like the last, only the Marks of the Flowers are much fairer and better than those of the other. The Elegant variegated *Danish* Rose has shorter and reddish Shoots than the former, Leaves smaller, and Flowers something double. The Damask *Provence* Rose, whose Shoots and Leaves are longer than any of the rest, and of a reddish green with very large Roses. The Monthly Rose bearing Flowers only three Months in *England*, viz. *June*, *August* and *September*. The Blush *Belgick* Rose that hath larger Branches, and is fuller of Thorns than any of the former, the Flowers growing very thick, sweet-scented, and the Water distilled therefrom is almost as good as that of the Damask. Thirdly, The yellow Rose, whereof the single Yellow Rose grows as high as the Damask, and whose young Shoots are full of small hairy Prickles of dark red Leaves, small, and Flowers single, and pale yellow. The Scarlet Rose of *Austria*, like unto the other, only the inside of the Leaves of the Flowers is a fine Scarlet, and the outside of a pale Brimstone Colour. The double yellow Rose, whose Shoots are small, and not so red as those of the single kind, the Flowers contain very many small pale yellow Leaves with a great Thrum in the middle. Fourthly, The White Rose, whereof the common one is well known; but there are two sorts thereof, the one being much doubler and fairer than the other. The Blush Rose that differs nothing from the other, but in the Colour of the Flowers, that at first opening are of a fine pleasant Blush Colour, and then grow somewhat white. The Double Musk Rose that rises high with many green Branches, and dark green shining Leaves armed with great sharp Thorns, the Flowers come forth together in a Tuft not very double ; but there is another of the kind that beareth single Roses, the scent of both Flowers is sweet like Musk. The Damask Rose, or the white Cinamon Rose grows not so high as the last, but the Leaves are larger and of a whiter green, and the Flowers bigger, whiter and more double, but not quite so sweet. The Double Dog Rose, that is in Leaves and Branches like the lesser White Rose. The Ever-green Rose, that grows like wild Eglantine, whose Leaves fall not away in Winter, as those of other Roses, from whence it took its Name ; and Flowers containing but five Leaves of a pure white Colour, stand four or five together at the end of the Branches. The Spanish Musk Rose, that hath great green Branches, and bigger green Leaves than the last, and single Flowers. The great Apple Rose, that hath a great Stock, and reddish Branches with green sharp Thorns, and single small Flowers

M 3   standing

standing on prickly Buttons. The Double Eglantine, whose Flowers are double made up of two or three rows of Leaves of a pretty red Colour.

But of all these varieties of Roses, the best and most esteem'd amongst the Red, are those call'd the Rose of the World, the Red Belgick, the Red Marble, the Rose without Thorns, and the Red Provence Rose. Among the Damask are the Chrystal Rose, the Elegant variegated Danish Rose, the Blush Belgick, the Monthly and the Damask Provence Rose. The Scarlet *Austrian*, and Double Yellow among the Yellow Roses; and of the White Roses, the Blush and Damask Musk Rose.

Now Roses are increased either by inoculating the Bud of them in other Shoots, or by laying down the Branches in the Earth; the best Stocks to inoculate upon, which must be done about *Midsummer*, are the Damask, the White, the *Frankfort*, and wild Eglantine; care must be had that all Stocks of budded Roses be kept from Suckers, and the Buds to be inoculated as near the Ground as may be, that the budded Launce may be laid in the Earth to Root after one Years growth. You may likewise prick many Holes with an Awl about a Joint that will lie in the Earth, and then cover the same with good Mould; this do in the Spring, and peg it down that it rise not again, and if water'd now and then in dry Seasons, it will be so rooted by Autumn, as to be remov'd and cut from its other part behind the Root, and becomes a natural Tree; one whereof is more valuable, than two of the other that are only budded, or ingrafted, because very many Suckers that come from them will be of the same kind. But all Roses being apt to yield Suckers, the fairest way to increase them is gently to bend down part of the Tree, or the whole in the Spring, to lay all the Branches in the Ground, and to apply unto them old and well rotted Dung about the Places where they are laid, which will make them root the sooner, and by Autumn there will be thereby

as

as many rooted Trees of the same kind as Branches, laid in the Earth, without prejudice to the old one, which when the new ones are cut off, may be eafily reduc'd to its place again, and the next Year bear as plentifully as ever : Neither will it prevent the bearing of Flowers, for the laid Branches will be as plentifully ftor'd, as if the Tree were erect, and not laid; fo that neither the profit nor pleafure of that Year is loft thereby ; they will alfo grow of Suckers, if they be never fo little rooted.

The double yellow Rofes bear not fo well when planted in the Sun, as other Rofes, but muft be plac'd in the fhade ; and for its better bearing, and having of the faireft Flowers, firft, in the Stock of a *Frankfort* Rofe, put in the Bud of a fingle yellow Rofe near the Ground, that will quickly fhoot a good length, put into it a Bud of double yellow Rofe of the beft kind at about a Foot higher in that Sprout; keep Suckers from the Root, as in all other inoculated Rofes, and rub off all Buds but of the defired kind, When big enough to bear, prune it very near the preceding Winter, cutting off all the fmall Shoots, only leaving the bigger, whofe tops are alfo to be cut off as far as they are fmall. When it Buds for Leaves in the Spring, rub off the fmalleft of them ; and when for Flowers, if too many, let the fmalleft be wip'd off, leaving as many of the faireft as you think the ftrength of the Tree may bring to Perfection, which fhould be a Standard, and rather fhaded than planted in too much heat of the Sun, and water'd fometimes in dry Weather, whereby fair and beautiful flowers may be expected.

Shearing off the Buds when they are put forth, for the retarding of the blowing of Rofes is practicable enough ; and a fecond fhearing of them may caufe them to be ftill later, and fo Rofes may be had when no other Flowers are in being ; but then care muft be taken that the whole Tree be ferv'd fo : For if one part of it be only fhear'd, the part unfhear'd will fpend

M 4   that

that Strength and Sap which you expected would have put forth new Buds in the places of those cut off, and frustrate your design. Monthly Roses, if you would have them bear in Winter, should be set in a Tub that they may be remov'd into the Conservatory.

As soon as the Roses have done blowing, they must be cut with Shears pretty close to the Wood, and each Branch ought to be cut again with the pruning Knife near the Spring, and that close to the Leaf; Bud, and all that is superfluous take away to bring the Tree into a handsome Form; they are hardy, and endure the severest Winters well enough; and they may be dispers'd up and down the Garden in Bushes, or to the Walls among the Fruit; or else set in Rows and Hedges, intermixing the several Colours in such a manner as to have no two alike. The well placing of them much advances their Prospect to the Eye. None of the Rose Trees should be left to grow too high; lower than a Yard and half in heighth is best; except the Musk Roses which will not bear well, except against a Wall, Pale or House-side, and must be suffer'd to grow eight or nine Foot, which is their usual height.

Rue.

Rue is multiply'd by Seed that is of a black Colour and rugged, but 'tis usually propagated rather by its Layers and Slips than by its Seed. It makes pretty Borders for Flowers, being kept clipt.

S.

Sage.

Sage, whereof there are several sorts, the red, green, small and variegated; but the first is the best, and the young Leaves thereof a very wholesome Sallet in the Spring. It is commonly a Border Plant whose Culture hath nothing particular; it is like that of other Border Plants, as Rosemary, Lavender, Wormwood, &c. It is raised by setting the Slips and Branches in the beginning of *April.* The tender tops of the Leaves, but especially the Flowers should be

sparingly

sparingly cropp'd, yet so as not to suffer it to be too predominant.

Salsifie.

Salsifie, or Goats-beard. The common sort is multiply'd only by Seed, which is of a very long oval Figure, as if it were so many Cods all over streak'd, and as it were engraven in the Spaces between the Streaks, which are pretty sharp pointed towards the end.

Spanish Salsifie.

Spanish Salsifie, or *Scorzanera*, is multiply'd by Seed as well as the other, and is very good boil'd, both for the pleasure of the Taste, and the health of the Body. It is sown in *March*, and must be sown very thin, whether it be in Beds or Borders, or else at least it must be thinned afterwards, that the Roots may grow the bigger. It runs up to Seed in *June* and *July*, and is gather'd as soon as it is ripe; it may also be raised of slips or by cutting of the Roots into small pieces, and planting of them in *March* or at other times; they are said to lie in the Ground all the Winter without any prejudice; and still to grow bigger and bigger, tho' they yearly run to Seed.

Samphire.

Samphire is one of our Sallet furniture that is multiply'd only by Seed, it should be planted by the sides of Wall, expos'd to the South or East. The open Air and great Colds are pernicious to it. It's usually sown in some Pot or Tub fill'd with Mould, or else on some side Bank towards the South or East, and that in *March* or *April*, and afterwards transplanted into those Places abovemention'd; but the *French* Seed is better than our *English*.

Savory.

Savory, Winter and Summer, the latter being annual, and rais'd of Seed; the other living over many Winters, and increas'd by Slips as well as Seed: They are both, as to the uses of them, well known

in

in the Kitchen, more particularly the Leaves are us'd to some Ragou's, and among Peas and Beans.

Scallions: See *Ciboules.*

Scurvy grass.

Scurvy-grass is rais'd of Seed. That of the Garden, but especially that of the Sea is a sharp biting and hot Herb of Nature, like unto *Nasturtium*, prevalent in the Scurvy, whereof a few of the tender Leaves may be admitted into our Composition of Sallet.

Selery.

Selery is only multiply'd by Seed, which is of a yellowish and longish oval Figure, and a little bunch'd; it is not good but at the end of Autumn and Winter Season. It is first sown in hot Beds the beginning of *April,* and because of the extreme smallness of its Seed, you cannot help sowing it too thick, so that without thinning of it seasonably before it be transplanted, it warps and flags its Head too much, and grows weak, shooting its Leaves outward in a straggling manner. In the transplanting of it, the Plants are to be plac'd two or three Inches one from another, for which we make holes in the Nursery Bed with our Fingers; only what comes from the first sowing, is transplanted the beginning of *June,* about which time the second sowing is sow'd, which is in open Beds at a Foot distance, and the same must be thin cropped and transplanted as the other, but more must be planted the second time than the first. The transplanting of them in hollow Beds is good only in dry Ground; so the second way of transplanting them is in plain Beds, not made hollow, but both must be extreamly water'd in Summer, which contributes to make them tender; and in order to whiten the same, begin at first to tie the Selery with two Bands when 'tis big enough in dry Weather, then Earth it quite up with Earth taken from the high rais'd Pathways, or else cover it all over with long dry Dung, or dry Leaves, and this whitens it in three Weeks or a Month; but because when it is whiten'd it rots as it stands,

stands, if not presently eaten, it is not to be so earth-
ed up, or cover'd with Dung, but in such Proportion
as you are able to spend it out of hand; hard Frosts
quite spoil it, and therefore upon the approach thereof, it must be quite cover'd over; in order to which,
after it is tied up with two or three Bands, it is taken
up with the Earth at the beginning of Winter, planted in another Bed, and the Plants set as close to one
another as may be, which will make them require
much less covering than before when more asunder.
To raise Seeds from them, some Plants must be transplanted into some by-place after Winter is past, which
will not fail to run to Seed in *August*. There is but
one sort of this Plant. The tender Leaves of the
blanch'd Stalk do very well in our Sallets, as likewise
the Slices of the white Stems, which being crimp and
short, and first peel'd and slit longwise, are eaten
with Oil, Vinegar, Salt and Pepper, and for its high
and grateful taste is ever plac'd in the middle of the
grand Sallet at great Men's Tables. Have a care of
a small red Worm that is often lurking in these Stalks.

Sives.

Sives are a diminutive kind of Leek, they are increased by parting and planting of them in single
heads early in the Spring; if planted in good Land,
they will multiply exceedingly.

Skirrets.

Skirrets are a sort of Roots propagated by Seed;
they require a rich Soil inclining to moisture rather
than drought, they should be sown very thin amongst
other things in *February* or *March*, but the surest way
is to set them of Slips; being parted as single as may
be. If you set them too thick, or above one Slip in a
place they will starve one another; they being also
apt to canker, they require fresh earth often.

Smallage.

Some use this Herb in their Pottage. It's rais'd either by Slips or Seed, which is reddish, and pretty
big, of a roundish oval Figure, a little more full and
rising on one side than the other, and streaked from
one end to the other.

Snap-

Snap-Dragon.

Snap-Dragon, *Antirrhinum*, has some pretty diversities. First, the white Snap-Dragon very common. Secondly, The white variegated one like the other, but broader leav'd, divided in the middle and turn'd up on the edges, with many small long purplish Lines on the inside. Thirdly, The red, which is of two or three sorts, the best flowered like the former of a deep red Rose Colour, but the other paler. Fourthly, The yellow distinguish'd only from the common white in the yellow Colour of its Flowers, they Flower from *May* to *July*, and the Seeds are ripe in *August*, they being all rais'd from Seed, bear Flowers the second Year, when the old Roots commonly perish; yet the Slips being taken off and set, will grow the best, being those that do not rise to Flower, and the best time of setting them is the end of *May*, or the beginning of *June*.

Solomon's Seal.

Solomon's Seal, is sometimes rais'd of Seed, but most commonly by the Tops or pieces of Roots; the Seed is ripe in *September*.

Sorrel.

Sorrel, of these are several sorts, of which the *French* Sorrel is the best; but of the common sort the largest is best for the Garden, and serves for many Uses in the Kitchen, being rais'd easily enough from Plants, which should not be set too near, the same being apt to grow large and spread abroad; but the usual way of propagating it, is by Seed, which is small, slick, and of a Triangular Figure, sharp pointed at the end, and of a dark Cinamon Colour. It may be sown (of whatsoever sort it be) in *March*, *April*, *May*, *June*, *July* and *August*, and the beginning of *September*, provided sufficient time be allow'd it to grow big enough to resist the vigour of the Winter, it's sown either in open Ground, or else in streight Rows or Furrows, in Beds or Borders; in all which cases it must be sown very thick, because many of its

Plants

Plants perish; the Ground it requires should be naturally good or well improved with Dung; it must be kept clean from Weeds, well watered, and once a Year covered with a little Mould after its first cut down to the Ground. The Mould serves to give it new Vigour, and the Seasons most proper for applying it are the hot Months of the Year.

Its Seed is gathered in *July*, by which it is propagated, tho' that called round Sorrel from the roundness of its Leaves (those of the other sort being sharp pointed) is multiplied by running Branches that take Root in the Earth as they run over it, which being taken off and transplanted produce thick Tufts, and these also other Runners.

Spinage.

Spinage is an excellent Herb crude, or boiled, being multiplied by Seed only, that is pretty big, horned and triangular on two Sides, having its corners very sharp pointed and prickly; and on that part which is opposite to those pointed Horns, it is like a Purse of a greenish Colour. This Plant requires the best Ground, and is planted either in open Ground, or in Furrows, in straight Rows upon well prepared Beds, and this several times in the Year, beginning about the middle of *August*, and finishing about a Month after; the first is fit to cut about the midst of *October*, the second in *Lent*, and the last in *May*. They may be also sown early in the Spring. Those that remain after Winter run up to Seed towards the end of *May*, and are gathered about the midst of the Month following. They must be well ordered; and if the Autumn prove very dry, it will not be amiss to water them sometimes. They are never transplanted.

Squashes.

Squashes are a small sort of *Pumpkin* lately brought into request, they are ordered like *Pumpkins* or *Cucumbers*.

Strawberries.

Strawberries deserve a place in the Orchard or Garden,

den, being humble and content with the shades and droppings of the more lofty Trees. There are various kinds of them, as the common *English Strawberry*, much improved by being transplanted from the Woods to the Garden, the white Wood *Strawberry* more delicate than the former, the long red *Strawberry*, the *Polonian*, and the green *Strawberry*, which is the sweetest of all, and latest ripe. But some esteem, that the best of all which hath not long since been brought from *New England* : It is the earliest ripe of all *English* Fruit, being ripe many Years, the first Week in *May*. They are of the best Scarlet Dye, and are propagated of Runners, which is a kind of Thread or String which grows out of the Body of the Plant, which easily takes Root at the Points or Knobs, and in two or three Months time are fit to transplant ; but the best to Plant are those that shoot first in Spring. They are planted either in Beds or Borders, and should be well watered. They thrive best in a moist Soil, or new broke up fresh Ground, or in such places as they have not grown in before, especially on the sides of Melon Banks, where the heat of the Sun is convenient to nourish them ; the time of planting them is in *May*, or *September*, in moist Weather. They bear well the Year after they are planted, especially if planted in single Rows, and thrive much better than if planted thick together according to the common way. But if you would have *Strawberries* in Autumn, the first Blossoms which they put forth may be cut away, and their bearing hindred in the Spring, which will make them afterwards blow a-new, and bear in their latter Seasons ; and in order to get some of these of a larger Size, as soon as they have done bearing, let them be cut down to the Ground, and cropt as soon as they spire, 'till towards the Spring : And when you would have them proceed towards bearing, now and then as you cut them, strew the Powder of dried Cow-dung, Pigeons-dung, Sheeps-dung, or fresh Mould, &c. upon

on them, and water them when there is occasion. Such
as are red, throughly ripe, large, and of a pleasant
Odour, are the best, being agreeable to the Taste;
they extinguish the heat and sharpness of the Blood,
by refreshing the Liver. They should be stringed
once in two or three Years, and transplanted once in
three or four Years. To preserve them over the
Winter, and cause them to come early, cover them
from the Frosts with a little Straw, and bestow some
new Mould on them.

Succory: See *Endive.*

Sweet Marjoram.

Sweet Marjoram is something tender, and therefore
if you will have it betimes, you must raise it on a
hot Bed, and in a warm Situation, sowing of it in a
warm dry Season, for if Moisture comes at the Seed
before it has lain some time in the Ground it will
turn to a Jelly, and never grow, it will do the same
if sown in a moist Soil.

T.

Tabacco.

Tabacco is raised of Seed, which must be sowed in
a good warm Soil sheltered from the Winds, in which
you sow the Seeds, mixing of them with Ashes, that
they may the more equally be sown: When they
begin to appear, they lay Boughs and other things
over them, to shelter them; and while they are
growing, they prepare another place to remove the
Plants into, where they plant them two or three
Foot distant; but with us, in these cold Countries,
they must be sown in a hot Bed. 'Tis fit to remove
when it puts forth four or five Leaves. When it
comes up, it must be carefully watched from the
Caterpillars; and once a Month the Weeds howed
up that are about about it: When the stock is got
to a due Heighth, it must be cut, except what you
design for Seed. When it loses any thing of its verdure, begins to bow down, or is come to a strong
scent, 'tis ripe, and when 'tis cut, they dry it in a
House

House upon Poles ; it must be often visited, and not hang too thick. The Roots left in the Ground produce another crop, but not so good as the first ; Mr. *Worlidge* says, it Tans Leather as well as Oak-bark.

Tansie.

Tansie is raised by Seeds, Slips, or parting of the Roots ; a Herb hot and cleansing, but in regard of its domineering Relish, must be sparingly used with our cold Sallets.

Tarragon.

Tarragon is one of the perfuming, or spicy Furnitures of our Sallets, being propagated both by Seeds and rooted Slips, and by setting of the tops which spring again several times after they are cut. It endures the Winter, and requires but little watering in the driest of Summers. When planted in Beds, it requires eight or nine Inches distance for each Plant one from another, and the best time for it is in *March* or *April*, which hinders not, but that it may be transplanted again in the Summer Season. The best for use, is that which is fresh and tender, and not the Leaves which hang on the Ground, but the Tops are to be preferred.

Thistle *Carduus :* See *Carduus.*

Thorn Apple.

Thorn Apple is of two sorts, the greater, which rises up with a strong round Stalk four or five Foot high, branched at the Joints with large, dark green Leaves, jagged about the Edges, having large white Flowers at the Joints, which are succeeded by great round thorny Heads, opening, when ripe, into three or four parts full of flat blackish Seed ; and the lesser differing from the other in the smalness of the Leaves that are smooth, indented on the Edges and Stalks, without Branches, the Flowers are not so big, but more beautiful ; the Heads rounder, less and harder than the other ; the Roots of both die in Winter. There are other sorts not worth mentioning.

Thyme.

Thyme.

Thyme is of several sorts which are multiplied by Seed, that is very small, and those Plants or Stems of it that produce several rooted Slips and Suckers are separated to replant into Borders, for *Thyme* is seldom planted otherwise; a Border of it is a considerable and necessary Ornament in a Kitchen-Garden.

Trip Madam.

Trip Madam is propagated of Seeds, Cuttings, or Slips; 'tis used in Sallads in Spring, while it is young and tender.

Turneps.

Turneps are of several sorts, as the round which is the most common, the long otherwise called narrow, and the yellow. These are usually nourished in Gardens, and are properly Garden Plants, yet they are very advantageous being sown in Fields, not only for culinary Uses, but for Food for Cattel, as Cows, Swine, and of late Years, Sheep. They delight in a warm, mellow, and light Ground, rather sandy than otherwise, not coveting a rich Mould. The Land must be finely plowed and harrowed, and the Seed sowed and raked with a Bush (as I have shewed already.) They are sown at two Seasons of the Year; in the Spring with other like Kitchen Trade, and also about Midsummer and after. Cows and Swine will eat them raw, if they are introduced into the Diet, by giving the *Turneps* first boiled to them, then only scalded, and last of all raw. It is a piece of great neglect amongst us, that the sowing of them is not more prosecuted, seeing the Land need not be very rich, and that they may be sown as a second Crop also, especially after early Peas. They supply the great want of Fodder that is usual in Winter, not only for fatning Beasts, Swine, &c. but also for Milch-Cows.

The Season for sowing this Plant for the Kitchen, is about Midsummer, that they may be ready to improve upon the autumnal Rains, which makes them much fweeter than the Vernal, yet you may fow in *April* to have *Turneps* in the Summer ; the fhallower you fow *Turneps, Onions*, or any of thofe forts of Roots that go but a little way into the Ground, the larger they will be. They muft not be fown too thick, for that will hinder the growth of the Root ; but if the over-fatnefs of the Ground, which is a very great fault for *Turneps*, or over-much wet caufes them to run out into Leaf more than in Root, then treading down the Leaves will make them Root the better. And if the Roots of them are ufeful and palatable, the Greens or Leaves of fuch as have been fown late, and lived over the Winter are fo too. They being frequently boiled and eaten with falt Meats, prove an excellent Condiment.

Turneps in Winter, before the great Froft prevents, may be taken up, and cutting off the green Tops, you may difpofe of them in fome cool place in Heaps, and they will keep a long time, but the beft way to keep them is to cover them with Sand.

V.

Valerian.

Valerian is of feveral forts, and is raifed of Seeds or Slips, the Seeds fhould be kept moift, and fown in *March* and *April*. It flowers in *March* and *April*, and moft of the Summer Months.

Violets.

Violet Plants, as well the double as fingle fort, and of what Colour foever they be, though they produce Seed in little reddifh Shells or Husks, yet they are multiplied only by Slips, each Plant or Stock of them growing infenfibly into a Tuft that is divided into feveral little ones ; which being replanted, grow in time big enough to be likewife divided into others. The double *Violets* more particularly ferve to make pretty Borders in our Kitchen-Gardens, their Flowers placed on the Superficies of Spring-Sallets making a very agreeable Figure.

W.

Winter Cherries.

Winter Cherries are increafed from the Roots by Sprouts or Runners.

Wormwood.

Wormwood is multiplied by Seed that is of a pretty odd Figure, as being a little bent inward in its fmalleft part, and on the other end which is bigger and rounder, a little open, and upon which laft end there is a little black fpot. Its Colour is yellowifh at the bigger end, and its fharper end inclines to black : Its Seed is feldom ufed, becaufe it is difficult to fan, it being very little ; and therefore when there is occafion of propagating *Wormwood*, its Cuttings, that are a little Rooted, are rather made ufe of. It's planted on Borders or Edges, in a Line, at two or three inches diftance, and five or fix deep in the Ground. It is good to flip them every Spring, to renew them every two Years, and to take away their oldeff and decay'd Stocks. The Seed is gathered about *Auguft*.

CHAP. II.

HAVING given an Account of the feveral Herbs, Plants, &c belonging to the Kitchen-Garden, I fhall, before I proceed to the Defcription of the Orchard and Fruit-trees, take Notice of feveral forts of Flower-trees, Winter-greens, and other Shrubs, that will bear the Froft, which are both convenient and ornamental for the making of Hedges, Walks, and the Partitions of the feveral Quarters of Gardens, Orchards, &c. whofe fhelter is of great Advantage to preferve your Gardens warm, as well as to afford a pleafant Profpect to the Eye ; I fhall obferve the fame Method with that of the Kitchen-Garden, and begin with

Acacia.

A.

Acacia.

The *French* do mightily adorn their Walks with the *Virginian Acacia* : It endures all sharp Seasons but high Winds; which, because of its brittle Nature it does not well resist. The Roots which run like Liquorice under Ground, are apt to emaciate the Soil, and therefore not fit for Gardens. It is increased by Suckers.

Alaternus.

The *Alaternus* thrives very well in *England*, and bears the severest Frost. It makes fine Hedges, and is a quick grower; the Seed ripens in *August*, the Blossoms of which afford an early Relief to the Bees : And the *Phyllyrea*, of which there are five or six sorts, are still more hardy, both which are raised of their own Seeds or Layers, only the *Phyllyrea* lies long in the Ground, and the *Alaternus* comes up in a Month after it is sowed. Being transplanted for Hedges or Standards, they are to be governed by the Shears, and transplanted at two Years growth; clip them in Spring after Rain, before they grow sticky, while the Shoots are tender ; thus it forms a fine Hedge planted in single Rows at two Foot distance, of a Yard thick, and twenty Foot high if you think fit, and furnished with Branches to the bottom : Only because of the Winds, it may be necessary to support it with some Wall or Frame, if you let it grow to such a Heighth.

Almonsdwarf.

Almonsdwarf is a very humble Shrub, bearing in *April* many fine Peach coloured Blossoms. 'Tis a very pleasant Plant, and yields plenty of Cions.

Althæa Fruticosa.

Althæa Fruticosa, or *Shrub-Mallow*, of which there are two sorts, the Purple and the White. They endure the Winter, and are usually planted Standards : They bring forth their Flowers in *August* and *September*, and last 'till the Wet or Cold spoils them ;

the

the Tree is increafed by Layers, and may alfo be raifed by Seed, which muft be fown in *February*, and kept well watered after they come up; they may be tranfplanted the fecond Year, and will blow the fourth; they are very fubject to be over-run with Mofs, which fhould be rubbed off.

Arbutus.

Arbutus, or *Strawberry* Tree, grows common in *Ireland*. It is difficult to be raifed from the Seeds, but may be propagated by Layers. It grows to a goodly Tree, endures our Climate, unlefs the Weather be very fevere, and makes beautiful Hedges.

B.

Bucks-horn Tree.

Bucks-horn Tree, or *Virginian Sumach*, grows in fome places fix Foot high, the young Branches being of a reddifh Brown, feeling like Velvet, and yielding Milk if cut or broken: The Leaves are fhiped about the Edges, and at the end of the Branches come forth long thick and brown Tufts, made of foft and woolly Thrums, among which appear many fmall Flowers: The Roots put forth many Suckers, whereby it is increafed.

Bays.

Bays are of feveral forts, and are propagated of Suckers, Layers and Seeds, or Berries, which fhould be dropping ripe e'er gathered. *Pliny* orders the Berries to be gathered in *February*, and fpread 'till their Sweet be over, and then to be put in Dung and fown. Some fteep them in Wine, but Water does as well; others wafh the Seeds from their Mucilage by breaking and bruifing the glutinous Berries: But the beft way is to interr them as you furrow Peas, or rather to fet them a-part. Defend them the firft two Years from piercing Winds. This aromatick Tree loves the fhade, but thrives beft in hotteft Gravel, on which Soil it beft endures the Froft; but if it is any way injured, cut it down to the Ground, and it will recover again, or elfe it will die. Having paffed the

firft

firſt Difficulties, culture about the Roots wonderfully augments its growth. They ſometimes grow thirty Foot high, and two in Diameter : They are fit
both for Arbours and Pallifado Work, if the Gardi-
ner underſtands when to prune and keep them from
growing too woody ; The Berries are emollient,
and ſovereign in Diſtempers of the Nerves, they are
uſed in Colicks, Gargariſms, Bathes, Salves, Perfumes ; and ſome uſe the Leaves inſtead of Cloves.

C.

Celaſtrus.

Celaſtrus, or Staff Tree, bears a few green Leaves
all Winter, and is fit to mix with the Pyracantha to
make an ever-green Hedge.

Celaſtius.

Celeſtins or Staff Tree, bears green Leaves all the
Winter, and does well to mix with Pyracantha for
making of ever-green Hedges ; 'tis raiſed of Seeds or
Layers, and is beſt removed in March or April.

There are Cherry Trees, Peach Trees, and alſo Apple
and Pear Trees, which bear double Bloſſoms.

Ciſtus.

Ciſtus, there are two ſorts of it ; Firſt, The ſmall,
which is a ſhrubby Plant, about a Yard high, with
two Leaves at every Joint, and Flowers coming forth
at the end of the Branches three or four together,
like a ſingle Row, of a fine reddiſh Purple, with
many yellow Threads in the middle, which are ſucceeded by round hairy Heads, containing ſmall brown
Seeds. Secondly, The Gum Ciſtus that riſes higher,
and ſpreads more than the former, and is bedewed
all over with a clammy Moiſture, which artificially
taken off, is the black ſweet Gum called Landanum.
Its Flowers are larger than thoſe of the former.
They are Plants which continue flowering from May
to September, and are raiſed from Seeds : But being
not able to endure Cold, they muſt be houſed in
Winter.

Cornel

Cornel Cherry Tree.

Cornel Cherry Tree grows to a good Heighth, in any fort of Ground, and may be raifed both of the Seed and Slips.

F.

Filberts.

Filberts are raifed of the Nuts fet in the Ground, or Suckers from the old Roots, or they may be grafted on the common Hazle-nut: They delight in a fine, light, mellow Ground, but will grow almoft on any Soil, efpecially if defended from cold Winds. 'Tis of two forts, the *White* and the *Red*. There is alfo another kind, called the *Filbert of Conftantinople,* the Leaves and Fruit whereof are bigger than either of the former; the beft of them are thofe with a thin Shell.

G.

Gelder-rofe.

Gelder-rofe is increafed by Suckers or Cuttings, being hardy, and what will grow almoft in any Soil that is not too dry.

Granade.

Granade, there are three forts of them: They differ little in Culture from the *Alaternus* ; Confiderable Hedges may be raifed of them in Southern Afpects: Their Flowers are a glorious Recompence for our Pains in pruning them. They muft be diligently purged of their Wood. If you plant them in Gardens to the beft Advantage, keep them to one Stem, and enrich the Mould with Hogs-dung well rotted. Plant them in a warm corner to have Flowers. If you plant them in Hedge-rows, loofen the Earth at the Roots, and enrich it Spring and Autumn, leaving but a few woody Branches. At the Tranfplantation of them they fhould be well watered.

H.

Hypericum Frutex.

Hypericum Frutex is a Shrub yielding abundance of fmall flender Shoots, which in *May* are very thick fet

N 4 with

with small white Blossoms, that the Tree seems to be all over hoary with Frost, are covered with Snow. It is increased by Suckers, and endures all Weathers.

I.

Jessamine.

Jessamine, there are several sorts of this Plant : *First*, The white *Jessamine*, that hath divers flexible Branches proceeding from the bigger Boughs that come from the Root ; at the end of white young Branches come forth divers Flowers together in a Tuft, opening into fine white pointed Leaves, and of a strong sweet Scent, which fall away with us without Seeding : *Secondly*, The *Catalonian* or *Spanish Jessamine*, that is not so high as the former, but bigger in Branches and Leaves as well as Flowers, which are white when opened, with blushed Edges, and sweeter than those of the former : *Thirdly*, the double *Spanish Jessamine*, it Flowers white like the first, but bigger and double, and consisting of two rows of Leaves that are sweet as the former : *Fourthly*, The yellow *Jessamine*, which upon long Stalks bears small long hollow Flowers, ends in fire, and sometimes six yellow Leaves, and are succeeded by black shining Berries : *Fifthly*, The *Indian* Scarlet *Jessamine*, whose Branches are so flexible as not to be able to sustain themselves without the help of something to support them : The Flowers come forth many together at the end of the Branches, being long like a Fox-glove, opening at the end into five fair broad Leaves, with a Stile in the middle of a Saffron colour. The *Jessamine* Flowers from *July* to the middle of *August*, the first white and common yellow being hardy, and able to endure our Winter and Colds, are increased by Suckers ; but the *Indian* yellow, or *Spanish*, must be planted in Boxes, or Pots, that they may be housed in Winter : They are usually increased by being grafted late in the Spring on the common white *Jessamine* by approach ; but they may be also propagated by vers or Suckers.

Jucca

Jucca Indian.

Jucca, Indian is increased by parting of the Roots, it must be secured in Winter from the Frosts.

Judas Tree.

The *Judas Tree* yields a fine purplish, bright, red Blossom in the Spring, and is increased by Layers or Suckers.

L.

Laurus Tinus.

Laurus Tinus is a Shrub yielding sweet-scented Tufts of white Blossoms in the Winter, as well as Summer, is easily propagated from Suckers or Layers, and makes a fine Hedge; but if 'tis injured by Frosts, cut it down to the Ground, and it will recover again.

Lentisc.

Lentisc is a beautiful ever-green, thrives abroad with us with a little Care and Shelter; it may be propagated by Suckers and Layers; it makes the best Tooth-pickers in the World, and the Mastick or Gum is of excellent use, especially for the Teeth and Gums.

Lilac.

Lilac, or *Pipe Tree,* which affords fine scented Flowers in *April* or *May,* and is a Tree yielding plenty of Suckers, by which 'tis propagated; or it may be increased by Slips put into the Ground in *March,* before the Sap begins to be in Motion.

M.

Maternus.

Maternus is a hardy Shrub, being something of the Species of the *Phyllyrea,* and doth as well for Hedges, being as easily managed.

Mezereon.

Mezereon, or *Dwarf-bay,* rises according to its Age from one to two, three, or four Foot high in a Bush full of Branches with whitish round pointed Leaves, that appear not 'till the Flowers are past, which are of a pale Peach colour, some others near red, and a third milk-white, and sweet-scented; they are succeeded by small Berries, when ripe of a delicate red:

The

The Berries and Seeds are to be sown in good light Earth in Boxes, as soon as they are ripe, or else such Earth laid under these fine Shrubs for the Seed as they ripen to fall into, and afterwards covered with the same Mould, not too thick.

Myrtles.

Of *Myrtles*, there is the broad leaf *Myrtle*, and the narrow leaf *Myrtle*, both very sweet smelling Shrubs; but the best is that which in Autumn affords plenty of double white Blossoms. They are not so tender, but small Defence will make them endure hard Winters. The Plants produced from Layers, are the most hardy; those from Seeds most tender. The same thing may be observed of most odoriferous Herbs, as *Thyme*, *Marjoram*, *Hyssop*, &c. There is a sort of *Myrtle* with a large Leaf, called *Spanish Myrtle*, that will endure all Weathers, without Shelter; but the hardiest sort of *Myrtle* of the other kind is that which comes from *Carolina* and *Virginia*; the Berries of which being boiled, yield a sweet or pinguid Substance, of a green Colour, which being scummed off, they make Candles of, which give not only a clear Light, but a very agreeable Scent; it thrives best near the Sea, and is raised of Seeds or Layers. They should not be planted too close together, because it will make them Mouldy, nor in too moist a place; the best transplanting of them is in Spring, that they may have time to get Root in Summer, to supply the Tree with Sap sufficient to support it in Winter, which is, what is necessary to be observed about all Winter greens. *Myrtles* must be well watered both Winter and Summer, or they will not shoot well.

N.

Night-shade Tree.

Night-shade Tree rises with a wooden Stem, a yard high, green leafed, and has Star-like Flowers, white with a yellow Pointel in the middle, succeeded by small green Leaves of a fine red, in *December*, wherein in are small white flat Seeds. It endures the Winter, and is raised by sowing of the Seeds in *March*, which are apt to come up and grow, especially if sown in a Pot, and housed in Winter.

O.

Oleander.

Oleander or *Rosebay* is a Plant bearing some of them bluish, and some of them white Flowers, and will prosper if secured from the most violent Cold. This Tree is commonly kept in Pots or Tubs, it blows all Summer, and when in Flower can hardly be watered too much. 'Tis increased by slitting of the Twigs in the place where you would have them take Root, and laying of them in the Ground, and keeping of them indifferent moist, and they will Root easily. They are also increased by Suckers that have Roots to them, they may be planted out in Summer, and taken up, and put into Pots again towards Winter to preserve them; and this will make them strong, the double sort are for the most part kept in Glass Cases.

P.

Periploca.

Periploca, or *Petisbea,* is a Plant that twists it self about a Pole like a Hop, and lives over the Winter, and yearly puts forth small blue Blossoms. 'Tis increased by Layers.

Phyllyrea.

Phyllyrea makes an excellent Hedge, and bears clipping into any Form, especially if supported by a Wall or Frame: There are five or six sorts of it (some of which are variegated.) 'Tis raised of Layers or Seeds which lie long in the Ground before they come up, and sometimes by Slips. The young Plants may best be removed at two Years old, in *March* or *April,* which may be clipped after Rain in Spring, before it grows sticky, and while the Shoots are tender. Thus it will form a fine Hedge, (though planted but in single Rows, at two Foot distance) a Yard in thickness.

nefs, and twenty Foot high. It will not well bear removing 'till the coldeft Seafons are over.

Privett.

Privett is a Plant that hath been in requeft for adorning Walks and Arbours ; but is of late difufed.

Pyracantha.

Pyracantha, this Tree deferves a principal place among thofe ufed for Fences, it yielding a very ftrong and firm prickly Branch, and ever-green Leaves. But it thrives beft in Standards, becaufe with often clipping it is apt to grow ſticky. It is quick of growth, and raifed either of the bright *Coralline-Berries,* which hang for the moft part of the Winter on the Trees, and lie as long in the Ground e'er they fpring as the *Hawthorn-berries ;* or elfe it is raifed of Suckers or Slips.

S.

Sabin.

Sabin or *Savin* will make fine Hedges, and may be brought into any fort of Form by clipping, much beyond any of the forts of Trees commonly made ufe of for that purpofe, efpecially fuch as are defigned not to grow of any Stature or Bulk, being eafily increafed by Layers, Cuttings, or Seeds.

Sena Tree.

Sena Tree is of two forts, the *Baftard Sena,* and the *Scorpion Sena,* both which yield a pleafant Leaf and Flower: They grow but flender, and fo need the fupport of a Wall or Pales ; but being tonfile, they may be reduced to any other Form ; and may be raifed by Layers or Seeds.

Southernwood.

Southernwood is raifed of Slips, planted any time in Winter.

Spanish Broom.

Spanish Broom is not much unlike the yellow *Jeffamine,* only the Flowers are larger. It flowers in *May,* and is increafed by Seeds or Suckers.

Spirea Frutex.

Spirea Frutex is a fmall Tree bearing fmall Peach-
coloured

coloured Blossoms about *August.* 'Tis a hardy Tree, and is increased by Layers.

Stone-crop Tree.

The greater *Stone-crop Tree* is a beautiful green, but not common; 'tis raised of Layers.

Sweet Bryer.

Sweet Bryer, that which bears a double Blossom is much the best, it makes very good Hedges, and will bear clipping: 'Tis easily raised by Layers, Slips, or Cuttings.

T.

Tamarisk.

Tamarisk is a Tree that grows tall and great, being increased by Suckers and Layers, and usually planted by those who respect Variety and Pleasure. Its Wood is also medicinal.

Tamarisk is a Tree grows to a considerable Heighth, which for its aptness to be shorn, and governed like the *Savin* and *Cyprus,* may be reckoned worth propagating; as also for its Physical Virtues: And though in some part of Winter it loses its Verdure, yet it quickly recovers it again; it may be raised of Layers or Slips.

V.

Virginian Climber.

Virginian Climber, or *Maraca,* comes out of the Ground in *May* with long round winding Stalks, more or less, and in Heighth according to the Age. From the Joints come the Leaves, and at each one, from the middle to the top, a Clasper like a Vine, and a Flower; also the Leaves are of a whitish Colour, having towards the bottom a Ring of a perfect Peach colour, and above and beneath it a white Circle; but the stronger part is the Umbrane, which rises in the middle, parting it self into four or five crooked spotted Horns, from the midst whereof rises another roundish Head that carries three Nails or Bars, biggest above, and small at the lower end. It bears Fruit like a Pomegranate. Its beautiful Flowers shew themselves

felves in *Auguft*, the Stalk dying to the Ground every Winter, fpringing again from the Roots in *May*,
which fhould be covered and defended from hard
Frofts in Winter. It fhould be planted in a large
Pot, to hinder the Roots from running; and for hou-
fing in Winter, and fetting in the hot Sun in Summer, it muft have the hotteft place that may be, or
it will not bear at all. The Pots may be fet in the
Spring in hot Beds to bring them forwards.

Double *Virgins Bower*.

Double Virgins Bower is a climbing Tree, fit to cover fome place of Repofe, or to be fupported by
Props for that purpofe; it bears many dark, blue,
double Flowers in *July* or *Auguft*, and 'till the Cold
prevents them. You may cut off moft of the fmalleft Branches in Winter; it fhoots early, and fpreads
very much in a Summer, and is eafily increafed by
Layers : There are of them fingle, both Purple and
Red; but the double is moft efteemed.

W.

Woodbine, or *Honey fuckle*.

Honey-fuckles bear a fine Flower, and efpecially thofe
of the double red fort, and may be brought to cover
Arbours, or to adorn other parts of the Orchard,
being to be clipped into any Form, and are eafily raifed of Layers.

Y.

Yucca.

Yucca is an *American* Plant, but hardier than we
take it to be : It will fuffer our fharpeft Winter without fetting in Cafes. When it comes to fome Age it
bears a Flower of admirable Beauty; and being eafily multiplied, might make one of the beft and moft
ornamental Fences in the World for Gardens.

Moft fort of Fruit Trees, as *Apples*, *Cherries*, *Codlings*, *Plumbs*, &c. make good Hedges, and afford a
good Shelter, being planted to divide Gardens, Orchards, &c.

Of such Plants as least endure the Cold.

THere being several Plants exotick to our Climate, which are brought out of hotter Countries, and are therefore too tender to indure our Colds: I shall, before I treat of them, describe to you the Green-house or Winter Conservatory to preserve them in.

Green-Houses are of late built as Ornaments to Gardens, as well as to preserve tender Plants; they ought to be open to the South, or very little declining to the East or West; the heighth or breadth about twelve foot, and the length according to the number of Plants you intend for it. It must by no means be plaister'd within with Lime and Hair, for dampness is observ'd to continue longer on such Plaister, than on Bricks or Wainscot. One part of it may have Trils made under the Floor to convey warmth from the Stoves made on the back side of the House, the better to preserve it from Cold or Dampness: This way of preserving for the most tender Plants, being thought much better, than Fire hung up or plac'd in holes on the Floor, as hath been practis'd; tho' in very hard Weather that way may be practis'd in other parts of the House. The Charcoals that are used in Pans, must be well burnt before they are put into the Houses. Coals of Wood Fires, or of Ovens, will serve very well: Some use Glass-doors, Casements, or Chases; but Canvas-doors are reckon'd best; Whatever it be, they are to be plac'd at such distance from the Wainscot-doors, that Mats may be set up before them in extream hard Weather. If Canvas-doors be used, they may be made to take off, and put on at pleasure.

But the cheapest sort of *Green-house* is to dig in dry Ground that is not annoyed with any Spring or soak of Water; as for a Cellar or Vault, about six foot deep, ten foot broad, and of such length as is necessary

to

to contain the Plants, to be reposited therein: Wall
up the sides with Brick, and at one end of the whole
breadth make a pair of Stairs, the better to carry
large Boxes or Cases up and down between them ; but
if a Crain be used, a Ladder will do. The Cover
must be made of Feather-edged Boards, in the Nature
of several Doors with Hinges fixed thereon, to be
put on Hooks fastened in a piece of Timber, lying on
the North-side, raised a Foot or more higher than the
South-side, that by a little shelving the Cover may
the better carry off the Rain ; and let there be a Joist
put between every pair of Doors for them to rest on ;
and unto the South-end or Fore part of each Door a
Rope or two must be fastened, and a Frame of two
Rails on the North-side of the Conservatory, that
the Ropes may be drawn over that one Rail to raise
the Doors, and be fastened to the other Rail, when
the Door is at the necessary Height. Whereby as the
Season is, the Doors may be raised and stand at what
height you please, and as few or as many may be o-
pened, to admit the Air or Sun-beams as are necessary.
Fern or some other kind of Straw, in very sharp
Weather, may be laid on the Top of the Boards, to
prevent the Winds piercing through. Range your
Pots and Cases, so as they may readily have the benefit both of the Sun and Air, and do not place them
so near as to touch one another ; neither water them
often, because 'tis apt to make them fading and sickly ;
but when you find a necessity for it, by the curling
and withering of the Leaves, warm the Water, and
mix a little Pigeons and Hen's Dung with it, pouring
it on moderately, and at some distance from the
Roots, that it may leisurely soak to them. Take off
such Leaves as wither and grow dry, and open the
Mould about your Plants, sprinkling a little fresh
Mould on them, and upon the Top of that some
warm Dung ; and if any Weeds grow, root them up.

Amomum Plinii, so called, being a Plant by him
esteemed, and by him reported to be naturally grow-

O 3   ing

ing in divers Parts of *Asia*, yielding a rich costly Berry used in Perfumes. This Plant is now nursed up in our Climate, by carefully preserving of it in Winter, in close Conservatories, where it requires the same care as the *Orange*.

Citron must be carefully planted. It always bears Fruit, some falling, some ripe, and some unripe. There are several kinds of them. The *Leaf* is like the *Bay-leaf*, except only that Prickles grow amongst them. The *Fruit* is yellow, wrinkled without, of a sweet Smell, and soure Taste: The *Kernels* are like the Kernels of a Pear. The *Tree* is planted four manner of ways; of the *Kernel*, the *Scion*, the *Branch*, and of the *Stock*: If you will set the Kernel, you must dig the Earth two Foot every way, and mingle it with Ashes: Let your Beds be short with Gutters on every side to drain off the Water. Set three Kernels together with the Tops downwards; and being covered, water them every Day; and when they spring, set them in good mellow Earth, and water them every fourth or fifth Day; and when they begin to grow, remove them again in the Spring to a gentle moist Ground, for they delight in a moist Soil. If you set the Branch, you must not set it above a Foot and a half deep, lest it rot: It must be well sheltered from the North. It delighteth to be often dug about. They are grafted in hot Countries, in *April*; and in cold, in *May*; not under the Bark, but cleaving to the Stock near the Root. They may be grafted both on the *Pear* and *Mulberry*; but when they are grafted, they must be fenced either with a weather Basket, or some earthen Vessel. Such Fruit as you mean to keep should be gathered in the Night with the Branches to them. If the Fruit grow too thick, they should be thinned, which would make the Remainder the larger.

Date-tree delights in a moist Gravel. They seldom Bear with us, and are only planted for a Rarity. 'Tis raised of Stones which must be planted in Trenches a Cubit deep and broad, and the Trench filled up again with any Dung but Goat's-dung. In the midst set your Stones, so as the sharper part stand upwards, upon which sprinkle a little Salt, and cover them with Earth well mingled with Dung; and every Day till they appear, water them; they may be removed when a Year old: but as they delight in Salt, every Year the Ground should be dressed with it.

The Sets are not presently to be put into the Ground, but first to be set in Earthen-pots, and when they have taken root, to be removed.

Limon-Trees are to be ordered the same way as the *Orange-Tree.*

Orange-Tree. These Trees preserved in strong Boxes may with ease be removed into the Conservatory, and thence in Summer placed in several places of the Garden, especially if the Boxes are set upon Wheels. They are raised of the Kernels sown in *March*, in cases of rich Earth. These Fruits were unknown in former Ages to the *Europeans*, and the Trees have not been long introduced; and not many Years has that more noble Kind, the *China Orange*, been propagated in *Portugal* and *Spain*, which annually furnisheth us with their Fruits; yet there they have in a few Years degenerated as to Size and Taste. The Fruit with us tho' it ripen not so well as in *Spain*, yet they serve for many Physical Uses, and the Flowers here are more valuable than the Fruit. They may be planted against a South-wall, where they must be well defended on all sides from the cold Winds, and the Top well secured from Rain, and against such a Wall they may stand without removing; only in Spring you must let the Sun and Air in to them by degrees, till they are left quite open, so as to have only the main stay standing till next Winter. In the Building of the Wall you may contrive Cavities through which the Heat of the Fire made in several places for that purpose may pass behind your Trees; or you may have other Fires in this Shed, as in a

green

green House. The most proper Earth to plant your *Orange-Tree* in, is that which is taken out of a Melon or Cucumber-bed, and equally mixed or tempered with a fine loamy Earth, and so to remain all the Winter to be sifted into the Cases. Instead of the Earth of a *Melon* or *Cucumber-bed*, you may use *Neats-dung*, and order it as the *Melon* Earth. Before you put your Earth into the Cases, lay on the bottom a good quantity of *Osier* or *Withy-*sticks, or such like, which will make it light; if they are in a small quantity mixed throughout, it will be the better. Place them in the green House before any Frost happen, and in hard Weather give them some warmth; and as the Spring appears, so acquaint them by degrees with the Air, opening of the Doors at Noon first, and shutting of them again, and so by degrees, till you can leave them open all Day. The same Discretion must be used at the setting of them into the Conservatory, that you do not shut them up too close, until Extremity of Weather require it. As the Trees grow large you may enlarge your Cases, and take out the Trees, Earth and all, and place them in new Cases. I know a Gentleman who annually makes a Shed or House over his *Orange-Tree*, and as the Tree encreaseth, he enlargeth his House, and his Trees are very large, and bear very well. You must gather the Flowers as they blow, leaving but few to knit into Fruit, that your Tree may not spend it self too much. You must carefully brush the Spiders-webbs off this Tree, for they delight to work on them, because of the fragrant Blossoms attracting of the Flies. The *Kernels* may be planted in hot Beds, and will produce fair Plants the sooner; moderate Heat will serve until the Frosts are very hard; then must you kindle greater Fires, but let not any Fire come too near your Trees, nor any Smoak annoy them. When you water them do it gently, and when 'tis needful, which may be discerned by the Leaf which will soon complain; give them rather too little than too much,

O 2   and

and wet not the Leaves. Renew and alter the Earth as tenderly as you can, by abating the Upper-part of it, and ftirring of it up with a Fork, taking great Care not to hurt the Roots, and applying the prepared Earth in the room of it, which may be done in *May* and *September*. If you kindle any Charcoal, when they have done fmoking, put them into a Hole funk a little into the Floor; about the middle of it is the beft Stove, and leaft annoys the Plants. The Water wherewith you water them ought to be prepared as well as the Earth : You may therefore mix it with *Sheeps* or *Neats-dung*, and let it ftand two or three Days in the open Air or Sun ; and it will be fit for Ufe.

Guinea *Pepper* has fome of it a long, and fome a round Fruit, 'tis fowed every Year, and therefore you muft let the forwardeft Pods grow 'till ripe for Seed, though others are pickled for Sauce; they muft be fown early, and great Care taken to keep them from the Frofts by fowing of them in a hot Bed; and afterwards tranfplanting of them into a Bed of good Earth.

Pomegranate. The double bloffomed *Pomegranate* Tree, is efteemed the rareft of all flowering Trees; they may be planted againft a warm Wall, being tender while young ; but afterwards are very hardy. They flower in *Auguft*, if they are pruned, they grow up high ; otherwife they grow into a thick Bufh, full of fmall Branches, which fhould be thinned : But to have them bear well in *England*, they muft be planted in a Box or Cafe made of Wood, that they may be houfed in Winter, and in Spring the young Sprouts cut off, that it fpend not it felf too much in Wood. 'Tis beft to keep it to a few or but one Branch. The Ground fhould be well enriched with Hogs-dung, for 'tis the plenty of Nourifhment that makes them apt to Bloffom ; if you do not houfe them, if you think your Wall ftands too open to the Wind, you may place a mat to fcreen them. They

are

are eafily propagated by Layers or Suckers. They
love both a hot Ground and a warm Air, and may
be grafted on their own Stock or Scion, that grow
from the Roots of the old Tree, they fhould be often
watered with water in which Hogs-dung has been in-
fufed, and it will make them bear.

BOOK XIV.

Chap. I. *Of Gardening.*

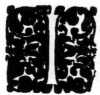 Have already given an Account of
the Orchard and Kitchen-Garden,
and fhall next proceed to the Garden of Pleafure or Flower Garden,
without which not only Courts and
Palaces are very imperfect, but even
the more retired Habitations are ve-
ry much wanting in one of the moft material Parts
of them.

How much Gardens have been admired by the Ancients: What Expences and Charges the *Romans* and
others have been at about them. What Variety of
Delights and pleafant Profpects they have been made
to afford, when laid out fo as to improve all the
Advantages of Nature and Art ; where the Groves,
Avenues, and Walks are noble and free, the Fruits,
Flowers, and Herbs ranged in due order, and every
thing made to confpire together to delight the Senfes,
being what is too numerous to particularize, and
what can be better manifefted by Experience and Ob-
fervation, than indicated by an imperfect Pen ; and

there-

therefore I shall choose rather to refer the Reader to his own Observation, with only this Caution, to suit things to the Circumstances of the place, the charge of making and maintaining of it, and the quantity of Land designed for this purpose.

Chap. II. *Of the Situation of the Flower-Garden.*

OF the Kitchen-Garden and Orchard, I have already given you an Account, and how 'tis to be managed and brought into Order; which being made fit for Trees and Herbs, cannot be improper for Flowers or other Rareties; but as your Garden is a necessary concomitant to your Habitation, from which, if 'tis remote, it can neither be pleasant nor useful; so if 'tis upon a barren Soil (which is commonly the wholsomest for your House) the more Charge and Labour must be bestowed upon it to make things thrive, which one would not chuse, except it be for the Advantage of some pleasant Grove, Prospect, or the Enjoyment of a good Air; and though Woods and Water are two of the best Ornaments of an Habitation, and what may be had in most places, together with a good Air, yet you will seldom meet with Water and a good Prospect near each other.

But if your Place and Situation is fixed, you must do as you can, and improve the Soil you have to the best Advantage, by considering the Nature, Product, Advantages, and Disadvantages of it.

If your Soil be dry and warm, a plain flat place is best for a Garden; but if it be cold and moist, then a declining or shelving piece of Ground that lies towards the Sun is best, because you can the better drain off the Water; and upon such Land Trees will thrive exceedingly, besides such Lands commonly afford good Prospects, and no form for a Garden can be pleasanter than the having of one Degree of it above another; but 'tis necessary in Affairs of this

kind

kind to remember to fuit things to the Circumftances of the place, the charge of making and maintaining of them, and the quantity of Land defigned for this purpofe.

Chap. III. *Of feveral Sorts of Soil for a Garden.*

A Deep rich black Mould is the beft for all forts of Garden-ware, if it is warm and eafy to dig, and is the moft fruitful, efpecially if well dunged and trenched in Winter.

Chalkey-Land is very fweet, and what it produces has a good Tafte. 'Tis very agreeable Land to moft Plants that are not too tender, it being cold in Winter, and backward in Summer, but it may be eafily corrected, and made fuitable to all forts of Plants and Trees, which commonly bear well upon it, efpecially when you have a good Depth of Mould before you come at it.

Marle is a very good Mixture for Garden Ground, being much of the Nature of Chalk.

Sandy-Land is very warm and forward, and agrees very well with moft forts of Fruits and Flowers where the Land is not too fandy, where 'tis, it will require a conftant fupply of proper Soils to enrich it, and fo will Gravel Soils.

Chap. IV. *Of the Improvement of Garden Ground.*

THE black Mould which is the beft for this Ufe is commonly found in bottoms, or near Rivers, or great Towns and Villages, where they have a great deal of Dung, Afhes, &c. to be conftantly mending of it with, or elfe where 'tis made Earth it will in time degenerate into a red-brick Earth, as may be feen in moft of the Fields about *London*.

Clay-Lands being cold and ftiff, are to be mended by Labour, and a Mixture of a contrary Nature, as

O 4   has

has been shewed before. If you dig them often, the
Sun, Rain, and Frosts will mellow them, so as to
cause them to shelder into Dust; these Lands retain
Manure the best of any, but they must be well drained; and tho' Chalk and Marle sweeten them at first,
yet in time they unite with the Clay, and are soon
converted into its own Nature.

Chalkey-Lands usually yield good rich Surface
where 'tis any thing deep; but their Surface being
commonly shallow, you must take care to plant on
them as shallow as you can, and where you can sink
your Walks, the Earth that comes out of them will
help to raise your Borders, and add to the Thickness of your Soil, which will be a great Improvement
of this or any other sort of Land that has but a shallow Soil; and as Chalkey-Lands are cold and binding,
warm Applications, and such things as may loosen
their binding Qualities, are the best Manure for them:
For Chalk, being of a heavy binding Nature, makes
a very great Improvement of light hot dry Grounds,
especially having suffered a Calcination.

Lands seated on Marle are commonly very rich,
altho' cold and heavy; and you need not doubt of the
Depth of the Soil, and the more you turn it up and
expose it to the Air, the better it converts into good
Earth, and any light, warm Mixture is of great Advantage to it.

Sandy-Lands are of several sorts, as I have shewed
already; but Land that hath a competent Mixture of
Sand in it is the warmest and lightest, and according
to its fatness is the best to produce Vegetables. Sandy and gravelly Grounds need a more constant supply
of Manure than other Lands, as I said before; but
they yield a good increase in moist Summers, or where
they may be well manured with Dung: Cow-dung,
Sheeps-dung, and Hogs-dung are the best.

Where you have any Trees, Plants or Flowers that
delight in a different Soil, from what your Garden
is composed of, your best way is in such places to
make

make a Mixture of some such sort of Soil with the natural Earth, as your Plants delight in.

Chap. V. *Of the Form of Gardens.*

AS to the Form I reckon that the best, that will allow of the most Uniformity and Regularity, and bring to view at once the greatest Variety the place will afford, which I think the Square, or rather the long square Figure will do the best; because in that Form your Walks will be straight, and your Trees and Plants stand every way in a direct Line. But where you are already limited, by reason of the Situation of your House, and other Boundaries, things must be formed and modelled according to such Limitations as are prescribed to you.

You should also observe to have one principal Walk in the Middle of your Garden, which is to lead from one of the Principal or most frequented Rooms of your House, and if possible you should cause it to terminate in the best Prospect your Situation will afford; and because of its affording of a Prospect do not make it too narrow, which is the fault of most Walks that I have seen; for the broader they are the more noble (especially where they have a length that will bear it) and to such Prospects let not your Inclosure bound your Sight, but rather leave it open with some thin Pallisadoes, or Iron-work.

If your House stand on the side of a Hill; if you must make your Garden either above or below it, chuse rather to make it below than above, because the Land below will be the richest, and best watered; and because you may have a Prospect of it from every Room of your House; besides 'tis much better to descend into a Garden, than to ascend.

Chap.

Chap. VI. *Of Fences and Inclosures of Gardens.*

OUR being obliged in thefe cold Countries to
have Walls for to ripen and fhelter our Fruit,
is a great prejudice to the Pleafantnefs of our Gardens ;
for in the hotter Climates where they do not need
Walls to ripen their Fruit, their Gardens lie all open,
where Profpects may be had, and Water-fences can
be made ; or elfe they bound their Gardens with
Groves in which are walks, Fountains, &c. which
are much pleafanter to bound the Eye with, than a
dead Wall : But Walls being neceffary with us for
Fruit, as well as to fecure our Plantations, I fhall begin with Brick-walls, as the warmeft and beft for
Fruit, of the price of which I have already given an
Account. Thofe Walls which are built Pannel-wife,
with fquare Pillars at equal diftance, fave a great deal
of Bricks in being thinner than other Walls, and look
much handfomer ; which Pannels may be only made
on one fide, or on both, as you think fit.

Next unto Brick, Stone-walls are preferred ; efpecially the fquare Hewen-ftone. The Rough alone is
very dry and warm, but its unevennefs is inconvenient to nail Trees to, unlefs you here and there lay
fome Timber to nail them to.

All Walls muft be well coaped ; efpecially thofe of
Stone ; left the wet get into them, which will quickly
deftroy them. I fhall not mention any thing of Pales,
becaufe I have already given an Account of the
Charge of cleft Pales ; and as for fawn Pales they are
as dear, confidering their lafting, as Brick or Stone ;
except where the one is very dear, and the other very
cheap ; befides they are too low for Fruit.

And for Earth-walls I fhall not perfwade any body
to make them. I think Pallifadoes the beft Fence,
where any thing of a Profpect is to be had, as I faid
before, and one of the cheapeft if made the common
way, and moft ornamental, though you may beftow
great

great charge on them, or may make them of Iron, which affords the best Prospect; or you may have wooden Bars set in the same Form, which are very handsome and lasting.

Quick Fences are also very fine, especially if kept well clipped; the best of which sort, is the *Holly*; concerning the raising of which I have already treated.

For parting of inward Gardens, or Quarters; Codlins, Cherries, Plumbs, Quinces, &c. are very handsom, kept clipped; and so are Hedges of *Piracantby*, &c.

Walks and Partitions may be made also of all sorts of Winter-greens, which may be planted about a Yard distance one from another; and you may plant one Plant of Laurel, and another of Yew; the one being light Green and a broad Leaf, and the other a dark Green and a narrow Leaf: If these are clipped square till they come to join each other in a Hedge, they will make a fine Chequer-work. There are also, several other Winter-greens that may be thus mixed, that will make very fine Walks.

Chap. VII. *Of several Sorts of Walks.*

THE Design of Gardens being for Recreation and Pleasure, they ought to be accommodated for all Weather, and be suited with such Walks and Places of Retirement, as may suit all Seasons, and all Occasions; that so when our Lassitude, the Rain or the scorching Beams of the Sun, render the open Walks unpleasant, we may have a Retirement till we are willing to repeat our Progress.

The best Walks in Winter wet Seasons are reckoned those paved with broad Stones; but such Stones being very apt to break and moulder with the Frost, I think gravel Walks much better where they are made of a good binding Gravel that will not poach; which if you find them to do, mix a good quantity
of

of Sand with it, and it will make it folid and firm, and make it the more beautiful. The loofeft biggeft grained Sand you can get is beft for this purpofe. Some grind or beat Sea-fhells, and therewith add a thin Coat on the Gravel, which by conftant rolling incorporates with the Gravel, and prevents its fticking to ones Shooes. Others make ufe of refufe Bricks, which they pulverize and ftrew on their Walks, which gives them a fine Colour and dries up the Moifture. Others pulverize Smith's Cinders, which are of a very drying Quality : Gravel-walks are alfo beft for Fruit-Trees, becaufe the Beams or Rays of the Sun reflect from them againft the Walls.

The great Inconvenience thofe Walks are fubject to, are Weeds and Moifture. To prevent the Weeds, you maft when you firft make them, dig the Earth away deep where you defign to lay the Gravel ; It 'tis clay Land, or a Soil apt to produce Weeds, you fhould dig it the deeper, and lay your Gravel the thicker. Some pave their Walks all over with large Pibbles or Flint-ftones, and lay their Gravel on the Top of them, the courfer Gravel underneath, and the fineft on the Top: You muft keep them well rolled, efpecially as foon after Rain as the Top will be dry enough not to ftick to the Roller, which will make them bind; and if they grow Moffy or difcoloured, you may ftir them with a Spade as deep as the fineft Gravel lies, and the watering of your Walks with the Brine your Meat is falted with, or which is better, with the Liquor the Salters call Bitterne, is very good to kill the Worms and Ants (which are commonly very pernicious to Walks) and alfo the Weeds.

And as for Moifture, efpecially after Frofts, which loofening of the Gravel, caufes long foaking Rains to make it ftick and hang to ones Feet: The beft cure of which, is to lay your Walks very round, and to make good Drains where you find the Water to fettle.

Walks of Grafs are much to be preferred in Summer, and in dry Weather, which may be made either

by

by laying of them with Turf, or by raking of them fine, and level, and sowing of them with Hay-feed, and keeping of them well rolled, and weeding of them of the larger fort of Weeds; the often mowing of them will make the Grafs fine. If thefe Walks prove moift you may lay them a little rounding, which will make them caft off the Water the better; and alfo after they are grazed, cover them with fine Gravel or Sand, which will dry up the Moifture on the Top of them, and make the Grafs finer in a little time, than it was before, when once the Stones or Sand is funk into the Ground; but till they are, which they will be fome time a doing, they will be but uneven and rough; however, if you lay them on againft Winter and roll them often, the Stones will quickly fettle, fo as you may be able to mow the Grafs, tho' not fo fhort as it ought to be at the firft: A Water-Table alfo on each fide of the Walk is very good to drain your Walks, and to keep your Grafs and Weeds from mixing with your Borders, and make your Walks the handfomer, and better to the Eye: Thefe Water-Tables fhould be new cut once or twice a Year, and be cut ftraight by a Line.

Terrafe-walks are very pleafant and ufeful, and alfo beneficial for the Air, efpecially where they raife you up to a Profpect, and where you have a great deal of fpare Earth, or Rubbifh, which would elfe coft a great deal to remove to another place, as where Water is near, and you make Canals or Ponds, &c. On the out-fide may be a Wall to fupport it, or on both fides, or the infide to the Garden may be made declining, and cloathed with Turf, and may be fet with Pallifadoes, or with a clipt Hedge.

Chap. VIII. *Of Arbours, Summer-Houfes,* &c.

ARbours have been much more in ufe than at prefent, becaufe their Seats are apt to be moift and wet, and fo unwholfome to fit on; and therefore I rather prefer cover'd Seats or fhady Walks, which are

are warm in Winter, as well as shady in Summer : The best Tree to plant, for Arbours, is the Hornbeam, as I have said before ; and for covered Seats they may be placed so as to face each Coast, that you may according as the Wind or Sun are, place your self so, as to be defended from them.

Summer-Houses may also be erected at each Corner, and made so as to let in the Air on all sides, or to exclude it, as you find it refreshing, or inconvenient by having of Windows, or Doors placed accordingly.

Chap. IX. *Of Water.*

A Good Soil may produce all Sorts of Plants proper for the Garden or Orchard, and they may be so ranged as to make it pleasant and delightfull, but a Garden cannot be said to be compleat nor convenient without Water ; not only for the pleasure that Ponds, Rivers, and Fountains afford, but also for the Necessity of having of Water at Hand on all occasions, especially in dry parching Times ; which Defect where Water is not at hand, may be supplied by Springs rising at a Distance, especially where they rise any thing higher than the place you want to have the Water conveighed to, and may be brought in Pipes of Elm, which if laid deep are not so liable to break with the Frosts, as Leaden or Earthen-pipes are.

Earthen-pipes may do well in some places: They are made about three Foot long, and to fit one into another ; the Joints of which may be closed with a Cement of Lime, Linseed-Oil, and Cotton-wooll ; but these Pipes are not to force Water in any height, they being apt to break, but to convey Water by a Descent ; they are cheap and lasting.

Though small Streams and Springs are ornamental, and necessary for watering your Garden, and supplying of your Fountains ; yet large Streams, Ponds,

or

or Canals, are more noble and pleasant, especially if made of regular Forms, and their Sides even and level with Rows of Trees, or Groves on the side of them, or near them.

Fountains are also a great Ornament to Gardens, being much more esteemed in *Italy* and *France*, than here: There they bestow very great Costs and Charges on them, and make them in very great Variety of Forms, some being made round, others square, and some oval, and some are flat in the bottom, others rounding like a Bason, some being made of Brick, others of Stone, or Lead, and adorned with Variety of Figures, from which by Pipes they cast Water from several Parts of them, according to the Contrivance of the Workman: From which, waste Pipes must be laid to carry off the waste Water, and to clear them; which Water running into lower Parts of the Garden, may be made use of for Cascades, and other Water-works.

But in dry places where neither Springs nor Rivers can be obtained, water may be procured for necessary Occasions from the Heavens, by preserving the Drips of the Houses, the Water of the declining Walks, and the Water-shoots of other adjacent Lands, which may be reserved in Cisterns or Ponds so as to be of use, and to add to the Ornament of the Garden.

Statues, Obelisks, Dials, and other Ornaments, are of use to adorn Gardens with, as they are a lasting Ornament for all Seasons of the Year, when vegetable Ornaments are out of Season, and afford good Variety to the Eye, especially if placed in Fountains, Ponds, or Groves, where they seem more surprizing than in open places.

BOOK

BOOK XV.

Chap. I. *Of Flowers.*

AVING given you an Account of the Site, Form, and other Ornaments of a Garden : I shall proceed to what remains for the beautifying of it, which is Flowers ; concerning which I shall observe the same Alphabetical Order, that I have already done about Flower Trees, and Kitchen-Garden Herbs.

A.

African Marygolds.

African Marygolds are of three forts, and are raifed of Seed fown in *April*, by fome, in a hot Bed ; but they will in a feafonable Spring thrive well enough without. The Seed fhould be faved of the largeft Flowers, only as they are a Flower that blows late, as in *August* and *September*, fo the more Care muft be taken of them, they being a Plant to be renewed every Year ; they require much Sun, and a light Mould, and when new fown, they fhould be watered.

Amaranth.

Amaranth Flowers gentle, or Princes Feathers, are of great Variety ; but the principal are, 1. The great purple Flower with a thick tall Stalk, and many Branches, large green Leaves, and long Spikes of
round

round hairy Tufts, of a reddiſh Purple, containing
many ſmall white Seeds, of which there are many
kinds. 2. The leſſer purple Flower with the yellow Leaves, a little reddiſh, broad at the Stock,
ſharp pointed, the Stock branched at Top, and bearing long, ſoft, and gentle hairy Tufts of a deep
ſhining murry Purple. The Seeds are ſmall, black,
and ſhining. 3. The Flower of divers Colours, which
differ little either in Leaves, Stalks, or Seed, only
that the Flowers are deeper or lighter coloured of
Purple Scarlet, or Gold Colour. The Soil wherein
they ſhould be ſowed muſt be light and rich ; the
Seeds ſhould be ſown about the middle of *March* in
a hot Bed, and when grown to any Strength, be removed into another new Bed, and taking of them up
with Earth about them, they ſet them about the beginning of *May*, where they may bear Flowers, which
by this means they will the ſooner do, and alſo ripen
their Seed the better. The Seed will grow, though
it be two or three Years old, which you may ſow,
except you deſire them forward without a hot Bed.

Anemonies.

Anemonies, or *Wind Flowers*, are diſtinguiſhed into
thoſe with broad and hard Leaves, and thoſe with
narrow and ſoft ones ; of both which ſorts there are
great Variety of Colours, ſome being double, and
others ſingle Flowered.

They muſt be ſet in a rich Mould, wherewith
Neats-dung and Lime ſhould be mingled, that has
lain together ſome time to rot : And the place ſhould
be rather ſhady, than have too much of the Sun.

The broad Leav'd *Anemone* Roots ſhould be planted
about the end of *September*, and the ſmall Eminences
which put forth the Leaves be ſet uppermoſt. Thoſe
with ſmall Leaves muſt be ſet after the ſame manner,
but not at the ſame time ; for being tender Plants,
they muſt not be put into the Ground 'till the end
of *October* at ſooneſt, for fear they ſhould come up

too early, and the Frosts destroy them, from which
they should be defended with Matts, Tyles, Peafe-
straw, &c. which once in two Days at farthest, when
fair Weather will allow of it, must be taken off for
an Hour, or so. If the Spring prove dry, they will
require often and gentle watering. They must not
be taken up 'till *July*, if they profper well; but if
their Leaves are few, Flowers small, and Stalks short,
'tis a sign that they like not the Soil: And therefore
in this Case they should be taken up as soon as the
green Leaves turn yellow, before they are quite dry,
and be put into Sand in some dry place for a Month,
they being wash'd clean and laid on a heap to dry,
and then taken out, and kept in Papers in some dry,
but cold place, 'till the time of their Planting.

As to the raising new Varietie of them, some double broad Leafed ones bear Seed, as the Orange-tawny,
which will soon yield Variety. To sow the Seeds,
take one or more of the best coloured Flowers, and
keep them a Year without Planting, and after that,
plant them again, and when the Seed is ripe or near
it, cover it with Glasses, or any thing that may prevent the Winds blowing of it away when ripe; gather it when the Wind is in the South, dry the Seed
in a Chamber, and lay it where it may not Mould,
sow it in *March* or later, steeping of it first six Hours
in some Wine and Water, which pour off, and dry
the Seed, that it may not stick together; sow it in
a Box filled with Horse-dung and Earth mixed together, the Seed must not be sowed above half an Inch
deep. The Seeds of these Flowers are commonly fit
to gather in *May*, earlier or later as they Flower,
which must be done as soon as ripe and not before,
which is known by the Seed with its Woollinefs, beginning a little to rife of it felf at the Lower-end of
the Head, at which time it must be gathered and laid
to dry, a Week or more, and then in a Bafon or
Earthern Veffel, rubbed with a little Sand or dry
Earth gently, to feparate the Seed from the Wool or

Down

Down that incompasses it. When your Seeds come up, take care to prevent their being scorched with the Sun, when you see the Leaves begin to grow dry. Leave them the first Year in the Box they were sown in, and preserve them from the Frosts, but let them have as much Air as you can, especially in Spring ; and if they stand too dry, water them a little, and the second Year take them up in a dry Season, and then lay them in a dry place 'till *September*, and the next Year, keep them all the Year out of the Ground, and then plant them in good Earth mixed with Horse-dung that it may be light, plant them about an Inch deep in the Ground, and strew over them some rotten Horse-dung about the same thickness.

Apples of Love.

Apples of Love, of which there are three sorts, the most common having long trailing Branches with rough Leaves and yellow Joints, succeeded by Apples (as they are called) at the Joints, not round, but bunched, of a pale Orange shining Pulp, and Seeds within : The Root dies in Winter, the Seeds are yearly sown about the beginning of *April*, and must be often watered to bring them forward before Winter.

Asphodils.

Asphodils are of no great beauty, but may be planted, and increased as other bulbous Roots are, for Varieties sake.

Auricula's.

Auricula's, or *Bears Ears*, is a Flower that affords a very great Variety of Form as well as of Colour, and is not only beautiful to the Eye, but also of pleasant Scent ; the double sort is the most rare, and the *Windsor Auricula* the most splendid. They blow in *April* and *May*, and some of them again about the end of *August*. If you crop off the Buds that offer to blow late in Autumn, it will cause them to yield you the fairer Flowers in Spring. They delight in a rich Soil, and shady, but not under the Drip of

Trees.

Trees. They muſt be often removed, once in two Years at leaſt, and the Ground inriched, elſe they will decay. The ſtriped and double muſt be removed oftner, or elſe they will degenerate. If you ſet them in Pots (which is the beſt way to preſerve them) fill the Pots almoſt half full with ſifted Cow-dung, and the reſt with a good light Mould enriched with the ſame Dung. In the Winter, place them in the Sun, but in Summer in the ſhade. Defend them in Winter from the wet, but for the Cold you need not regard them. You may raiſe them from the Seed by careful gathering of them, and preſerving of them in their Umbels 'till about *Auguſt* or *September*, when you muſt ſow them in Boxes almoſt filled with the Mixture you made for the Plants, and about a Finger thick at the Top with fine ſifted Mellow Earth, or dried Cow-dung beaten ſmall, and mixed with Earth ; in which ſow your Seeds mixt with Wood-Aſhes : Then cover them with the ſame Mixture of Earth ſifted thereon, and about *April* following they will come up, and you may plant them out where you deſign them. It will do well to water them once a Week with the ſame Water which you water Orange Trees with, ſome of them will bear Flowers in *April* following. When they have done blowing, wait for a ſhower of Rain, and then take up your Plants, cut off ſome of the Roots and re-plant them if it doth not Rain, water them very well. If the Leaves diſcolour, cover them from the Sun all the heat of the Day.

B.

Bacchus-bole.

Bacchus-bole, is a Flower that is not tall, but a very full large broad-leafed Flower, being of a ſad light Purple, and a proper white, divided equally, having the three utmoſt Leaves edged with a crimſon Colour, bluiſh Bottom, and dark Purple.

Batchelors-buttons.

Batchelors-buttons are raiſed or increaſed by Slips. They are of three or four ſorts.

Bears-

Bears-Ears Sanicle.

Bears-Ears Sanicle is almost of the Form of an *Auricula*; 'tis usually raised of Seed planted in Pots, and preserved as other tender Plants.

Bean-trefoil.

Bean-trefoil so termed from the likeness of its Leaves to the Herb *Trefoil*, and its Pods to Beans. It affords many fine yellow Blossoms, and is a very pleasant Flower. It is increased by Seeds, Cuttings, and Layers; it requires some Artificial Help to support its weak Branches. There are three kinds of it; the smallest is called *Cytisus Secundus Clusii.*

Bee.

Bee or *Gnat-Flowers* are of several sorts, and are very beautiful, but tender, and therefore are cautiously to be removed: They are to be taken up, Earth and all, and you must endeavour to plant them in such Ground as you removed them from.

Bell-Flowers.

Bell-Flowers are of several sorts, and are both double and single; they are increased by parting of the Roots, and of Seeds sown in *April.*

Bindweed

Bindweed is of two sorts, the larger and the smaller; the first sort Flowers in *September*, and the last in *June* and *July*: The Roots die in Winter, and so they must be annually raised of Seed; the first sort requires a hot Bed, but the latter will thrive without any trouble.

Bladder-nut.

Bladder-nut grows low, if neglected to be pruned up, and kept from Suckers; the Bark is whitish, and the Leaf like Elder-leaves, white and sweet, hanging many on a Stalk, after which come greenish Bladders with a Nut in them; 'tis increased of Suckers.

Blue-borage.

Blue borage leaved, *Auricula* is leaved like Borage, yields fine blue Flowers. It is a tender Plant, and

set

set in Pots, and must be preserved in the Conservatory from the Extremity of the Winter.

Bastard-bittany.

Bastard-bittany is of two sorts the white and the red, which grows about two Foot high bearing a reddish Flower, having many brown woodish Stalks, and on the Lower part of them many winged Leaves like those of the Ash, but larger, longer, and purpled about the Edges, being of a sad green Colour. The white is hard to be increased, and must stand in a rich Soil, they are increased by parting of the Root, which may be done every year to the red ; but to white not above once in two or three years.

C.

Camomile.

Double *Camomile* is like the common sort, only the Leaves are larger and greener, and the Flowers bigger and brighter. 'Tis more tender than the common sort, and must yearly be renewed by setting of young Slips thereof in Spring.

Candy-Tufts.

Candy-Tufts must yearly be increased by Seed, the Roots perishing in Winter.

Cardinal-flowers.

Cardinal-flowers have large Leaves, from whence arise tall hollow Stalks set with Leaves, which are smaller by degrees as they come nearer to the Top, from which come forth three Flowers consisting of five Leaves, three standing close together hanging down-right, and two turned up. The rest which is composed of many white strings lasts many Years : They must be planted in a Pot in good rich light Earth, and the same in Winter; set in the Ground under a South-wall three Inches deeper than the Top, and cloathed about on the Top with dry Moss, and covered with a Glass, which may be taken off in warm Days, and gentle Showers, to refresh it, which must be observed especially at the first setting of it out in April. They may be increased by parting of the

Roots in *Auguſt*, in rainy weather, or by planting of the Stalks that have flowered, which muſt be cut off from the plant the length of three joints, including the button you put into the Ground.

Ciſtus.

Ciſtus is of two ſorts, the ſmall ſort, and the Gum-Ciſtus, which is bedewed all over with a clammy ſweet Moiſture, which artificially taken off is the black Gum called *Laudanum*. They are Plants which flower from *May* to *September*, and are raiſed of Seeds; but they being tender muſt be houſed in the Winter.

Columbines.

Columbines are of ſeveral ſorts and colours. They flower in the End of *May*, when few other Flowers ſhew themſelves: They all bear Seeds, but thoſe that come of a ſingle Colour ſhould be nipped off, and only the variegated ones left for Seed, or thoſe which come of the beſt double Flowers; which being ſown in *April*, will bear the ſecond Year. The Roots will continue three or four Years; but they are apt to degenerate, unleſs the Seed be changed.

Cornflag.

Cornflag is a Plant fit for our Borders, becauſe of its rambling broad long ſtiff Leaves, the Stalk riſing from among them bears many Flowers, one above another. They are of ſeveral ſorts, of which the moſt remarkable are firſt that of *Conſtantinople*, having deep red Flowers, with two white Spots in the Mouth of each Flower; their Roots afford many Off-ſets if they grow long unremoved. Secondly, The *Cornflag* with a bright red Flower; and Thirdly, The Aſh-colour'd ones.

They flower in *June* and the beginning of *July*, the *Byzantine* being the lateſt: Some of them have there Colours intermixed, they all of them loſe their Fibres as ſoon as the Stalks are dry, and may then be taken up and kept out of the Ground free from their many Off-ſets, and ſet again in *September*.

Corn-

Corn-flowers.

Corn-flowers or *Blue-bottles*, whereof there are many Sorts, being raised from Seeds differing in Colours; the Seeds should be saved in Spring, because the Roots perish every Winter; some of them flower in *June* and *July*, and others in *August*, the seeds should be sown in *March*; they require a good Soil, or else they will not come up.

Cowslips.

Cowslips are of various kinds, as those that have Hose within a Hose, as the double Cowslip: The double green ones, the single green, the tufted, the red, the orange, &c. and some of a fine scarlet and very double; whose Flowers must often change their Earth, or they will degenerate and become single.

The Seeds are to be sown in a Bed of good Earth in *September*, and they will come up in the Spring; they may likewise be increased by parting of the Roots.

Crains-bill.

Crains-bill is of several sorts, but the sort most used in Gardens is the musked kind; 'tis raised by Seeds or by parting of the Roots.

Crocus.

Crocus are of divers sorts, whereof some flower in *February* and *March*, and others in *September* and *October*; They are also of great Variety of Colours; when they loose their Leaves they may be taken up and kept dry: Those of Autumn till *August*, and those of the Spring till *October*; they are hardy and will prosper any where. But the best place to plant the Spring *Crocus*'s is close to a Wall or Pail, or on the Edge of boarded Borders round the Garden, mingling the Colour of those of a Season together; as the White with the Purple, the Gold with the Royal, &c. The Seed must be kept in the Husk till sown, and a light rich Ground should be chosen for them: They must not be placed too thick, they may be increased also by Off-sets.

Crown-Imperial.

Crown-Imperial hath a great round Root, a long Stalk and long green Leaves, with a Tuft of small ones on the Top, and under that eight or ten Flowers of an Orange-colour. There are other sorts also as the double, and the yellow ones. They flower in *March* or the beginning of *April*, being propagated by Off-sets that yearly come from the old Roots, which may be taken up after the Stalks are dry, which will be in *June*, and kept out of the Ground till *August*. The double Orange-coloured and Yellow shew finely intermixed. The double ones bear Seeds from which, and from the Yellow, when attainable, you may expect, if sown, some new Varieties, but from the common ones there is but little hopes.

D.

Daffodils-Narcissus.

Daffodils-Narcissus are of great Variety, of different Colours, and some are single, others double; and some bear many Flowers on a Stalk, others but one. They flower from the End of *March* to the beginning of *May*. They are hardy and will grow most of them in any shady place; most sorts of them should be taken up in *June*, and kept dry till *September*, and then set again. To make Varieties of them, the Seeds of the best single ones, for the double bear none, are to be sown in *September*, in such places as they may stand two or three Year before they be remov'd, and then taken up in *June*, and set presently again in good Ground; the seeds are ripe when they look black; at which time pull them off, and clearing of them from the Husk; lay them by till *August* and then sow them an Inch deep in the Earth, which should not be too strong nor stiff; they should be sowed when you expect rain, which if you want, lay a Straw mat over them, and pour water upon it, which soaking through, will moisten the Ground enough; for too much moisture is not good; which mat should always be left on them, except when it rains, or the

morning-

morning dews fall, which are better for feeds than
watering. When they come up, they muſt not be
medled with the firſt year, except to clear them of
weeds; and the ground ſhould be covered with dung
if the Froſt in Winter be very hard; you may cover
them, but then they ſhould not be uncovered again
till *March*. The Seedlings ſhould be taken up about
the middle of *July*, when they are two years old.
They will be four or five years old before they bear;
and the firſt year of their blooming they bear not many Flowers, but the next year they come to perfection.
This Plant requires a good fat Soil, and to ſtand in
a warm place, becauſe it flowers early.

Dazies.

Dazies are of various ſorts, as the great white,
the all red, the great red and white Daſie; the abor-
tive, naked, green, &c. They all flower in *April*,
and may eaſily be increaſed by parting of the Roots
in Spring, or Autumn; but they muſt be well watered, eſpecially if they ſtand too much in the Sun.

Dittany.

Dittany is a hardy Plant, and of ſeveral ſorts, which
Flowers in *June* and *July*, their Seeds being ready to
gather in *Auguſt*, which will be all loſt without care
taken to prevent it. It endures long without removing, and yields many new Roots which ſhould be taken from the old the beginning of *March*; various
kinds are raiſed of them from their Seeds ſown in rich
Earth, ſoon as ripe, as the deep red, white, Aſh-
colour, &c.

Dog-Fennel.

Dog-Fennel has deep dark green Leaves, and broad
ſpread double white Flowers at the Top of the Branches: The Root is only many ſmall ſtrings, which is
increaſed by parting of it in the End of *Auguſt*, nipping off the Buds for Flowers as ſoon as they appear.

Dogs-Tooth.

Dogs-Tooth, or Dogs-Tooth-violet is a kind of Satyrion, grows about half a Foot high with one Flow-

er

er at the Top. Of which there are these following
sorts. 1. Those with white Flowers, 2. with Purple, 3. with Red, 4. with Yellow; all of them Flower about the End of *March*, or beginning of *April*.
They love a good fresh Earth, but not a dunged Soil;
They should be planted in *August* e'er they put forth
new Fibres; for tho' they lose the old, they quickly
recover new ones, wherefore they must not be kept
out of the Ground; and when set, must be defended
from Rain for a Fortnight, for much Wet will ret
them.

Dorothea.

Dorothea is a fine Flower of a deep brown Purple,
curiously edged and dapled with red and lighter Purple, with a white bottom.

E.

Æthiopian Star-flower.

Æthiopian Star-flower is a beautiful Flower in *August*, but it must in Winter be preserved from the
rigorous Colds, by some shelter; or the removing
of it into some Garden-house.

F.

Double Featherfew.

Double Featherfew is like the single, only the Flowers are thick and double, being white and somewhat
yellow in the Middle. It's increased by slips that run
to Flowers in *August*.

Flower-de-Luce.

Flower-de-Luce, of which there are two sorts; one
with Bulbous roots, of which there are several kinds,
and colours, and the Tuberous rooted ones, of which
there are as great Variety as of the former. They
flower in *May* and *June*. The Roots of those whose
Leaves die, should be taken up and replanted in *September*, which is the best time to remove them. They
are increased by parting of the Roots: They delight
in a good Soil. And some sorts of the Bulbous roots
being apt to shoot forth green Leaves in Winter,
they should be a little defended from the Cold.

Indian

Indian Fig.

The *Indian Fig* is preserved for the Rarety of it, there being no Plant in Nature that puts forth its Leaves like it: It must be housed in Winter, and is increased by laying down of the Leaves.

Fox-gloves.

Fox-gloves are of various sorts and colours, and flower in *June* and *July*, and some sorts of them in *August.* They are raised of Seeds which should be sown in *April*, in good rich Earth: The best time to remove them is in *September.* They do not bear Flowers till the second Year.

Fritillaries.

Fritillaries have small round Roots made of two pieces, as if joined together, from whence springs a Stalk bearing the Flower. There are great Varieties of them of divers Colours, and some single, and other double, of which some flower in *March*, some in *April*, *May* and *August.* The Roots lose their Fibres as soon as the Stalks are dry; and may then, or at any time, before the Middle of *August* be taken up, and kept dry some time; though they should not be taken up too soon, nor kept too long out of the Ground, both being apt to weaken them.

G.

Gentianella.

Gentianella is a low Plant, yielding many blue Flowers in *April* and *May.*

Glastenbury-Thorn.

Glastenbury-Thorn which blossoms in *December*, as other Thorns do in Summer. Its first Original coming, as they say, from *Glastenbury* Monastery: 'Tis raised and increased as the common Thorn.

Gilly-flowers.

Gilly-flowers, or rather *July* Flowers, from the Month they blow in, are of very great Variety; but they may be reduced to these four sorts; Red and White, Crimson and White, Purple and White, Scarlet and White, the various Kinds of which are

too,

too many to enumerate, and therefore I shall rather proceed to their Propagation and Culture. The chief thing to be considered to make them produce fair large Flowers, and many Layers, is the Soil in which you plant them, which should be neither too stiff nor over light. In order to which provide a quantity of good fresh Earth, and mix it with a Third-part of Cow-dung, or Sheeps-dung, that hath been long made: To which add a small quantity of Lime; lay this mixture on a heap, and make the Top of it round that it may not take in wet: Let it lie so long as to ferment well together, turning of it up often that it may be mellow before you put it into your Pots, or Beds, for planting of your Layers in. Take your Layers off in *September* or *March*, which last is always best, and cut from them all the dead Leaves, and the Tops of all that are too long, which take up with as much Earth as you can, and set them in Pots, or in a warm place, where you may cover them in Winter, in the afore prepared Earth. They must be well watered in dry Seasons, and not too much exposed to the Noon-sun; the Morning-sun being esteemed the best for them.

Some have used another sort of Earth for them, and that is the Rubbish of a Tan-pit, that by long lying is turned into Earth. To one Barrow-full of which they add four of the Earth of Wood-stack: But what is esteemed the best, especially for Layers, is the Earth in an old rotten Willow. When the Flowers begin to spindle, all but one or two of the biggest, at each Root, should be nipped off, leaving them only to bear Flowers, and the same thing must be observed about the Bud, which will make the Flowers the fairer, and gain the more Layers, by which the Kinds are continued and increased. The Spindles must be often tyed up, and as they grow in height to small Rods, set on purpose by them, lest by their bending they should break, and their Flowers be lost.

The

The chief Time of laying Gilly-flowers, is in *July*, when the Flowers are gone which muſt be done thus: The ſtrongeſt Side-ſhoots having Joints ſufficient for laying, are to be choſen, whoſe ſides and end of the Top-leaves are to be cut off; the undermoſt part of the middlemoſt Joints are to be cut off half through, and the Stalk from thence ſlit through the Middle up-wards to the next Joint; the Earth is to be opened underneath to receive it, and the Layer to be gently bent down into it, with a ſmall Hoop-ſtick to keep it down, the End of the Layer being bent upwards, that ſo the ſlit may be kept open when covered with Earth, which muſt be kept well watered, eſpecially if the Seaſon be dry. It will make them root the ſoon-er, ſo as that you may remove them with the Earth, about the beginning of *September* following into Pots or Beds of the aforeſaid Earth, which muſt be ſhaded and gently watered; but too much Water may be apt to rot them, wherefore they are to be ſheltered from Rain with Boards, ſupported with Forks and Sticks laid on them, but not too near leſt they ſhould periſh for want of Air. Care alſo is to be had in tranſplanting of them, that the Layers are not ſet too deep, which is what hath occaſioned the loſs of many. And if any ſhould not have taken root, you may lay them a new, and make the Cut a little deeper, and ſo let them remain till Spring, and you may then plant them out as you ſee fit.

Some of theſe Flowers, in Summer, ſhoot up but with one Stem or Stalk, without any Layers; which if ſuffered to blow, the Root dies: Wherefore the Spindle muſt be cut off in time, that it may ſprout anew, which will preſerve the Root; but when any of them die in Pots, they are to be emptied of the old Earth, and new put in before another Flower is planted therein. If Roots produce too many Layers, in good Flowers, three or four are enough to be laid. When your Gilliflowers blow, if they break the Pod, open it with a Penknife or Lancet at each Diviſion, as low as the Flower has burst it, and bind it about with a narrow slip of Goldbeaters's Skin, which moisten with your Tongue and it will stick together. The first Flowers are to be kept for Seeds, and their Pods left to stand as long as may be for danger of Frost, and kept as much as possible from Wet : The Stems, with the Pods, must be cut off and dried so as not to lose the Seed, which is ripe when black and the Cod dry.

The best time to sow the Seed is the beginning of *April*, if the Frosts are over, on indifferent good Ground mix'd with Ashes in a place which has only the Morning Sun ; the Seeds must be taken from the best double Flowers, and be daily watered till they come up sometimes ; they must not be sown too thick, and the same Earth should be sifted over them a quarter of an Inch thick that you prepared for the planting of them in ; the Seedlings, when grown to a considerable heighth, may in *August* or *September* following be removed into Beds, where they must stand till they flower. If you have any Gilliflowers that are broken, small, or single, you may graft other Gilliflowers on them that are more choice, but graft them on the most woody part of the Stalk, the best way is Whip-grafting. The Earth about your Gilliflowers ought to be renewed once in two Years at least ; for by that time they will have exhausted the better and more appropriated part of the Earth and Soil. They cannot be kept too dry in Winter, except you find them begin to wither, then you may water them sparingly.

H.

Hearts-ease.

Hearts-ease is a sort of Violet that blows all Summer, and often in Winter ; it sows it self, and may also be increased by parting of the Roots ; it requires a good Soil.

Hellebore.

Hellebore is of several sorts ; as the Black, which flowers at *Christmas*, and the White and Red, which
flowers

flowers in *June*: Their Roots are compofed of divers long brown Strings running deep in the Ground, from whofe big End the Leaves and Flowers arife. They may be increafed by parting the Roots in Spring; and as they are hardy and abide long without removing, they fhould at firft be planted in good Ground.

Hepatica or Liverwort.

Hepatica or *Liverwort*, is a very fine Flower, 'tis of two forts, the fingle and the double ; it never rifeth high, yet yields variety of Bloffoms in *March*, of which the White is moft valued ; they are raifed of the Seed of the fingle ones, which muft be very ripe before you gather it. Keep the Ground moift till they come up, and then houfe them, for they are very tender when young; and when they are houfed, they muft be watered once in five or fix days, and that not too much. You may tranfplant them when they are a year old into a fandy foil, a little dunged ; they may alfo be increafed by parting of the Roots ; they deferve your Labour and Care, which is not much to plant and propagate.

Honey-Suckle.

Honey-Suckle : Of this Plant there are three forts; *firft*, the Common One, 2dly, that called the Double One, producing a multitude of fweet Flowers growing feveral Stories one above another. 3. The Red *Italian Honey-Suckle*, which grows fomewhat like the Wild or Common One, but has redder Branches, the Flower longer and better formed than thofe of the former, being of a fine red colour before they are fully grown, but afterwards more yellow about the Edges. They flower in *May* and *June*, and are eafily increafed of Layers or Cuttings.

Hollihocks.

Hollihocks far exceed the *Poppies* for their Durablenefs, and are very ornamental, efpecially the Double, whereof there are various Colours, they fhould be fown in good Earth an Inch deep, and kept moift till they are fome heighth ; They are fown one Year

and

and flower the next; they may be removed in *August* or *September* from your Seminary, they being raised of Seeds, into their proper places of growth, which should be near some Shelter from the Wind because of their Height; they may also be increased by Shoots cut off of the Stem.

Hollow Roots.

Hollow Roots are of several forts, they come up in *March*, and flower in *April*, and fade away again in *May*; they may be kept out of the Ground two or three Months; they increase very much in any Soil, but they like the sandy best, if not too much exposed to the Sun.

Humble Plant.

The *Humble Plant*, so called, because so soon as you touch it, it prostrates it self on the Ground, and in short time elevates it self again; 'tis raised in a hot Bed, being preserved with great Care, it being one of the most tender Exoticks we have.

Hyacinths.

Hyacinths, or *Jacinths*, are all bulbous Rooted, except the tuberous Rooted *Indian Hyacinth*, which we reserve for the Conservatory. The forts of them which are Muscaries or Grape-flowers, whereof there are great Diversities, as Yellow, Ash-coloured, Red, White, Blue, and Sky-coloured, &c. and also the fair haired branched *Jacinth*, the fair curled haired *Jacinth*, the Blue, White, and Bluish, starry *Hyacinth* of *Peru*, and the blue silly leaved starry *Hyacinth*, that yields fair Flowers on large Stalks. These flower in *May*, and may be removed in *August*; they lose not their Fibres, and so are not to be kept long out of the Ground.

But there are several forts of them that lose their Fibres, and may be kept longer out of the Ground, and are to be preferred to the other, for that they come early in the Year, as from *February* to *April*, being very sweet and well coloured.

Some are more double, as well White as Blue; and therefore are to be esteemed because of their Party-flowering. They may be raised of Seed, which must be sown in *August* in very rich Ground: They come up in Spring, and sometimes in Winter; they should be covered with Straw to preserve them from the Frost; they are also increased by slitting the Bulb with a Penknife on four Sides, but not so deep as the Heart. The Seedlings should be taken up about *Midsummer* when the Seed is black, and layed in a dry place; and as soon as they are quite dry they should be replanted. They begin to blow the fourth Year; they should be planted in a lean Soil, without any Dung.

I.
Indian.

Indian or *Garden-Cresses*, may be raised of Seed in hot Beds, if you desire them forward; else if you sow them in *April* in good Garden Ground, they will thrive very well; they are from a Flower become a good Sallad.

Indian Reed.

Indian Reed has fair large green Leaves coming from the Joints of the Stalks, which bear divers Flowers on the Top, like the *Corn-Flag*, of a bright Crimson Colour, being succeeded by three square Heads, containing Seeds. It hath a white tuberous Root, by parting of which 'tis increased. There is also a sort with yellow Flowers, they must be set in Boxes of good Earth, often watered, and housed in Winter, for one Night's Frost destroys them.

Irises.

Irises are both bulbous and tuberous Rooted: The Bulbous afford very great Variety, some of them (as the *Persian*) flowering in *February* or *March*, others in *April*, *May*, *June*, and *July*. There are some of them very fair and beautiful, their Colours are either Blue, Purple, Ash-coloured, Peach-coloured, Yellow, White, or Variegated. Their Roots may be be taken up as soon as the Leaves begin to wither, for soon after they are quite withered, the Bulbs will issue out more Fibres, and then it will be too late to remove them, otherwise you may keep them dry 'till *August* or *September*. They delight in good Ground, but not too rich, on a sunny Bank, but not too hot : The Eastern Aspect is the best.

Those with tuberous Roots are not altogether so various as the Bulbous, yet they afford many curious Flowers, the best whereof is the *Chalcedonian Iris*, vulgarly called *The Toad Flag* from its dark marbled Flower. This Species of them ought to be carefully ordered, else it will not thrive well ; it requires warm and rich Soil to be planted in, and because 'tis apt to shoot forth green Leaves before Winter, it expects to be a little sheltered from the Cold. These may be taken up when the Leaves begin to dry, and kept some time in the House, and then replanted in *September* or *October*, which will make them thrive the better. The other sorts of Tuberose Rooted ones are much more hardy, and increase exceedingly in good Ground.

Jucca-India.

Jucca-India hath a large Tuberous Root and Fibres, whence springs a great round Tuft of hard, long, hollow, green Leaves, with Points as sharp as Thorns. Its Flowers consist of six Leaves, being of a reddish blush Colour. It must be set in Boxes that are large and deep, and housed in Winter, though some say it will endure our Climate.

Junquils.

Junquils are a kind of Daffodils, and are of several sorts, as the single and the double, and those that have many Flowers on a Stalk, and must be ordered like them ; they Flower about *April* and *May*, and are increased by parting the Roots. They should be taken up about *Midsummer* once in two or three Years whether they be green or not, you must lay

them

them in a dry place without cutting off the Leaves, and plant them again in *August*.

L.

Ladies Slipper.

Ladies Slipper is a Flower valued by moſt Floriſts, although wild in many Places of the North of *England :* It yieldeth its Flowers early in Summer, is a hardy Plant in reſpect of cold, but not very apt to be increaſed.

Ladies Smock.

Ladies Smock has ſmall ſtringy Roots that run in the Ground, and comes up in divers places, by part-ing of which it may be increaſed.

Larks Heels.

Larks Heels are of ſeveral ſorts, as well double as ſingle, though but one kind is worth preſerving, and that is the double upright *Larks Heel* with jagged Leaves, tall upright ſtalked, branched at top, and bearing many double Flowers, ſome Purple, and ſome Blue, *&c.* and ſome Roots now and then produce ſtriped and variegated Flowers with Blue, White, *&c.* The Seeds ſucceed the Flowers in ſmall hard Pods that are black and round, which being ſowed, will produce ſome ſingle, but moſtly double Flowers : The Roots in Winter periſh, they flower ſooner or later, according as they are ſown. The uſual time of ſowing them is the beginning of *April* ; but to get good Seed ſome may be ſown as ſoon as ripe in places defended from long Froſts, and one of theſe Winter Plants is worth ten of thoſe planted in the Spring.

Lentiſe.

Lentiſe is a beautiful Ever-green, that thrives a-broad with us with a little Care and Shelter ; it may be propagated by Suckers and Layers.

Lily.

Lily Of this Plant there are divers kinds, as, I. The *Fiery Red Lily,* that bears many fair Flowers on an high Stalk, of a fiery Red at the top, but at the bot-

tom

tom declining to an Orange-colour, with small black Specks. 2. The *Double Red Lily,* having Orange-coloured single Flowers, with little brown Specks on the sides, and sometimes but one fair double Flower. 3. The *Yellow Lily,* which is the most esteemed of any, being a fine Gold-colour. 4. The common White one, which is like the Red. 5. The *White Lily* of *Constantinople,* smaller every way than the last, but bears a great many more Flowers. 6. The *Double White Lily,* which is like the common sort except in Flowers, which are constantly Double, and seldom open but in fair Weather. 7. The *Persian Lily,* rooted like the Crown Imperial, beset with Leaves to the middle of the Stalk, and from thence to the top with many small Flowers hanging their Heads; These (except the last, which Flower in *May*) put forth their Flowers in *June :* They are increased by parting of the Roots, which hold their Fibres, and therefore should not be often removed, but when there is occasion; the best time is when the Stalks are quite dried down, for then the Roots have fewest Fibres. They may be raised of Seed, but then they are long before they blow. They ought to be set five Inches deep, and their Roots uncovered every Year without stirring the Fibres of the old Roots; the young ones may be parted from them, with only an addition of new rich Earth put to them, which will much advance the fairness of their Flowers. They should be taken up once in four or five Years.

Lilies of the Valley.

Lilies of the Valley, though wild in some places are very much valued for their rich Scent and Usefulness in Physick : They are increased by parting of their Roots, and delight in a moist rich Soil, and to grow in the Shade. They Flower in *May.*

Lupines.

Lupines are here annually sown in Gardens for the sake of their Flowers, but in *Italy* 'tis an ordinary Pulse sown in the Field for their Cattle. They are of two forts, the greater and the leffer, fome of which are Blue, White, and Yellow, which laft is moft valued for its Sweetnefs. They are annually to be fown in *April*.

Lychnis.

Lychnis or *Calcedon* are fingle and double, the fingle only bear Seed, but the double may be increafed by dividing of the Root in *Auguft*, or by fetting of the Stalk, as is faid of the *Cardinal* Flower. They delight in a good Soil.

M.

Mallows of the Garden.

Mallows of the Garden, is a fair large Flower, much diverfify'd in their Form and Colour ; the time of its Flowering is in *Auguft* and *September*, when the Flowers are paft ; the Seeds are contained in round flat Heads, and as they Flower late, the firft Flowers muft be preferved for Seed ; for though the Plant is of fome continuance, it is chiefly raifed from Seed fown the beginning of *April*, which will bear Flowers the fecond Year.

Martagon : See *Lilies*.

Marvel of Peru.

Marvel of Peru, fo termed from its wonderful Variety of Flowers on the fame Root : They Flower from the beginning of *Auguft* 'till Winter, being deftroyed by the Frofts : The Seeds fhould be fet the beginning of *April*, and from their hot Beds removed into rich Earth, where they may have the benefit of the Sun : Upon their failure to Flower the firft Year, Horfe-dung and Litter muft be laid on them before the Frofts, and fo be covered all Winter, and they will Flower the fooner the fucceeding Year, and the Roots of the beft kind, when they have done Flowering, may be taken up and dried, and wrapped up in Woollen Rags, and fo kept from Moifture all Winter, and being fet the beginning of *March* will profper and bear.

Maracoc.

Maracoc.

Maracoc, ufually termed the *Paſſion-Flower:* This Plant is encreaſed by Suckers naturally coming from it, and if the Root be preſerved from the Extremity of Froſts, it will Flower the better, it ſhould elſe be planted againſt a South Wall. They Flower in *Auguſt.* *Snails* as naturally affect this Plant as they do the Fruit of the *Nectarine* Tree, and as Cats do *Marum Syriacum,* and therefore care muſt be taken to defend them.

Meadow Saffron.

Meadow Saffron is of ſeveral Sorts and Colours, and is both double and ſingle. Their Roots ſhould be taken up about the middle of *July,* and ſet about the end of *Auguſt* or beginning of *September,* will ſuddenly put forth Fibres, and ſoon after Flowers, being firſt blown from the time of the ſetting of the dry Roots of all others; they are eaſily planted, the Roots loſing their Fibres, which may be taken up as ſoon as the green Leaves are dried down, and kept out of Ground, 'till the time of planting: They will thrive almoſt in any Soil, tho' they affect moiſt beſt.

Mezereon.

Mezereon grows about three or four Foot high, the Leaves of which appear not 'till the Flowers are paſt, which are of a pale Peach Colour, ſome near Red, and others quite White. They are ſucceeded by ſmall Berries that are of a fine Red when ripe. The Berries and Seeds are to be ſown in good light Earth in Boxes as ſoon as ripe, or elſe ſuch Earth laid under the Shrubs for the Seeds, as they ripen to fall into; and afterwards to be covered with the ſame Mould, but not too thick.

Moly.

Moly, or *Wild Garlick,* is of ſeveral Sorts or Kinds, as the *Great Moly of Homer,* the *Indian Moly,* the *Moly of Hungary, Serpents Moly,* the *Yellow Moly, Spaniſh Purple Moly, Spaniſh Silver-capped Moly, Dioſcorides's Moly,* the *Sweet Moly of Montpelier,* &c. The

Roots

Roots are tender, and must be carefully defended from Frosts.

As for the time of their flowering, the *Moly of Homer* flowers in *May*, and continues 'till *July*, and so do all the rest except the last, which is late in *September*. They lose their Fibres, and may be taken up when the Stalks are dry, and the biggest Roots preserved to set again, casting away all the small Off-sets, wherewith many of them are apt to be pestered, especially if they stand long unremoved. They are hardy, and will thrive in any Soil.

N.

Shrub Nightshade.

Shrub Nightshade has a woody Stock and Branches, dark sad green Leaves, and flowers like that of the common *Nightshade* ; it is increased by Layers, and flowers the end of *May*.

Nonsuch.

Nonsuch is distinguished into two sorts, the single *Nonsuch Flower of Constantinople*, or *Flower of Bristol*, which bears a great Head of many Scarlet Flowers, whereof there is another, which only differs in the Colour of the Flowers, that at first are of a Blush Colour, but grow paler, and a third with white Flowers. And the double rich Scarlet *Nonsuch*, which is a large double-headed Flower of the richest Scarlet Colour : They flower the latter end of *June* ; they are a hardy Plant, but prosper the worst in hot or too rank Ground. They continue long, and are increased by taking young Roots from the old at the end of *March*, when they come up with many Heads ; each of which divided with some share of Roots will grow, and soon come to bear Flowers.

Noli me tangere.

Noli me tangere may be planted among your Flowers, for the Rarity of it ; because its Pods, though not fully ripe, if you offer to take them between your Fingers, will fly to pieces, and cause the un-

wary

wary to ftartle at the Snap. This Plant is annually raifed of Seeds, and only propagated for Fancy fake.
P.

Peafe.

Everlafting *Peafe* is fo called; becaufe, although it be firft raifed of Seed, yet it annually produces new Branches, which furnifhes you with many Bloffoms. 'Tis eafily propagated by the Seed which muft be fown early in the Spring: For the Seed is long in coming up, and muft be kept well weeded. Some think it would be a great Improvement to fow Land with it, becaufe of the great Bulk one Seed produces.

Peony.

Peony is a Plant of feveral forts, as the fingle and the double. The firft is very ufeful in Phyfick, and bears a Flower of a purple red Colour. The fecond are of feveral forts; as, *Firft*, The double purple *Peony*, which is fomething fmaller than the common red one, and the Leaves of a whiter green. *Secondly*, The double Carnation *Peony* of a bright fhining Carnation Colour, at the firft opening, but daily waxing paler, 'till almoft white, but never drops the Leaves, which wither on the Stalk. *Thirdly*, The double blufh or white *Peony* large flowered, and at firft opening, tinctured with a light blufh, but in a few Days turns perfectly white, and continues fo, long before it decays, and then withers on the Stalk: 'Tis the beft fort yet come to our Knowledge. *Fourthly*, The double ftriped *Peony*, that is fmaller than the laft, and the Flowers of a fine red, ftriped with white, lafts long, and drops no Leaf.

All the feveral forts of them flower in *May*, are hardy Plants, and endure long in the Ground without ftirring. The only time to remove them is *October*; none of thefe Roots will grow but fuch as have Sprouts or Buds to them, which you may part from the main Root, which fpreads in the Ground. The double ones fome Years bring Seeds to Perfection; which

which being sown very thin in *September*, and let stand two Years where sown, will produce them, and sometimes Varieties.

Perriwincle.

Perriwincle is a low creeping Plant, some bearing white, some blue Flowers; it will grow in any shady place, or under the dropping of Trees, and its Leaves are always green. There is also a sort of it, whose Leaves are very finely gilded, which will make a fine show under gilded Trees.

Pinks.

Pinks, though a mean Flower, yet the common red sort, or the double ones planted on the Edges of your Walks, against the sides of your Banks; will not only preserve your Banks from mouldring down, but when in Blossom are a great Ornament, and an excellent Perfume; and when out of Blossom they may be clipped by a Line, which will keep your Borders even and straight. They flower in *June*, and are commonly raised of Seed, sowed in *March*, or planted out by parting of the Roots which are apt to spread.

Poppies.

Poppies are a fine Flower for Colour, only of an ill Scent, and not lasting: They are of divers Colours, and very double, as Red, Purple, White, and some Striped; but the most esteemed is the fine Golden coloured one, which flowers in *May*. They yield much Seed, by which you may increase them, which should be saved in *March*, or they will encrease themselves by the falling of the Seed.

London Pride.

London Pride is a pretty fancy, and does well for Borders; 'tis increased by parting of the Roots.

Primrose.

Primrose is an early springing Flower, of which there are great Varieties; as the double Pale, Yellow, single Green, single Yellow, the Red, the Scarlet, the red Hose in Hose, the double Red, &c. Their
Seed

Seed sown in *September*, or in a Bed of good Earth, will come up at Spring; or you may increase them by parting of the Roots.

Q.

Queen's Gilliflower.

Queen's Gilliflower, or *Dame's Violets*, are of several sorts; as the single with a pale Blush, the single White, the double White. Like the single, only there are many Flowers on a Branch, standing thick on a long Stalk, of a pure white and sweet Scent. The Purplish differing only in the Colour of the Flowers, that are of a fine pleasant, light, reddish Purple; and the double striped, which is most esteemed.

These Plants flower the beginning of *June*, and blow 'till the end of *July*; being easily raised from any Slip or Branch, which set in the Ground at Spring, and shaded, and watered, will grow: But the Buds of it must be nipped off as soon as they appear, for Flowers; otherwise they will blow and kill the Root.

R.

Ranunculus.

Ranunculus's are to be ordered like *Anemonies*. They excel all Flowers in the Richness of their Colours; nor is there any Flowers so fine and fair, as the larger sorts of them, of which there are great Variety; as the double White, Crow-foot of *Candia*, the Cloth of Silver Crow-foot, the double yellow Crow-foot or *Asian Ranunculus*, the double Red one of *Asia*, the striped ones, the Monster of *Rome* very rich and double, the Monster striped, the Puroin of *Rome*, the *Morvila*, the *Ferius*, the *Ferius Trache*, the *Ranunculus* of *Aleppo*. The best of the single ones are the Golden *Ranunculus* striped with Scarlet. The *Rosa Frize*, the *Roman*, the *African*, the *Besanon*, the *Melidore*, the *Pannisan*, the *Didonian*, &c.

They must be planted in a very rich dry Earth, well dunged, and about *Midsummer* taken up and kept dry in Papers or Boxes 'till they are set again, which should be done in *December*, for they come up

too

too soon if set earlier, and are destroyed by the Frosts, unless they be daily covered and carefully aired ; and then in such case you may plant them in *October.* They should be planted about two Inches deep. Some commend the mixing of Human dung with the Soil you set them in. When they come up in *March* or *April*, they should be often watered. Their Leaves once nipt by the ,Frost, which their brown Colour will discover, often kills them to the Root ; but covering of them often recovers them. If you would keep them long in Flower, cover them from the heat of the Sun only, when they have almost done Flowering uncover them that they may dry in the open Air, and according as the Weather is, dry or moist, they should be taken up sooner or later ; and when they are taken up, put them in a dry place, so as to let them be thorow dry before you put them into Boxes, least they grow Mouldy, which will cause them to rot when you replant them. :

Rockets.

Rockets delight in a dry Soil, and are increased by Seed, which often sows it self, or by Slips or Cuttings.

S.

Scabious.

Scabious, the common sort grows wild, but those planted in Gardens are of several sorts, the *White Flowered Scabious*, the *Red Scabious* of *Austria*, and the *Indian Scabious* : These Plants commonly die after their Seeding. The two first Flower about *July*, and the other in *September*, so that to get good Seed from them, the best way is the beginning of *June* to remove the young Plants, to keep them from running into Flower the first Year, which will cause them to bring Flowers sooner the next, and so have time to ripen the Seeds. The Seeds must be gathered from the first Flowers, and sowed in *March*. In dry Weather you may water them ; they live sometimes the Winter over when 'tis not too severe.

Sensible

Sensible Plant.

The *Sensible Plant*, so called, by reason that as soon as you touch it, the Leaf shrinks up together, and in a little time dilates it self again : And the *Humble Plant*, so called, because as soon as you touch it, it prostrates it self on the Ground, and in a short time elevates it self again : They are both of them raised in hot Beds, and preserved with great Care, being the most tender Exotick we have.

Snap-Dragon.

Snap-Dragon or *Calves-Snout*, so called, from the form of its Blossom.   It has some pretty Diversities, as the *White Snap-Dragon*, which is very common ; the White variegated one, the Red, which is of two or three sorts, and the Yellow : They flower from *May* to *July*, and the Seeds are ripe in *August*, they being all raised of Seeds, and bear Flowers the second Year, when the old Roots commonly perish, yet the Slips being taken off and set, will grow ; the best being those that do not rise to Flower, and the best time of setting them is the end of *May* or beginning of *June*.

Snow-Drops.

Snow-Drops, so called, because they shew their Snow White Flowers sometimes in *January*, for which early blowing they are esteemed.   They are increased by parting the Roots.

Sow-Bread.

Sow-Bread is of several sorts, as the bright shining Purple, the Vernal one, the pale Purple, and those that Flower in Spring, also White ones single, &c. so that some of the sorts are always in Flower from *April* to *October*.   They are increased by dividing of the Root, which you may do in *April*, or about the middle of *July*, by cutting of them into three or four pieces ; each piece will grow, only you must mind not to cut them too often ; they should be planted, and on good Ground, if they are kept in the Conservatory, they will blow in Winter, if not, they should be covered a little from the Frosts. *Spring*

Spring Cyclamens.

The *Spring Cyclamens* are the best, especially the double ones: They seldom increase by Roots, and therefore are raised by Seeds; the Head of the Vessel that contains them, after the Flowers are past, shrink down and wind the Stalks in a Scrowl about them, and lieth on the Ground hid under the Leaves, where they grow great and round, containing some small Seeds, which as soon as ripe must be sown in Pots or Boxes in good light Earth, and covered near a Finger thick. When they are sprung up, and the small Leaves dried down, some more of the same Earth is to be put upon them; and after the second Year they must be removed, where they may stand and bear Flowers; set them about nine Inches asunder.

Spider-Wort.

Spider-Wort. The *Italian* and the *Savoy* are the only ones fit for your choice. They Flower about the beginning of *June*. They are hardy Plants, and live and thrive in any Soil, but best in a moist one.

Star-Flowers.

Star-Flowers are of several sorts, as the *Star-Flower* of *Arabia*, the great white *Star-Flower* of *Bethlehem*, the *Star-Flower* of *Naples*, the yellow *Star-Flower* of *Bethlehem*, &c.

The *Arabian* Flowers in *May*, that of *Naples* and the *Yellow* in *April*, and some sorts of them not 'till *August*. They lose their Fibres, so that the Roots may be taken up as soon as the Stalks are dry, and kept out of the Ground until the end of *September*, except the Yellow, which will keep out but a little time; and the *Arabian* and *Æthiopian* are so tender as not to endure the Severity of a long Frost, for which reason they should be planted in Boxes, and set in rich Earth; but the rest are hardy.

Stock-Gilliflowers.

Stock-Gilliflowers are usually distinguished into single and double ones: The single are only valuable for their bearing of Seeds; the double ones may be diftinguifhed into, 1. The double *Stock-Gilliflower* with divers Colours. 2. The double ftriped with White. 3. Another double one not raifed from Seed. 4. The Yellow, whofe Seeds produce double Yellow ones.

They begin to flower in *April*, and flourifh in *May*, and fo continue to do 'till the Frofts check them. To raife them, procure good Seed, and of the right kind, which the Seed of thofe fingle ones that bear fine Leaves in a Bloffom are faid to afford ; and alfo the White fingle ones, and the Yellow double ones, and the older you take your Seed from, the better ; thefe are to be fown at the Full of the Moon in *A- pril* or in *Augufl*, provided you keep them from the cold and wet, they muft not be fown too thick, they de- light in a good light Earth, and when grown three or four Inches high, muft be removed into other Earth, or they may be fet again in the fame Earth, after turn- ing of it, and mixing fome Sand with it ; which muft be done fpeedily upon their taking up, that they may be fet again prefently at convenient Diftances, and in fome time ferve them fo again to prevent their growing high ; by which means they will grow more hardy, grow low, and fpread in Branches ; which will make them endure the Winter, and be better to remove in Spring, than fuch as run up with long Stalks, which feldom efcape the Winter Frofts. It may be feen in Spring by the Buds, which will be double and which fingle, for the former will have their Buds rounder and bigger than the others ; them remove with care, not breaking off the Roots, but taking up a Clod of Earth with them, and fet them in your Flower Garden, where they fhall abide all Summer in good Earth ; where being fhaded, they will grow and bear Flowers as well as if not removed at all. Thofe that are fingle may ftand to bear Seed, that muft be yearly fown to preferve the kinds ; for after they have born Flowers they are apt to die, but may be preferved by Slips or Cuttings, that willgrow and bear Flowers the next Spring. The manner of

doing

doing of which, is in *March* to choose such Branches
as do not bear Flowers, which being cut some distance from the Stock, slit at the end of the Slip,
about half an Inch, in three or four places; then
peel the Rind back as far as it is slit, and take away
the inward Wood, turning up the Bark, which must
be set three Inches in the Ground, by making of a
round Hole that depth, and putting the Slip into it,
with the Bark spread out on each side or end thereof;
which cover up, and shade and water for some time,
and the Ground being good it will grow and bear,
some prefer *June* for setting the Slips in. If you take
away the blowing Sprigs the preceding *Autumn*, it
will much further their Duration. If you set them
in Pots they should be housed in Winter, and kept
as dry as you can, except you find them drooping,
which if you do, you may moisten them a little, just
to keep them alive and no more.

Sun Flowers.

Sun Flowers, some of them must be sown every
Year, and others will keep always, and are increased
by dividing of the Root, those raised of Seed are
sown the beginning of *April* in a good Soil, they love
the Sun and Air.

Sweet Williams.

Sweet Williams, or *Sweet Johns*, are of several sorts,
but the Double and the Velvet are chiefly worth
your propagating; every Slip of them set in Spring
will grow. They flower in *July*, and if their Seed
be kept and sown, other Varieties of them may be
gained, which must be sown in *April*; they do not
Flower 'till the second Year: They do very well to
sow in Borders, and make a fine shew.

T.

Thorn Apple.

The *Thorn Apple* is of two sorts, the greater, which
rises up with a strong round Stalk, four or five Foot
high, branched at the Joints with large dark green
Leaves, jagged about the Edges, and having large
bell-

bell-fashioned white Flowers at the Joints, are succeeded by great round prickly green Heads, opening when ripe into three or four Parts, and full of blackish flat Seeds. And the lesser, differing from the other in the smallness of the Leaves that are smooth, rent at the Edges, and Stalks without Branches; the Flowers are not so big but more beautiful, the Heads rounder, less and harder than the other. The Roots of both die in Winter. They are common, and will grow any where, being raised of Seeds.

Toad Flax.

Toad Flax are of several sorts, as, the *Wild Flax* with a white Flower, broader leaved than the common Flax, whose Root will abide many Years. The *Yellow Flowered Flax,* whose Roots are durable, for though the Branches die in Winter, the Root will send up new ones next Spring. The *Toad Flax,* whose Root dies as soon as the Seed is ripe. The *Sweet Purple Flower,* whose Root perisheth. The *Toad Flax* of *Valentia* is yellow Flowered; and brown *Toad Flax* with reddish Flowers. They Flower in *July* and *August,* and the Seed is ripe soon after: Such whose Roots abide the Winter are fit to be set together, the rest to be set with Seedlings in some place open to the Sun. They come up dry and need but little Attendance.

Tube Rose.

Tube Rose, the Stalks run up four Foot high more or less, the common way of planting of them is in Pots in *March,* in good Earth, well mixed with rotten Dung, they should be set in the House 'till *April* is over, and kept dry 'till you see them spindle, and then you must water them, and set them in the open Air, or you may put them in a hot Bed, which is the surest way to have them early; when they have done blowing, lay the Pot on its side that no moisture may come to it, that the Plant may grow dry, and when the Leaves are dry take them out, and hang them up in a dry place.

Tulips.

Tulips afford such great Variety, that it would be endless to make a Catalogue of their several sorts, since every Year produces new kinds of them. Their Colours being from the deepest Dye to the purest White, intermixed with the brightest Yellow, the transcendentest Scarlet, the gravest Purple, with other Compounds inclining to Green and Blue, &c. I shall therefore only name a few of the principal and best of them, and begin with the *Precoces*, or *early blowing Tulips*, of which sorts may be reckoned

The *Florisaine*, which is low Flowered, of a pale flesh Colour, marked with some Crimson and pale Yellow, and the bottom Bluish. The *Rhodenburg*, middle sized, the Tops of whose Leaves are of a Pease-blossom-colour, the sides White, Yellow, and Bluish. The *General Malmilk*, well marked with Carnation White and Yellow. The *Morillion Cramafine* of a bright Crimson or Scarlet and pure White. The *Perishot* of a shining bluish Colour often marked with White. The *Fair Ann* of a Claret Colour with Flakes of White. The *Omen*, a fair large well-formed Flower, of a pale Rose-colour, with many Veins of Crimson with great Stripes of White. The *Galatea*, the *Superintendant*, the *Aurora*, &c.

The next are those called the *Medias*, or *middle Flowering Tulips*, as the *General Essex*, which is Orange-coloured striped with Yellow, and the bottom of a Purple. The *Pluto* of a sooty Orange-colour, variably marked with light and dark Yellow. The *Agat Robbin Paragon* of a sullen Red, marked with Dun-colour, Crimson, and White. The *Royal Tudeal*, of a sad Red-colour about the Edges, whipped with Crimson, and striped with pale Yellow. *Cardinal Elambiant*, a pale Scarlet marked with White. *Morillion* of *Antwerp*, a pale Scarlet and pale Yellow. *Bell Branc*, a dark brown Crimson well marked and striped with White; with many others.

_Only

Only 'tis to be noted, That the white Yellow and Red never change, but there are *Tulips* of different Reds, some deeper, some lighter, and some are glittering, and others of a dull Colour; if any of these have a good bottom, let them stand for Seed, the best Colours being raised from them.

The times of their Flowering is the latter end of *March*, *April*, and *May*; and to continue them the longer, pretty strong Hazle Rods bended Archways are stuck in the Allys, of such an height that the Flowers may not reach them; over which a Tilt, made of Cap-paper, is laid, so starched together, that it may be wide enough to reach the middle of each side, with Rods pasted along the sides of this Tilt, as in Maps, to roll it up, and to each Rod a String in the middle to tie to the Bows over the Flowers to keep the Wind from raising or blowing it off.

As to the planting *Tulips*, make your Beds for them of fresh light Earth, a Foot deep, and a Yard square, which will contain thirty Roots placed at about three or four Inches distance; but such as are designed for Seed must be sunk two Inches lower than the others, lest their Stalks dry before the Seed ripen; and do not set two Flowers of the same Colour together. When the others put forth their Leaves, if any of them do not appear, or their Leaves fade, the Earth is to be opened to the bottom to find its Distemper; and if the Root be moist and squashy there is no hopes of it; but if hard, 'tis recoverable by applying dry Sand and Soot to it, but not to blow that Year: And when 'tis taken up, which must be done as soon as the Fibres are gone, care must be had to keep it free from Moisture, 'till the Season require it to be set again. *Tulips* may be increased by parting of the Off-sets, when you take them up, which being tender and small, should be replanted about the latter end of *August*, or a Fortnight after your taking of them up; you may leave them two Year in the Ground without taking of them up, pro-

vided

vided you Weed them well, and keep the Beds clean.

Tulip.

Tulip Roots need no watering, but when they begin to Flower you may shelter them with Tilts, especially in the Night, to keep off the sharpness of the Frosts; such as hang their Heads must be tied up to small Rods that will just reach the Flowers. When they drop their Leaves, break off the Pods of all but what are intended for Seed, which must be clean, three edged ones, and of such Flowers as are large and strong, and the bottoms either Blue, Dark, or Purple, these must stand longer than the rest because of the Seeds ripening. As soon as the Stalks of the *Tulips* are dry and withered, the Roots will lose their Fibres, at which time they must be taken up every Year, but not when the Sun shines too hot, especially those of any value, and every sort put by themselves, that you may know how to set them again without Confusion; which you may know how to do by laying them on distinct Papers on which their Names are writ. They must be laid in the Sun to dry, and then put into Boxes and kept in a dry Room, and once in a Fortnight or three Weeks be looked on to see that they do not Mould, which if they are not gently aided and aired in the Sun sometimes they will be apt to do, which will spoil the Roots: If any of them are shriveled or crumpled, and feel soft, it is a sign of their decaying; to prevent which, wrap them up in Wooll dipped in Sallet Oyl, and place them where the Warmth of the Sun may just reach them : About the end of *August* set them in the Earth, and mix Wood-Ashes and Soot with some fine Mould and place about them. They should be covered with a Pot that no Wet may hurt them 'till the Fibres are put forth, which will be at the end of *September* or not at all, about which time the other Roots should be set in the same manner and form; and if they are all the Winter kept from too much Moisture it will do well, the Sun then not being

ing able to dry them, and Frosts coming on may occasion their rotting and spoil them. *Tulips* are subject to the Canker, which you may know by their Leaves lying down, rolling up, and wrinkling, which when you find you must cut them off to prevent the ill Effects of it falling into the Root, and if you take up the Root too soon they will grow withered in two or three Days; which, when you observe, lay them in the Ground in a place where no Rain can fall upon them; let them lie seven or eight Days, and when you see them recovered, and grow close and firm, lay them up, and put some dry Earth over them, and they will keep well. But if you take up the Bulbs in dry Weather, put them into Boxes, and cover them with dry Earth that they may not dry too soon, but by Degrees, and let the Box stand in a dry Room, and in about three Weeks or a Months time the Earth that is about them will dry, at which time it should be taken away, and the Bulbs left bare, and by that time they will be fit to keep 'till the time of transplanting of them. If you dung the Earth you plant them in, it should be with Neat's Dung that has lain so long as to be rotten. The best Composition, if the Earth be not naturally light enough for them, is two parts of Neat's Dung, two parts of fresh Earth, and two parts of Sea Sand where 'tis to be had, for want of which, Brook Sand will do, which should be well mixed and turned up together, before it be laid into the *Tulip* Bed. To make them continue long in Flower, cover them from the Sun and Rain; but you must mind, that when they shed their Flowers they be uncovered again, that the Rain and Air may come at them, because at that time the Roots grow moist, and stand in need of the greatest Refreshing.

To raise Variety of *Tulips*, the best way is by saving of the Seed, which are ripe, when the Pods begin to open at the Top, which cut off with the Stalks from the Root, and keep the Pods upright that the

Seed do not fall out of them, which you may set in a sunny Window, to perfect the ripening of them, and so let them remain 'till *September,* or thereabouts, and then separate the Seed from the Chaff, which sow in Boxes of about six Inches deep, some time in *September* or *October* at farthest : Four Inches of which fill with the finest sifted Mould that can be got, which must be light and rich, and rather ridled in than pressed down of an equal Thickness; upon which sow your Seeds about half an Inch asunder ; then let more of the same Earth be ridled over them, not above half an Inch thick if you sow it in Boxes; but if you sow the Seed in Beds, empty them four Inches deep of their old Earth, laying Tiles flat on the place you emptied ; on which Tiles sift some of your finest Earth, and order your Seed as before is directed in Boxes : When *March* comes it will be convenient to water the Seeds a little. The Violet coloured *Tulip* striped with White, is by many e-steemed the best Colour to raise Seed of, which must not be sown on heavy Lands.

The Seeds being thus managed, the Roots of each should be taken up every Year 'till they Flower ; as soon as their Leaves are dry, and kept free from moisture, or being too dry, 'till the latter end of *August,* and then be set again at wider Distances; the third Year they may produce two Leaves, and when they do they will Flower, and after the first Year they may be set in a deeper Soil, and richer Earth ; a rich Soil being what will make them thrive best, and a barren one Flower best; the Change of Soil is what must be observed for them some time.

Tyme.

Mastick Tyme is a Plant of a curious Scent, and vulgarly known, apt to be increased by Slips, and as apt to be destroyed by Cold ; but 'tis worth your Care to preserve it.

Violet

V.
Violet Morian.

Violet Morian, or *Canterbury Bells*, come up the first Year, the whole Plant dies as soon as the Seeds are ripe, of which 'tis raised, which should be sown in *April*, and afterwards removed where the Plants may stand to bear Flowers.

W.
Wall Flowers.

Wall Flowers are of several sorts, as the common Ones, the great single Ones, the great double Ones, the single White, the double White, the double Red, and the pale Yellow, all which sorts flower about the latter end of *March*, and in *April* and *May*. They are increased or continued by Slips planted in *March*, which should be set against a South Wall, whereunto they should be fastened and defended from Frosts and hard Weather, especially the double Ones.

Wolfs-Bane.

Wolfs-Bane is an early Flower, and may be removed at any time. 'Tis increased by parting of the Roots.

BOOK XIV.

Chap. I. *Of Fruit-Trees.*

H AVING treated of the Kitchen-Garden, I shall next consider the Orchard, there being nothing more profitable than the Planting of Fruit Trees; of which *Worcestershire*, *Herefordshire*, *Gloucestershire*, *Kent*, and many other Places can give us ample Instances: And therefore it will be necessary, more particularly, to consider the Improvement that is to be made of this part of Husbandry,

bandry, and the Advantages of it, which confift in many Particulars : As,

Firft, In the Univerfality of it, there being hardly any Soil, but one fort of Fruit Trees or other may be raifed on them, efpecially if judicioufly managed.

Secondly, The ufe of Fruit is alfo Univerfal, both for eating and drinking, there being hardly any Places, where of late Years, Fruit is not much made ufe of, efpecially the Juice for Cyder; which being made of good Fruit, and well prepared, is a moft delicious wholefome Liquor, and moft natural to our *Englifh* Bodies, there being no County in *England* that hath afforded longer lived People than the Cyder Counties. The greateft Inconvenience that attends it, is, that it is a very ticklifh Liquor, and requires a great deal of Art and Skill to manage, as I fhall have occafion to fhew hereafter.

Thirdly, In the Charges and Expences of it, which are very fmall, efpecially if compared with that of other parts of Husbandry, there being hardly any more required than the trouble of gathering them, after a few Years at firft, which is a very inconfiderable Charge to the Profit and Advantage that accrues to the Owner afterwards. Mr. *Hartlib*, in his Legacy, telling us of the Benefit of Fruit Trees, fays, " That they afford curious Walks, Food for Cattel " in Spring, Summer, and Winter, Fuel for the Fire, " Shade from the Heat, Phyfick for the Sick, Refrefh- " ment for the Sound, Plenty of Food for Man, and " that not of the worft, and Drink alfo of the beft, " and all this without much Labour, Care, or Coft.

So that confidering the great Expence of the other parts of Husbandry ; as alfo the Charge of Plowing, Sowing, Reaping, Inning, and Thrafhing of Corn, it will come much fhort of the Profit of Fruit Trees ; nay, many times Fruit amounts to more than Corn will yield, though the Charges were not deducted.

And I cannot but think Fruit-trees a great Improvement of the Land where they are planted, in that the

Grafs

Grafs which grows underneath them will be forwarder
in Spring than any other, and if mow'd will yield
twice the quantity of Hay: But the mowing of Orchards being prejudicial to the Trees, I shall rather advise the keeping of it short fed with Cattle the fore-
part of the Summer (especially if your Plantation is
so large that you cannot keep it constantly dug) which
the Cattel will then eat as well as other Grafs ; and
if it should grow rank, and get a head of you the latter part of the Year, let it but stand till the Frost nips
it, and the Cattle will be glad of it : But I think
Trees may be so planted both in Orchards and Fields
as only to shade the Grafs, and prevent the burning
of it in Summer, and to drop on the Grafs but very
little, which is the only occasion of its Sournefs, as
I shall endeavour to shew afterwards. And that the
Leaves of Trees are a very great Improvement of
Land, may be seen by small Inclofures, which are
commonly richer and more fruitful than large Fields
adjoining to them, tho' of the fame Soil; of which
also Woods are an evidence, by the Improvement
they make of any fort of Soil they are planted on.
And that Fruit-trees are a great Improvement of
Land, is the Opinion not only of the Ancients, who
have composed many large Volumes to encourage
this Work, giving it the greatest Encomiums, and
preferring it before moft other Employments ; but
likewife of Mr. *Blith*, Mr. *Austen*, and all others that
have writ lately on this Subject, in that they caufe
Land to yield a double Crop, and increafe the Advantage of its common Produce of Grafs too. And befides the Advantage that accrues to the private Owner,
it would be of benefit to the Publick to have Fruit-
trees much more propagated than they are, in that
it would hinder the vaft Confumption of *French*
Wines, which is the enriching of a Foreigner, by a
Trade very prejudicial to this Nation, and inftead of
it might procure po us a confiderable foreign Trade,
of no lefs Advantage than the other has been prejudicial.

dicial. In order therefore to advance and promote this useful part of Husbandry, I shall first begin with the Seminary and Nursery, as what is the first Work to be taken care of where you have not the opportunity of buying Trees, or that you design the raising of them your self.

Chap. II. *Of the Seminary and Nursery for Fruit-Trees.*

THE Seminary and Nursery of Fruit Trees is to be order'd much after the same way as is before described for Forest-trees: As first, You must towards *October* cleanse the Ground of Weeds, Roots, &c. which you design for this purpose: And note, that wet or very stiff Clay, and Land rich with Dung, is not good for this Use. Make the Mould very fine, and where you can get Crab-stocks enough in the Woods, you may plant your Nursery with them; but if your Nursery be large, and they are hard to get, your dependency must be upon those you raise in your Seminary, which are esteem'd the best. The way of doing of which, is to keep the Stones of such Fruit as are early ripe in Sand till *October*; and then stretching of a Line cross your Beds, if you make Beds for them, prick Holes by it about a hand's breadth distant one from another, setting of the Stones about three Inches deep; and having finished one Row, remove your Line to another, which must be about a Foot distance from the former: And so you may go on with your setting of them if you raise your Seeds on Beds; but if not, your Rows must be two Foot or more distant from one another, that so you may have liberty to go between to weed them, observing to keep each sort by themselves. All kind of Nuts, &c. may be set in the same manner: And for Stocks raised from the Seeds of Kernels, of Apples, Pears, or Crabs, some propose this Method; which is, after having made any Cyder, Verjuice, or Perry, to take

the

the Muft, which is the Subftance of the Fruit, after the Juice is preffed out, and the fame Day, or next after, before it heats, have the Seeds fifted out of it with a Riddle, on a clean Floor or Cloth, which fow as foon as may be upon Beds of fine Earth very thick, for fome being bruis'd in the grinding or pounding, and others not being ripe, many will never come up; upon them fift fine Mould about two Fingers thick, laying White-thorns or Furze on them till the Ground is fettled, to prevent the Birds or Fowls from fcraping of them up: And, to keep them warm in the Winter, lay fome Fern or Straw on them, which you muft be fure to remove in Spring, before the Seeds begin to fhoot, which is commonly in *May*, and likewife to keep them well weeded; and if the Summer happen to be dry, they may be fometimes watered. Be careful likewife to fet Traps for the Moles and Mice, which are very greedy of them; or you may, as fome fay, poifon the Mice with pounded Glafs mixed with Butter and Oatmeal and caft in bits upon the Beds.

The beft Stocks to graft on are thofe that are raifed *Crab*of the Kernels of Wildings, and Crabs of the moft *ftocks* thriving Trees, tho' in *Herefordfhire* they reckon the Gennet-Moyl, or Cydodine Stock (as they call it) to be the beft Stock to preferve the Guft of any delicate Apple, it being obfervable, that the Wild-ftock enlivens the dull Apple, and the Gennet-Moyl fweetens and improves the over-tart Apple, but that the Tree lafts not fo long as if grafted on a Crab-ftock; and tho' the Fruit doth always take after the Graft, yet it is fomething altered by the Stock, either for the better or worfe.

To be furnifhed with fuch variety of Stocks as is neceffary for the feveral forts of Fruit-trees that you are to raife, the Seminary ought to be filled with fuch as are raifed of Peach-ftones, Plumb-ftones, Cherryftones, Quince-ftocks, &c. or of fuch as are raifed of Suckers from the fame, which are as good according
ing

ing to what each fort of Tree requires ; of which I
fhall give an Account hereafter.

Quince-
ftocks.

But the beft and moft expeditious way to raife a
great quantity of Quince-ftocks for your Nurfery, is
to cut down an old Quince-tree in *March* within two
Inches of the Ground, this will caufe a multitude of
Suckers to rife from the Roots. When they are
grown half a Yard high, cover them a Foot thick with
good Earth, which in dry times muft be watered ;
and as foon as they have put forth Roots in Winter,
remove them into the Nurfery ; where in a Year or
two, they will be ready to graft with Pears.

Plumb-
ftocks.

Plumb-ftocks and Cherry-ftocks may be raifed
from Suckers as well as from Stones, and alfo the fame
way, as is above directed for Quinces, only you muft
have regard to the kinds from whence they proceed,
becaufe of the forts you graft or inoculate on.

Cherries may be grafted on Plumbs, and Plumbs
on Cherries, or Apples, or Pears, and they will take
and grow the firft year ; but they are apt to die the
next ; and therefore if a *Scion* were taken from the firft
years Shoot, and grafted again on a proper ftock,
I am apt to think fome Improvement might be made
by it.

Pear-
ftocks.

Pear-ftocks may alfo be raifed of Suckers, and
tranfplanted like the former ; but thofe that are raifed
of Seeds or Stones are fteemed much better than thofe
raifed from Suckers or Roots.

Removing
of Stocks.

Thefe Stocks when they are two Years old, or
one Year, according to fome, are beft to be removed
into the Nurfery, tho' they are never fo fmall, provided they make but large Shoots ; where after they
come to make ftrong Shoots, they may be grafted,
inoculated, *&c.* according to their Nature, and the
Ufe you defign them for, obferving to cut off the
down-right Roots, and the Tops and fide Branches
of the Plants, leaving of them about a Foot above the
Ground, and letting neither the Roots be too long,
nor fet too deep, becaufe they will be removed afterwards

wards with the more eafe: And it it neceffary to remove Seed-plants often as well as Foreft-trees, becaufe
by that means they get good Roots, which otherwife
they thruft down only with one fingle Root, nor will
they bear well without it: And alfo all ftone-Fruit
fhould be fown quickly after gathering; for if you
keep them long after, they will be two Years before
they come up; if they have not all the moifture of
the Winter to rot the fhells, the Kernel will not come
up. And obferve to fet the biggeft and leaft by
themfelves in different places.

Fruit-Trees being of feveral kinds, and raifed and
increafed feveral ways, as by grafting, inoculating,
or budding, fome by Seeds or Nuts, and others by
Layers, Cuttings, Suckers, Slips, &c. according to
the nature of them; and you having furnifhed your
felf with feveral forts of Stocks for thefe purpofes:
I fhall in the next place endeavour to fhew the manner of ufing them, and the particular Ways and Methods ufed for the raifing of each feveral fort of
Trees, and begin with Grafting.

Chap. III. *Of Grafting of Trees.*

CHOOSE your Grafts from a good bearing
Branch, and from an old Tree rather than a
young one, and covet not one that is too flender, left
the Sun and Wind dry it too much, and caufe it to
wither. The beft Scions are reckon'd fuch as are of
the laft Year's Wood, and that have fome of the former Year's Wood to them, which is ftronger to put
into the Stock than the laft Year's Wood, and is reckon'd to advance the bearing of the Graft; but a Graft
only of the laft Year's Shoot will do very well; tho'
in *Herefordfhire* they commonly chufe a large Graft:
However, thofe Scions are efteemed the beft, whofe
Buds are not far afunder, which ufually determines the
length of the Graft.

And as the Stock is more or lefs thriving, and is
capable of yielding of more or lefs Sap, fo let the
Graft

Graft have more Buds ; but ordinarily three or four are sufficient. I am told that a Scion grafted with the top on, will bear sooner than if the Top be cut off.

And tho' you may graft and inoculate at moft times of the Year, either by beginning early in the Autumn, or by Budding in the Summer ; yet the principal time for Grafting is the Month of *February*, for Cherries, Pears, Plumbs, and forward Fruits, and *March* for Apples. Mild open Weather is beft, and moft propitious for this Work, but by no means graft in wet Weather ; and if you ftay till you can be pretty certain of the Frofts being over, tho' it be to the beginning or middle of *April*, if 'tis a late Spring, it will be the better.

Obferve that a Graft cut fome time before, and ftuck in the Ground, and then grafted at the rifing of the Sap, takes better than thofe that are grafted fo foon as cut, and Grafts from an old Tree fhould be taken fooner than from a young one.

Note alfo, that the Scions or Grafts of Plumbs or Cherries are not to be cut fo thin as thofe of Apples or Pears.

As to the Succefs of Grafting, the main point is to join the infide of the Bark of the Scion and the infide of the Bark of the Stock together, that fo the Sap that runs between the Bark and the Wood, may be communicated from the one to the other, efpecially towards the bottom of the Scion.

Choofe the ftraiteft and fmootheft part of the Stock for the place where you intend to graft, but if the Stock be all knotty or crooked (which fome efteem no Impediment) rectifie it with the fitteft pofture of the Graft you can ; and if your Stock be fmall, Graft it about fix Inches above the Ground ; but if it is large, and where Cattle come, it is beft to place it above their reach : In which way of Grafting there is a great Advantage to fome fort of Apples, in that it caufes them to partake more of the Sap of the Crab,

which

which makes the Fruit of a sharp brisk Taste, and
much helps sweet Apples, and is a particular Advantage to Golden-Pippens ; tho' for young Trees for
Standards it's not so practicable, because Trees so
grafted cannot be so well removed : But if your Stock be
removed it should stand at least three Years before 'tis
grafted , except it makes very large Shoots the second
Year.

Graft your Scions on the South-west side of the
Stock, because that is the most boisterous Wind in Sum-
mer; by which means the Wind will blow it to the
Stock, and not from it, which is the way that the
Graft will best bear the Force of it : but as to this
Point, the Shelter that the Grafts have in their stand-
ing is chiefly to be regarded.

Be careful that the Rain get not into the Cleft of
your young grafted Stocks, but keep them clay'd till
the Bark is grown over them, and leave not the Grafts
above four or five Inches in length above the Stock,
because its being long occasions its drawing more fee-
bly, and exposes it more to the shocks of the Wind,
and the hurt of the Birds.

Only the Gennet-Moyl is commonly propagated by
cutting off the Branch a little below a Bur-knot, and
setting of it without any more Ceremony ; but if they
are grafted first as they grow on the Tree, and when
they have cover'd the Head, are cut off below the
Bur, and set, it is much the best way ; only in the
Separation you should cut a little below the Bur, and
peel off or prick the Bark almost to the Knot. And
thus if the Branch have more Knots than one, you may
graft and cut off yearly till within half a Foot of the
Stem, which you may graft likewise, and so let it stand.

To perform this Work well, it is necessary to be
provided with a good strong Knife, with a thick Back,
to cleave the Stocks with ; a neat small Hand-saw,
to cut off the Head of the large Stocks ; a little Mallet
and a grafting Chisel, and a sharp Pen-knife to cut
the Grafts. You must likewise have a stock of Clay
well

well mix'd with Horse-dung, to prevent its freezing, and with Tanners Hair to prevent its cracking, Bass-strings or Woollen Yarn to tie Grafts with, and a small Hand-basket to carry them in, with such other Instruments and Materials as you judge necessary, according to the way and manner of grafting that you design to use, which is perform'd several ways, as,

Cleft grafting.   First, By grafting in the Cleft, which is the most known and ancient way, and the most used for the middle-siz'd Stocks: The way of doing of which is, first to saw off the Head of the Stock in a smooth place, and for Wall-trees or Dwarf-trees to graft them within four Fingers of the Ground, and for tall Standards higher, as you think convenient, or your Stocks will give way to do. Pare away with your Knife the roughness the Saw hath left on the Head of the Stock, and cleave the Head a little on one side of the Pith, and put therein your grafting Chisel, or a Wedge, to keep the Cleft open; which cut smooth with your sharp Knife, that the top may be level and even, except on one side, which must be cut a little sloping; then cut the Graft on both sides smooth and even from some Knot or Bud in form of a Wedge, suitable to the Cleft, with a small Shoulder on each side; which Graft so cut, place exactly in the Cleft, so as that the inward Bark of the Scion may joyn to the inward part of the Bark, or Rind of the Stock closely, wherein lies the principal skill and care of the Grafter if he expects an answerable success of his Labour, as was said before: Then draw out your grafting Chisel, or Wedge; but if the Stock pinch hard, left it should endanger the dividing of the Rind of the Graft from the Wood, to the utter spoiling of it, let the inner side of the Graft that is within the Wood of the Stock be left the thicker, that so the woody part of the Graft may bear the stress, or rather leave a small Wedge in the Stock to keep it from pinching the Graft too hard, and then you may leave the out-side of the Graft a little thicker, especially in smaller

Stocks. Cover the Head of the Stock with temper'd Clay, or with foft Wax, to preferve it not only from the extreamity of the cold and dying Winds, but principally from the Wet.

The fecond way of grafting, and much like unto *Rind* the former, is grafting in the Rind or Bark of the *grafting.* greater Stocks, and differs only in this, that where you cleave the Stock and faften the Grafts within the Cleft in the other way; here you with a fmall Wedge of a flat half-round Form, cut tapering to a thin point, made of Ivory or Box, or other hard Wood, you only force the Wedge in between the Rind and the Stock, till you have made a Paffage wide enough for the Graft; after the Head thereof is fawn off, and the roughnefs pared away, then you are to take the Graft, and at the fhoulder or grofs part of it cut it round with your fmall Grafting-knife, and take off the Rind wholly downwards, preferving as much of the outward Rind as you can; then cut the Wood of the Graft about an Inch long, and take away half thereof to the Pith, and the other half taper away like to the Form of the Wedge, fet it in the place you made with your Wedge between the Bark and the Stock, that the fhouldering of the Graft may join clofely to the Wood and Rind of the Stock, and then with Clay and Horfe-dung cover it as you do the other.

This way is with moft conveniency to be ufed when the Stock is too big to be cleft, and where the Bark is thick. Here you may fet many Grafts in the fame Stock with good fuccefs; and the more you put in, the fooner the Bark will cover the Wound.

The third way of Grafting that is made ufe of is to *Whip* be performed fomewhat later than the other, and is *grafting* to be done two ways; Firft, By cutting off the Head of the Stock, and fmoothing of it, as in Cleft-grafting; then cut the Graft from a Knot or Bud on one fide floping, about an Inch and a half long, with a fhoulder, but not deep, that it may reft on the top of the Stock; the Graft muſt be cut from the ſhoul-dering ſmooth and even, ſloping by degrees that the lower end be thin; place the ſhoulder on the head of the Stock, and mark the length of the cut part of the Graft, and with your Knife cut away ſo much of the Stock as the Graft did cover (but not any of the Wood of the Stock:) place both together, that the cut part of both may join, and the Sap unite the one in the o-ther; and bind them cloſe together, and defend them from the Rain with tempered Clay or Wax, as before.

The other way of Whip-grafting is, where the Grafts and the Stocks are of one equal ſize: The Stock muſt be cut ſloping upwards from one ſide to the o-ther, and the Graft after the ſame manner from the ſhoulder downwards, that the Graft may exactly join with the Stock in every part, and ſo bind, and Clay or Wax them, as before.

Theſe ways of Whip-grafting (eſpecially the firſt) are accounted the beſt; firſt, becauſe you need not wait the growing of your Stocks, for Cleft-grafting requires greater Stocks than Whip-grafting doth: Secondly, This way injureth leſs the Stock and Graft than the other: Thirdly, the Wound is ſooner heal-ed, and upon that account better defended from the nipping of the Weather, which the cleft Stock is in-cident unto: Fourthly, this way is more facile both to be learned and performed.

Grafting by Ap-proach. The fourth way of Grafting is by Approach or Ablactation: And this is performed later than the former ways; to wit, about the Month of *April*, ac-cording to the Forwardneſs of the Spring. It's to be done where the Stock you intend to graft on, and the Tree from which you take your Graft, ſtand ſo near together that they may be joyned: Upon which ac-count it is commonly practiſed upon Orange and Li-mon Trees, and other rare Plants that are preſerved in Caſes, and may upon that account be joyned with the more facility. Take the Sprig or Branch you in-tend

tend to graft, and pare away about three Inches in length of the Rind and Wood near unto the very Pith; cut alſo the Stock or Branch on which you in-tend to graft the ſame, after the ſame manner, that they may evenly joyn each to other, and that the Saps may meet; and ſo bind and cover them with Clay or Wax, as before.

As ſoon as you find the Graft and Stock to unite, and to be incorporated together, cut off the Head of the Stock (hitherto left on) four Inches above the binding, and in *March* following the remaining Stub alſo, and the Cion or Graft underneath, and cloſe to the grafted place, that it may ſubſiſt by the Stock only.

Some uſe to cut off the Head of the Stock at firſt, and then to joyn the Cion thereunto after the manner of Shoulder-grafting, differing only in not ſevering the Cion from its Stock. Both ways are good, but the firſt moſt ſucceſsful. This manner of Grafting is principally uſed for ſuch Plants as are not apt to take any other way; as, for Oranges, Limons, Pome-granates, Vines, Jeſſamines, Althæa Frutex, and ſuch like. By this way alſo may Attempts be made to graft Trees of different kinds one on another, as Fruit-bearing Trees on thoſe that bear not, and Flower-trees on Fruit-trees, and ſuch like.

Theſe are the moſt uſual ways of Grafting; ſome others there are, but they differ very little from the former; and where they differ, it's rather for the worſe, and therefore not worth the mentioning.

Thoſe Grafts that are bound you muſt obſerve to unbind towards Midſummer, leſt the Band injure them.

Where their Heads are ſo great that they are ſub-ject to the Violence of the Winds, it is good to pre-ſerve them by tying a Stick to the Stock which may ex-tend to the top of the Graft, to which you may bind the Graft the firſt Year; The beſt thriving Grafts are moſt in danger; afterwards they rarely ſuffer by the Winds.

S 2   Grafts

Grafts are alfo fubject to be injured by Birds, which may be prevented by binding fome fmall Bufhes about the tops of the Stocks.

Root grafting. There is another way of Grafting that hath not been fo long in ufe as the former, which is, to take a Graft or Sprig of the Tree you defign to propagate, and a fmall piece of the Root of another Tree of the fame kind, or very near it, or the pieces of Roots that you cut off of fuch Trees as you tranfplant, and Whip-graft them together ; and binding of them well, you may plant this Tree where you have a mind it fhall ftand, or in a Nurfery, which piece of Root, will draw Sap and feed the Graft, as doth the Stock the other way.

Only you muft obferve to unite the two But-ends of the Graft and Root, and that the Rind of the Root join to the Rind of the Graft.

By this means the Roots of one Crab-ftock or Apple-ftock will ferve for twenty or thirty Apple-grafts; and in like manner of a Cherry or Merry Stock, for as many Cherry-grafts; and fo of Pears, Plumbs, &c.

Thus may you alfo raife a Nurfery of Fruit-trees inftead of Stocks, by planting of them there while they are too fmall to be planted abroad.

This is alfo the beft way to raife tender Trees that will hardly endure grafting in the Stock, becaufe this way they are not expofed to the injuries of the Sun, Wind and Rain.

It is alfo probable, that Fruits may be meliorated by grafting them on Roots of a different kind, becaufe they are more apt to take this way than any other.

The Trees thus grafted are reckoned to bear fooner, and to be more eafily dwarfed than any other, becaufe part of the very Graft is within the Ground, which being taken off from a bearing Sprig or Branch will bloffom and bear fuddenly, in cafe the Root be able to maintain it.

The only Objection againft this way is this, that the young Tree grow flowly at the firft, which is occafioned eafioned by the fmallnefs of the Root that feeds the Graft ; for in all Trees the Head muft attend the increafe of the Roots, from whence it hath its Nou-rifhment.

Neverthelefs this Work is eafily perform'd, Roots being more plentiful thǝn Stocks, and may be done in great quantities in a little time within Doors, and then planted very eafily with a flender Dibble in your Nur-fery, and will in time fufficiently recompence your pains. If you tranfplant a Tree that the grafting place is not well grown over, place the back of the Graft to the South, and the cut to the North, and it will caufe it to heal the better. Slips are not fo good to plant or graft on as feedlings, becaufe they often mifs taking of Root; and if they do grow, the Root being part of the old Wood, it often doats, and rots in the Ground, and grows but flowly.

But for the Encouragement of thofe that defire to be furnifh'd with good Fruit, and good Bearers, Grafts may be carried with a little care either by Sea or Land, and will keep good from *October* to *March*, provided they are buried in Earth, and kept moift, tho' without any thing put to them : They will keep a great while, fo as that I my felf had fome Pear-Grafts which grew well that were fent from *Paris*: Only if you find them any thing dry, lay them as foon as you receive them, in Water twenty four hours, and afterwards in moift Earth; or you may ftick the large ends of them in Clay tying a Rag about them to keep it from falling off, and to wrap the other end in Hay or Straw Bands, which will fecure them from the Wind and Bruifes, and is a good way for tranfporting of them.

Chap. IV. *Of Inoculation.*

INoculation is by fome prefer'd before any of the ways of Grafting before treated of, and differs from the other ways, in that it is performed when

the Sap is at the fulleſt in Summer, and the other
ſorts of grafting are before the Sap aſcends, or at leaſt
in any great quantity.   Alſo by this way of inoculating ſeveral ſorts of Fruits and Trees may be propagated and meliorated, which by grafting cannot be
done, as the Apricock, Peach and Nectarine, which
rarely thrive any other way, becauſe few Stocks can
feed the Graft with Sap ſo early in the Spring as the
Graft requires, which makes it fruſtrate your Expectation; but being rightly inoculated in the fulneſs
of the Sap rarely fails.

The Stocks on which you are to inoculate are to
be of the ſame kind as before, and in the next Chapter
is directed to graft on.

The time for this Work is uſually from Midſummer to the middle of *July*, when the Sap is moſt in
the Stock.   Some Trees, in ſome Places, and in ſome
Years you may inoculate from Mid-*May* to Mid-*Auguſt*.
As to the time of the Day, it is beſt in the Evening
of a fair Day, in a dry Seaſon; for Rain falling on the
Buds before they have taken will deſtroy moſt of them.

The Buds you intend to inoculate ſhould neither
be too young or tender, nor yet too old, but young
ones are beſt; the Apricock Buds are ready ſooneſt,
they muſt be taken from ſtrong and well-grown Shoots
of the ſame Year, and from the ſtrongeſt and biggeſt
end of the ſame Shoots.

If Buds be not at hand, the Stalks containing them
may be carried many Miles, and kept two or three
Days, being wrap'd in freſh and moiſt Leaves and
Graſs to keep them cool : If you think they are a little wither'd, lay the Stalks in cold Water two or
three hours, and that, if any thing, will revive them
and make them come clean off the Branches.

Having your Buds and Inſtruments ready for your
Work, *viz*. a ſharp pointed Knife or Penknife, a Quill
cut half way and made ſharp and ſmooth at the end,
to divide the Bud and Rind from the Stalk, Wollen-
Yarn, Baſs-matt, or ſuch like, to bind them withal;

then

then on some smooth part of the Stock, either near or farther from the Ground, according as you intend it, either for a Dwarf-Tree, or for the Wall, or a Standard, cut the Rind of the Stock overthwart, and from the middle thereof gently slit the Bark or Rind about an Inch long, in form of a T, not wounding of the Stock; then nimbly prepare the Bud by cutting off the Leaf and leaving of the Bark about half an Inch above and below the Bud, and sharpen that end of the Bark below the Bud, like a Shield or Escutcheon, that it may the more easily go down and unite between the Bark and the Stock : Then with your Quill take off the Bark and Bud dextroufly, that you leave not the Root behind; for if you fee a hole under the Bud on the infide, the Root is left behind, therefore cast it away and prepare another; when your Bud is ready, raife the Bark of the Stock on each fide of the slit (preferving as carefully as you can the inner thin Rind of the Stock) put in with care the Shield or Bud between the Bark and Stock, thrusting it down until the top join to the crofs-cut, then bind it clofe with Yarn, &c. but not on the Bud it felf.

There is another way of Inoculation more ready than this, and more fuccefsful, and differs from the former only in that the Bark is slit upwards from the crofs-cut and the Shield or Bud put upwards, leaving the lower end longer than may ferve; and when it is in its place, cut off that which is fuperfluous, and join the Bark of the Bud to the Bark of the Stock, and bind it as before, which fooner and more fuccefsfully takes than the other, In Inoculating you may put two or three buds on a Stock.

You may alfo cut the edges of the Bark about the Bud fquare, and bind it faft, and it will fucceed well; which is the readier way, and eafier.

About three Weeks or a Month's time after your Inoculation you may unbind the Buds, left the binding injure the Bud and Stock.

When

When you unbind them you may difcern which are good and have taken, and which not, the good will appear verdant, and well colour'd, and the other dead and wither'd.

In *March* following cut off the Stock three Fingers above the Bud, and the next Year cut it clofe, that the Bud may cover the Stock as Grafts ufually do.

Chap. V. *Of the Ways and Methods ufed for the raifing of Fruit-Trees.*

HAVING already fhew'd how to furnifh your Seminary with Stocks, and the method of Grafting, Inoculating, &c. of them, I fhall in the next place confider the feveral forts of Fruit-Trees and what Ways and Methods are the beft and moft proper for the propagation of each of them: I fhall begin with the Apple, as it deferves the preheminence, both upon the account of its Univerfality and Ufefulnefs, being both Meat and Drink, it may be had at all times of the Year. Of Apples there are great Variety, and for feveral Ufes; but as Apples and Pears and other forts of Fruit are of feveral kinds, I fhall have occafion to mention a great number of them hereafter, and at prefent, only take notice of fome pe-culiar Apples and Pears proper for Cyder and Perry.

Apples, how rais'd.

Apples are commonly raifed by grafting upon the Crab, that being efteemed the beft and hardieft Stock for them, and the leaft fubjeft to Canker; they take beft when grafted in the Cleft or Bark, except the Gennet Moyle, which will grow of Burrs, as I have mentioned before, and the Codlin, which may be raifed of Suckers or Slips, tho' indeed moft of our beft Fruit has been produc'd from the Kernel, being a kind of Wilding, as the Red-ftreak, the Golden Pippin, &c. are reported to be; but it being very rare that any can be raifed this way, moft Kernels producing Fruit of a wild, auftere, fharp Tafte, tending rather to the wildnefs of the Stock on which the Tree was grafted, than that of the Graft; although many of them may feem fair, yet they want that

briknefs

briskness of Spirit, and are more woody than the grafted Fruit, being also much longer before they bear, and
are more unfruitful; for the often grafting of Trees upon the same kind, is esteem'd a meliorating of them,
which hath occasion'd many to endeavour the raising of
new kinds of Apples, and other Fruits, by grafting of
them upon different Stocks, which way I think deserves
Incouragement; if any should attain to any sort of Graft
of this kind, I think it would be convenient at the second, or at farthest, at the third Year's growth, to graft
it upon a Stock of a more natural kind to it, for Nature
delights more in an advance than going backwards. By
a failure in this Point I lost several Grafts that I believe
would have been of Advantage had I been aware of the
sudden Blast that took them the third Year. I think Experiments of this kind more likely to succeed by try-
ing a dry or an insipid Fruit upon a pungent vigorous
Sap, than the contrary; but for the gaining of a new
Species of Apples, tho' tis rare to have them take, yet
I think the doing of it by Seeds and Kernels the most
likely way, because many have been procur'd that way,
I have obtained four several sorts of this kind my self.
I am told that Apples, may also be raised of Layers; if
you cut the Layer when you put it in the Earth, as you
do the Layer of a Gilli-flower; and let them be two
Years in the Ground before you take them up, and that
such will bear sooner than those that are grafted.

   Next unto Apples I think Pears may come in the *Pears,how*
second place, and would be much more useful than *raised.*
they are, were such care taken to improve the Juice of
them as might be; in one thing they are to be preferr'd, that they prosper in cold, moist, hungry, stony and gravelly Land, where Apples will not bear so
well; they are best grafted the same way as the Apple, and upon the wild Pear-stock raised of their Kernels for Standards; but for Dwarf or Wall-Trees the
Quince-stock is esteem'd the best, but then they should
be planted in moist Ground; they will grow likewise
when grafted on the White Thorn, but not so well

as on the former ; they may likewise be buded on their own kind ; and if Winter Pears, or any sort of latter Fruit, be grafted on the earliest Pears, or earliest Fruit of the kind, it will forward and help them very much. They may be grafted when the blossom is on them.

Cherries, how raised. Cherries are a fine Summer-Fruit, and are of several sorts; they do best grafted on the Black-Cherry-stock, or the Merry-stock, which may be raised in great quantities from Cherry-stones ; Suckers also from the Roots of the Wild or Red-Cherry, will do well; they are commonly grafted about a Yard from the Ground by whip-grafting ; they may likewise be budded or inoculated on their own kind.

Plumbs, how raised. Plumbs are of several sorts, and commonly cleft-grafted on any Stocks of their own kind, except the Damascen ; but one of the best sorts to graft them on is the Pear-Plumb ; tho' I have often found them to prove well raised only from the Stones, especially Damascens.

Medlars. Medlars may be cleft, or Stock-grafted, on the White Thorn, but prove best on the Pear or Quince-stock.

Filberts. Filberts may be cleft-grafted on the common Nut, and Servises on their own kind, or propagated by Suckers, Layers or Seeds.

Quinces, how rais'd Quinces may be cleft-grafted on their own kind, or raised of Slips or Layers, and of Cuttings ; they delight in a moist Soil. If you have a Quince-Tree which grows so low that you can by plashing, or otherwise bring it to the Ground, do it the beginning of the Winter, and cover it with Earth, except the two ends of the Boughs ; and every twig will put forth Roots, which being cut off and transplanted will make a Tree ; they may be also Inoculated which will make them bear the sooner. If they are planted on dry Ground, they should be planted in *October*, They are better grafted in the stock than in the bark,

Apricocks

Apricocks are ufually inoculated in Plumb-ftocks, *Apricocks,* raifed either from Suckers that have not been grafted *how rai-* before, or Stones; thofe of the White Pear-Plumb *fed.* are efteemed the beft, and thofe of any other great White or Red Plumb that hath large Leaves and Shoots are very good for the budding of Apricocks or Peaches. I am told that an Apricock, inoculated on a Peach, mightily improves the Fruit. I have known Apricocks bear very well that were only raifed of Stones, but their Roots are reckon'd very fpungy, and fo not apt to continue long. Apricocks do well alfo in Standards; but as they bloffom very early againft a Wall, fo they are very much in danger from the Frofts, againft which they fhould be fhelter'd with Matts, or other fhelter, when they are in Bloom. You may alfo plant them in feveral Afpects, and by that means have them ripe in feveral Seafons.

They thrive beft on a light foil if inoculated on a Plumb-ftock; but the beft for a heavy Soil are thofe inoculated on an Apricock-ftock, raifed of the ftone.

Peaches are of feveral forts, and are raifed the fame *Peaches,* way with Apricocks, and delight in the fame Soil, and do beft grafted on Apricock-ftocks.

Nectarines delight in the fame Soil as Apricocks *Necta-* and Peaches; they are raifed by Inoculation, the beft *rines.* Stock for which is that raifed of the Peach-ftone.

Figs are of feveral forts, they are multiply'd of *Figs.* Suckers, and delight in a warm moift Soil.

Figs may be increafed by Cuttings, Layers or Suckers, and are to be planted againft a Houfe or Wall, where they may have Sun to ripen them. They may be raifed to the Wall like other Trees, but muft not be pruned more than needs muft.

If you plant fome fmall Fig-Trees in Pots or large Boxes, as you do Orange-Trees, and put them in fome houfe (from the beginning of *November* until *April*) without any Fire, or any other Curiofity; you may have early Figs if when you fet them out, you fet them under a South-wall, and keep them watered once a Week.

Week. I am told that if you lay a quantity of Walnut-shells round the Roots of a Fig-Tree, and cover them with good Earth till they rot, they will cause the Fig-tree not only to bear, but to produce large Fruits.

Oranges and Limons.   Orange and Limon Trees, in hot Countries, are raised of Slips, but they will not grow so here; they are commonly grafted or inoculated, or are raised by sowing their Pippins or Seeds in Boxes, and when they are two Years old transplant them in Cases, every one in a Case by it self fill'd with rich Melon-bed Mould mingled with Loam, refined and matured by one Winter Season; and when they can well support it, you may either inoculate or graft them by approach in the Spring of the Year. Be diligent to secure them from Cold, and commit them early to their shelter, where they may intirely be preserved from the Frost; you may give them a gentle Stow, and attemper the Air with a Fire of Charcoal during the extream rigour of the Winter, in case you suspect the Frost has at all invaded them.

But so soon as the Spring appears, and the Frosts are intirely past, you may acquaint them with the Air by degrees, beginning first to open the Doors of the Conservatory in the heat of the Day, and shutting them again at Night; and so by little and little you may set open the Windows, and shut them again in the Evening till all danger is past, and then you may bring them forth and expose them boldly to the Air during all the Summer following.

As these Trees grow big you may change and enlarge their Cases, but be sure to take them out Earth and all, raising the stringy and fibrous Roots a little with a Knife before you replace them, supplying what their new Cases may want with the fore-described Mould. Some when they alter their Cases denude them of all the Earth, conceiving it exhausted and insipid, but it is to the extream prejudice of the Tree, and doth set it so far back that a Year or two will hardly recover it.

You

You may gather the Flowers every Day to prevent their knotting into Fruit, sparing only some of the fairest and best placed for Fruit, and as many as you conceive the Tree can well nourish.

The Spiders do extreamly affect to spread their Toils among the Branches and Leaves of this Tree, because the Flies so much frequent their Flowers and Leaves, which attract them with their Redolency and Juice; to remedy this, use such a Brush as is made to cleanse Pictures withal from Dust, but treat them tenderly.

Mr. *Hartlib* condemns us much for neglecting the *Almonds.* propagating of Almond-Trees, which (saith he) grow very well and bear good Fruit, he having, as he says, seen divers Bushels on one Tree in his Brother's Orchard; they grow large and upright, and need not the help of a Wall. The Almonds are in some sweet, and in others a little bitter; but I am told, that often removing of them makes them grow sweeter; the Tree is chiefly received for the Beauty of its Flowers, which being early, and of a fair, pale, reddish Colour, make a fine show in a Garden.

They are raised by setting of the Nut in the Shell in the Month of *October* or *February*, which should be laid in soft dung, a day or two before you plant them; and when they come up they should be watered once a Week, till they grow pretty large: they may also be raised by slips from the Roots; they delight in the Sun, and a dry Soil; and grow barren in a fat one.

You may increase several Sorts of *Fruits*, by making choice of a convenient Branch, or Shoot of an indifferent size; which about *Midsummer*, sometimes a little sooner, and sometimes later, according as the Weather proves, when the Sap is very high, which in some Trees is sooner and in some later; a little above the place you think best, apply a pretty quantity of well-tempered Mortar round about it, and make such provision with convenient tying of a Cloath about it as may cause the Mortar not to fall off, either by the washing of the Rain or otherwise, and

cover

cover it with Clay, which form so on the Top, that it may a little receive the Moisture in case of Rain, and then cut the Bark off round about the Branch under the place where the Clay is, about two or three Inches wide, and in the Clay or Mortar, it will either put forth root, or prepare it so for rooting, that being cut off about the beginning of Planting-season, it will grow, which sometimes may be done about the latter End of *September*, or beginning of *October*. You must observe in planting of it to proportion the Top to the Root, and not to leave too much for the young Root to feed, and plant it in good Ground with the Mortar on, which is to be made of Clay, fine Earth, and a little Dung, which must be clapped to the bared place as well as to the Bark that is about it, and the Ball made as big as a Foot-ball. It must be set pretty deep, and kept often watering.

Pear-Trees have commonly more brittle Roots than Apple-trees, and therefore more care should be had in taken of them up, and the Roots of such Apple-trees, or stocks as have been raised of kernels, are likewise more brittle than those raised of the Crab or Wilding.

Medlars are raised best by grafting on a servise Tree; the great Dutch Medlar is the best, and a good bearer; they are also raised of the stone and some times from Suckers, but when sowed they require a great deal of time, before they bear. Neither is the Fruit so large as when grafted; they are commonly planted near the Water, but they thrive and bear well in other places.

Chap. VI. *Of Dwarf, Wall and Standard Trees.*

YOUR Trees being grafted or inoculated on proper Stocks, the next thing to be consider'd, is which are to be for Dwarfs, Walls and Standards, that a proper Method may be used about each sort, the ordering of which consists mostly in the pruning

of

of them : Dwarf Trees muft be pruned fo as to make them hollow, and to branch low into as many Branches as you can : The fame care muft be taken of fuch as are to plant againft Walls ; only as the one are to grow round, the others are, as much as you can, to be made to grow flat, that they may fpread the better on the Wall; but for the Standards, as foon as you can, you are to reduce them to one Branch, which when it comes to the intended heighth you defire, you muft cut the top of it to caufe it to head at that place, always confidering the ftrength of the Body, which if it have ftrength enongh you may prune it clear up; but if weak, leave according to its ftrength, more or lefs Branches to grow out of the fides, and check the Sap which ftrengthens the Body: and fo will the pruning of the Head, when once the Body is of a good ftrength.

By this means being provided with all forts of Trees, it will in the next place be convenient to confider the Methods to be ufed for the Planting of them out, in doing of which it will be neceffary to know, firft, how to order the Ground, and fecondly, how to tranfplant the Trees.

Chap. VII. *Of the manner of cultivating Ground for an Orchard.*

THE natural Soil of an Orchard is more to be regarded than that of a Garden, becaufe the Garden-Produ￼ rooteth but fhallow, and fo may eafily be manur'd to the depth that is required for Garden-Commodities; but Fruit-trees growing large and rooting deep, ought to have a deep and rich Soil, where your conveniency will allow of it.

And if the Land that you intend to plant be a Turf, or Green-fwerd, you will do well to plow it two Years before you fet the Trees in it, to make it mellow and loofe ; and the deeper you plow it the better, becaufe the Trees will have the better Opportunity

nity

nity to Root; and if you lay dung or Manure on its
the plowing will mix it the better with the natural
Soil, and it will be much the better to dig if you defign
to fet Beans, Peas or other Commodities with your
Trees, which is the beft way of advancing the Growth
of your Trees; and if you would have your Trees
to thrive, you muft take care that your Trees be not
too near together, and that no fort of Plants be near
them, which may deprive them of their Nourifhment,
or any way hinder thofe Refrefhings and Helps that
they might otherwife receive by the Rain or Dew.

Take care to keep the Earth about your Trees always light and clean, and often cultivated, fo as to
mend and clean it as often as it requires.

Earth that is hot or dry muft be dug or tilled in
Summer-time, either a little before, or whilft it rains,
or foon after; at which time you can neither Till it
too often nor too deep, becaufe the doing of it in
hot Weather will kill fuch Herbs or Flowers as grow
in it, except they are water'd; but cold, ftrong and
moift Earth is beft to be tilled in dry Weather, only
there are fome Grounds that will not work till Rain
comes.

The frequent ftirring of the Earth prevents its
goodnefs from being wafted by the growth and nourifhment of ill Plants; but fuch ftirrings are not e-
nough without pulling of the Weeds up: For ill Weeds
that ufually grow in Summer and Autumn, multiply
without end if they are fuffered to run to Seed. At
the time that the Trees bloffom and the Vines fhoot,
the Earth is not to be meddled with.

To dry Earth a large Culture or Tillage may be
allow'd the beginning of Winter, and the like, as
foon as it is paft, that the Snow, and Rain of the
Winter and Spring may eafily fink into the Earth; but
to ftrong and moift Earth a fmall Tillage in *October*
only, to remove the Weeds, is beft, and to give a
large Tillage in Spring when the greateft Rains are
over, and if you Trench heavy moift ground againft
Winter,

Winter, your firſt breaking of it up ought to be ſlight.

But if your Orchard is ſituated in a ſandy or dry Ground, endeavour by the help of ſome Gutters to carry off all the Water that falls in haſty Showers to thoſe Places that are manur'd, that none of it may be unprofitably waſted in the Walks or Allies, but if your Soil is ſtrong and fat, drain it off from the Orchard as much as you can. And if your Land lie flat, that wet is apt to ſtand upon it, or if 'tis a ſhallow Soil, you may ſomething help it in plowing, by gathering of the Land always up in and near the Place where you intend the rows of Trees ſhall afterwards ſtand; which will make the Soil deep where they are to ſtand, and draw off the Moiſture, as was ſaid before.

Rain-water ſinks not ſo deep into Land as Snow, and therefore in moiſt Land it is a good way to remove the Snow as much as you can from about your Trees.

Chap. VIII. *Of the Tranſplanting of Trees.*

I Have already given ſeveral Directions for the Tranſplanting of Foreſt-Trees, which will ſerve for Fruit-Trees alſo, and therefore what I ſhall conſider about Fruit-Trees at preſent, is the diſtance they are to be planted at, and what may tend to the making of them the more fruitful, the ſeldom bearing of Fruit being one of the greateſt Diſcouragements that attends Planting.

As for Standards of Apples or Pears, I am not for planting of them nearer than forty Foot; which diſtance, if any one think too far for an Orchard, and that by the thriving of their Trees (eſpecially while they are young) they ſhall ſuſtain loſs, I ſhould rather adviſe them to plant Cherry-Trees, Codlins, Plumbs, &c. between, becauſe in about thirty or forty Years time they will be decaying, and ſo by cutting them down will make way for the other Trees, the largeſt.

of Apple and Pear-Trees, aad the room that they have being what I think doth much contribute to their bearing and thriving, in that they receive the more Benefit and Refreſhment of the Sun and Air, and have the more room for their Roots to ſpread by their diſtance from each other, and the Fruit is much the better. But if you plant them in Fields or Paſtures, fifty or ſixty Foot diſtance will do well, becauſe they will be leſs prejudicial to the Graſs; but for the preventing of any Inconveniency that may come to Paſture-Lands from the dropping of the Trees, I ſhall propoſe one thing that I have already hinted at, which is, to plant all your Trees in rows from Eaſt to Weſt. Let the Trees and rows be forty Foot one from another, and let all the Trees, by pruning of them, be made to grow like a Fan, or to ſpread in the ſame Form as if they were to grow againſt a Wall, only in that the Stem muſt be taller: This I think will prevent the prejudice that the dropping of Trees will do to the Graſs, and will ſhade the Land from burning, and improve it by the falling of the Leaves, and will alſo cauſe the Fruit to ripen much better than if they grew in the common Form, of which laſt Particular I have had the Experience of two or three Trees that I kept trimm'd after the ſame manner on purpoſe for a tryal.

Cherries, Plumbs, Quinces, and ſuch like Trees, may be planted at fifteen or twenty Foot diſtance, which is ſufficient.

It will be neceſſary at every three or four Years end to lay about aged Trees ſome Soil, eſpecially Lime or Chalk, which is done by uncovering the Mould within a little of the Roots, and applying of it, and then covering of them again with Earth; the beſt Seaſon for which Work is the beginning of Winter, that the Rains way waſh it to the Roots before the heat of the Summer invade it.

Wall- *Trees.* But Wall-trees ſhould be planted at ſuch diſtances as the heighth or breadth of the Wall, the Nature of

the

the Tree, and the Nature of the Ground requires; the higher the Wall the nearer the Trees, may be together; and the lower the Wall the farther diftance, that they may have room to fpread in breadth where they want it in heighth; efpecially Vines, which require a more fpacious and ample place to fpread againft than other Fruit, it being certain, that the more they fpread the better they bear and thrive, which is contrary to the Opinion of all Foreign Parts: And the fame may be obferv'd of moft other Wall-Fruit, efpecially Pears and Apricocks, upon which account I cannot but think moft of our Walls too low, and our Trees commonly planted too thick. And,

Having occafion to find fault with the common fort *Walls for* of Walls for Fruits, it gives me an opportunity of *Fruit.* recommending the Propofal made by Mr. *Fatio* for floping Walls, that fo what is planted againft them may lie expos'd to the direct Beams of the Sun: This fort of Walls, breaking the Wind and reflecting the Sun-beams from one Wall to another, muft be of great advantage for the ripening of Fruit in our cold Climate; to which I fhall only add an accidental Experiment made by a Friend of mine, who had a Wall, the Foundation of which being bad, oblig'd him on the planted fide, between each Tree, to make Buttreffes of about a Yard from the Wall, which caus'd his Fruit to ripen much fooner than it did before: His Wall ftood a little facing to the Eaftward of the South.

But to proceed to what I think the great Point to *Of the* be taken care of about Fruit-trees, which is the Un- *Unfruit-* fruitfulnefs of them, it being often many Years toge- *fulnefs of* ther, that both Apples, Pears and other Fruits fail, *Trees.* as we have of late experienc'd for feven or eight Years together; which Inconveniency being one of the greateft Difcouragements that attends Planting, as I faid before, it may not be amifs a little te confider the occafion of this Unfruitfulnefs, and likewife to propofe fome Remedies for the fame. Now the Unfruitfulnefs of thefe Trees commonly proceeds,

Firft,

Occasion'd by Blasts. Firſt, From Blaſts occaſion'd by the Winds, as many times from the Eaſt-wind in the Spring, which coming after Rain, when the Bloſſoms are wet, and bringing of Froſts with it, ſhrivels up the Leaves of the Bloſſoms and ſpoils them. If a Weſt-wind ſucceeds it commonly brings Caterpillars; the beſt Remedies againſt which are good Shelter, to prevent the Froſts; and the burning of Straw, Hawm, &c. to kill the other, but a great means to have Fruit all Years, is where you have the Conveniency of different Situations, that ſo when a Blaſt comes by one Wind, you may have another under ſhelter from it; and tho' the South Aſpect is the beſt for ripening of Fruit, yet the moſt conſtant bearing Orchard that I have met with, is an Orchard belonging to a ſmall Farm I have in *Hertfordſhire* that is not above a Rood of Ground which is ſituated on the ſide of an Hill that faces the North-Eaſt: The Soil is a kind of a yellow Tile-Clay, and the Weſt end of it is ſhelter'd by the Houſe, and the South ſide by high Trees, but the North ſide and Eaſt end is wholly expoſed to thoſe Winds. But having a conveniency of ſeveral Situations where I now live, I may, when the Trees grow up, which I have lately planted, be able to give a better account of this particular; for Experiments of this kind muſt rather be try'd in the South Parts of *England* than North, eſpecially for North or Weſt Situations.

Want of Rain. Secondly, The want of Rain juſt at Bloſſoming-time often occaſions the dropping off of the Bloſſoms for want of Sap to nouriſh them, eſpecially in dry Grounds; and therefore I have heard of ſome in *Eſſex,* whoſe Orchards were upon a dry Soil, that have had great quantities of Fruit when all their Neighbours about them have fail'd, only upon this account, that they kept their Trees water'd at Bloſſoming-time; and of a Gentleman who had an Orchard planted on the ſide of a Hill that he could Water when he would, which hardly ever fail'd.

The

The third occafion of Unfruitfulnefs is the not fuit- *Suiting* ing of your Fruit and Soil together, a Point that de- *Fruit to* ferves more particular Obfervations than I have been *the Soil.* yet able to make, and what I would defire the Affift- ance of fuch in as are willing to promote this Work, being fatisfied that not only fome fort of Fruits bear better on fome Soils than others, but alfo that they thrive much better; however in the mean time I fhall advife them that plant an Orchard, to enquire what fort of Soil their Neighbours Orchards are of, and what fort of Fruits bear beft in thofe Soils, and ac- cordingly to ftock themfelves with that fort of Fruit.

But a great hindrance to a due enquiry into this ufeful part of Husbandry is the variety of Names that are in many places given to one and the fame Ap- ple or Pear, and therefore where any one hath a good bearing Fruit, I would advife them to be fure of the Name of them from fome experienc'd Gardeners, and then a certain Judgment may be made of them. And if you find the Ground wherein you plant your Fruit-trees not fuitable to the Nature of the Tree, it may be alter'd by applying of Earth, Clay or Sand of a different Nature to your Soil.

A fourth Reafon of the Unfruitfulnefs of Fruit- *From bar-* trees is the Barrennefs of the Soil they are planted on, *renness of* for I cannot but think it as neceffary to dung Orchards *the Soil.* as Plowed Lands, that fo the Dung may wafh to the Roots of the Trees to nourifh them; this I reckon was the reafon of the fruitfulnefs of the Orchard men- tion'd by Mr. *Hartlib* in his Legacy, when he advifed the turning of the Wafh of a Sheep-common to the Roots of the Trees, which, he fays, occafioned fuch a fruitfulnefs to an Orchard that belong'd to a Farm that an Acquaintance of his held, that the Occupier got an Eftate out of the Farm by it, which before was fo dear rented, that it had likely to have ruined him. And I knew a Farmer in *Kent* that ufed to fay, that he had often obferv'd it, that whenever he let his Hogs go into his Orchard unrung, to root about the

T 3 Trees,

Trees, and to dung them, he had always a Crop of Fruit; and it is certain, for Fruit-trees no Dung is so good as Hogs-dung, which Mr. *Worlidge* confirms in his *Vinetum Britannicum*; and tho' they may spoil some Grass in Winter, it is easily levell'd in Spring. Hogs-dung is likewise an excellent Medicine for a Canker.

From Moss.

Fifthly, Moss is very prejudicial to the bearing of Fruit-trees; and is commonly occasion'd by the coldness of the Land the Trees grow on, whether it is wet or dry, or their being planted too deep; and if it proceed from the coldness of the Land, lay Sea-coal-ashes, Horse dung, &c. If from Moisture, drain the Land well; but if it proceed from deep planting, if they are small, it is the best way in very moist Weather to draw them up higher; but if they are too large, for that there is no remedy but replanting of them, or to plant new in their places. To cure the Moss, in *Staffordshire*, I am told, they burn off the Moss of their Trees about *December* with a wisp of Straw; but the common way is, to rub it off of young Trees with a Hair-Cloth, or to scrape it off with a wooden Instrument that may not hurt the Bark of the Trees. I knew one that had an Apple-tree very much run over with Moss, and he made a Stye under it, in which he fatted Hogs, and it cured it. But as Moss is sometimes caused from the want of Sap, which is the reason that old Trees are commonly more Mossy than young, it is good to lop off several of the Branches of such Trees, which will make them prosper the better, and be less Mossy; especially where Trees are Mossy that grow on dry Ground.

Sixthly, Many Trees run altogether to Wood and Branches, and seldom bear any Fruit; to remedy which Inconveniency, some propose hacking of the Tree, or to cut Crosses or other Stroaks upon the Bark, to give some check to the Sap; others propose to bore a Hole through the Body of the Tree; this way carries some probability with it, because hollow

Trees,

Trees, or fuch as are hurt or decay'd in the Body or
Stem, are more apt to bear than found ones. The
fame reafon may be for the cleaving of the Roots of
Trees, and for the putting of Stones or Wedges in
them; for Trees blown afide by the Wind, or by
fome other Accident, do ufually bear great Quantities,
and fometimes more than when they ftood firm and
upright: The reafon of which may be the Check that
is thereby given to the Sap running into the Branches,
when lefs Sap might do to produce Fruit; but the
beft way is to prune off fome of the Branches in Summer time, when the Sap is in them, which is the beft
way to abate their Luxuriancy; but this muft be
done with Judgment, the beft time of doing which is
in *June*.

Seventhly, It is good to have variety of Fruit in an *Variety*
Orchard, becaufe fome Apples and Pears that bear in *good.*
one Year will not do fo in another; and therefore
where Variety is, you will feldom fail of having fome
take.

Eightly, Another thing to be confider'd, is, to plant *To plant*
fuch Fruit as will hang on the Trees, till ripe, there *Fruit that*
being many forts of Fruit that will fhake off almoft *will hang*
with any Wind; upon which account it is good always *Trees.*
to have the S. W. fide of your Orchard well fhelter'd
with high Trees, that Wind being, as I faid before
the moft troublefome about the latter end of Summer.

I fhall here add fomething concerning the fencing of
Trees out of Mr. *Evelyn*, communicated unto him from
Dr. *Beal*, which will be of advantage to the planting
of Trees in Fields for fecuring of them from Cattel.

The fencing of fingle Trees ufeth to be done by
Rails at great Charges, or by Hedges and Bufhes,
which every other Year muft be renew'd, and the
Materials not to be had in all places neither; I there-
fore prefer and commend to you the enfuing form of
Planting and Fencing, which is more cheap and eafie,
and hath other Advantages it it not commonly known;
I never faw it but once, and that imperfectly per-

form'd, but have practis'd it my self with success. Take it thus,

Set your Tree on the green Swarth, or rather five or six Inches under it, if the Soil be very healthy; if moist or weeping half a foot above it: Then cut a Trench round that Tree two Foot, or more, in the clear from it. Lay a rank of the Tufts, with the Grass outward, upon the inner side of the Trench toward your Plant, and then a second rank upon your former, and so a third and fourth, all orderly plac'd (as in a Fortification) and leaning towards the Tree after the form of a Pyramid or larger Hop-hill: Always as you place a row of Tufts in compass, you must fill up the inner part of the Circle with the loose Earth of the second Spit, which you dig out of your Trench, and which is to be two foot and half wide, or more, as you design to mount the Hillock; which by this means you will have raised about your Plant near three foot in heighth. At the Point it needs not be above two Foot or eighteen Inches diameter, where you may leave the Earth in form of a Dish, to convey the Rain towards the Body of the Tree, and upon the top of this Hillock prick up five or six small Briars or Thorns, binding them lightly to the Body of the Plant, and so you have finished the work.

The Conveniencies of this kind of Planting are,

First, That neither Swine, nor Sheep, nor any other sort of Cattel can annoy your Trees.

Secondly, You may adventure to set the smaller Plants, being thus rais'd and secur'd from the reach of Cattle.

Thirdly, Your Trees are fasten'd in the Hillock against the violence of Winds without Stakes to fret and canker them.

Fourthly, If the Soil be wet, it is hereby made healthy.

Fifthly, If very dry, the Hillock defends them from the outward Heat.

Sixthly, it prevents the Couch-grass, which for the

firſt Years inſenſibly robs moſt Plants in ſandy Grounds apt to Graze. And,

Laſtly, The grazing Bank will recompence the niggardly Farmer for the waſte of his Ditch, which otherwiſe he will ſorely bethink.

In the ſecond or third Year (by what time your Roots ſpread) the Trench, if the Ground be moiſt or Seaſon wet, will be near fill'd up again by the treading of Cattle, for it needs not be cleanſed; but then you muſt renew your Thorns: Yet if the Planter be curious, I ſhould adviſe a caſting of ſome ſmall quantity of rich Mould into the bottom of the Trench the ſecond Year, which may improve the Growth, and invite the Roots to ſpread.

In this manner of Planting, where the Soil is not rich, the exact Planter ſhould add a little quantity to each Root of Earth from a frequented High-way, or Yard where Cattle are kept. One Load will ſuffice for ſix or ſeven Trees, this being much more proper than rotten Soil or looſe Earth, the fat Mould beſt agreeing with the Apple-Tree.

The broader and deeper your Ditch is, the higher will be your Bank, and the ſecurer your Fence; but then you muſt add ſome good Earth in the ſecond Year, as before.

I muſt ſubjoin, that only Trees of an upright growth be thus planted in open Grounds, becauſe ſpreading Trees will be ſtill within reach of Cattel as they increaſe. Nor have I met with any Inconvenience in this kind of Tranſplanting (which is applicable to all ſorts of Trees) but that the Mole and the Ant may find ready Entertainment the firſt Year, and ſometimes impair a weak rooted Plant, otherwiſe it rarely miſcarries. In ſum,

This manner of Fencing is ſoon executed by an indifferent Workman, who will eaſily ſet and guard ſix Trees in a Winter's Day.

I ſhall conclude this Point with giving a ſhort account of ſome of the beſt or moſt common ſorts of

Apples,

Apples, Pears and other Fruits that I have met with, with some Remarks upon the several sorts, in order to procure a more exact account of them, which is a Work that will need more assistance exactly to perform than I am able to procure at present, every Country, and many parts of each Country, having some sort of Fruit or other not known in the next; and therefore in the mean time the Rule that I shall advise most to observe is, to take an account of the best Bearers, and most useful sorts of Fruits, and the particular Soils that they thrive and bear best in.

Chap. IX. *Some further Observations concerning the Unfruitfulness of Trees.*

GOOD Dunging, Chalking, or Liming of Orchards, are very beneficial to Trees, which when large draw a great deal of the Heart of the Ground, and cannot maintain themselves with vigour without it.

Sir *Hugh Plat* says, That if you mix green Cowdung and Urine together, and wash your Trees with it, with a Brush, once in two or three Months, it will keep *Conies*, *Hares*, &c. from barking of them; and he says likewise, that it will destroy the Canker.

A Gentleman near *Hereford* assures me, that he fed several Hogs about some old Apple-Trees that he thought had done bearing, and that the next Year he had thirty Bushels of Apples, a piece, off of several of them; and therefore he was proposing to me to have a moveable Sty, and about it to make a Yard with Hurdles, to remove from one Tree to another; which I cannot but think a very good way, and what will be a very great Improvement of all sorts of Fruit-Trees, not only to promote their bearing, but likewise to cure the Moss, Canker, and other Infirmities of them, especially since this way will save the Urine, which I prefer for Trees much before Dung, because it penetrates better to the Roots, and is much to be preferr'd for the curing of any of the afore-mentioned

oned Infirminies; which I am confirmed in by a Letter I received from——concerning the preserving of Fruit-Trees in *Kent*, which I shall give in his own words.

" It was formerly (as all Ancient Graziers know) " a Custom to keep up and fatten their Oxen in Stalls, " in which Earthen-vessels were placed under the " Planks to save the Urine that came from them: With " which Urine so saved, they washed two or three " times in the Month of *March*, their mossy, canker- " ed, worm-eaten and unsound Trees, and poured " some of it to the Roots; and if plenty of it were now " to be had, I do not doubt but that Pippin-Trees " might be raised, made thrive, and flourish, as well as " heretofore.

" Happening lately to mention this to a Carpenter, " he said he had several times seen at the pulling up of " such old Stalls, some that have had a well steen'd " Channel under the Planks, leading to a large steined " Receptacle without the Stall, at which he had of- " ten wondred, but could never think the reason of " it till I mentioned the aforesaid——————

This useful Observation I had from——near *Tunbridge*, in which part of the Country Apple-Trees formerly flourished in such abundance, that *Pembury* a- lias *Pippenbury*, a Parish not far distant, is said to have its Name from the plenty of that Fruit once growing there.

That Urine is a very beneficial Manure, is confirm- ed by Sir *Hugh Plat*, Dr. *Plot*, and others, and may as well be applied to the Body and Branches of the Tree to cure Cankers, kill Moss and Worms, as at the Root (that is, in moderate quantities) to warm, invi- gorate, and quicken the Circulation of the Sap, espe- cially in a cold barren Soil.

Nor may it be improper here to observe, that ge- nerally speaking, all Manures differ in goodness and strength, according to the different kinds or classes of Bodies they are made of; as that vegetable Substances, *viz. Rotten-straw, Beans, Grass*, &c. are better (quan- tity for quantity) than *Marle, Lime, Mud, Sea-Sand,*

&c.

&c. and that Animal Substances, as *Urine*, *Dung* of *Men*, *Beasts*, and *Fowls*, *Bones*, and *Horns* of *Beasts* burnt or putrefied, *Wollen-rags*, and the like, are of greater strength and nourishment than either.

I shall be glad if this Account may put any upon the Trial of raising that excellent Fruit the *Kentish Pippen*, which else, I fear, will be lost : For I find in several Orchards both in *Kent*, *Essex*, and *Hertfordshire*, old Trees of that sort, but I can find no young ones to prosper. A Friend of mine tried a great many Experiments in *Hertfordshire*, about raising of them, and could never get them to thrive, tho' he had old Trees in the same Orchard that grew and bore very will. I likewise tried several Experiments my self, and have had young Trees thrive so well, as to make many Shoots of a Yard-long in a Year, but these young Shoots were always blasted the next Year, or cankered ; which makes me think that the Ancients had some particular way of raising of them, that we have lost the Knowledge of, so that what is proposed seems very probable to be the way they did make use of : Which the Circumstances of the place not only confirms, but the Usefulness of the Matter proposed, there being nothing of greater Advantage, as I have found by Experience, than Urine, for the Improvement of all sorts of Vegetables.

And therefore I am sorry to find the ancient Husbandry out-do the present; and that so useful a Material as *Urine* should be so much neglected, and generally let run away to waste, of which so great Advantage is made in *Flanders*, and other parts beyond Sea : concerning which I have in several places given an Account of its Advantages.

Blood, *Soot*, and the *Dirt* of *Sinks* is good to lay to the *Roots* of *Trees*, *Vines*, &c. to make them bear.

If your *Orchard* be on dry Ground, to make your Trees bear about the beginning of *May*, when the Trees are in the height of their Blossom, dig a little about them; especially if you'll mix Sheeps-dung with
it,

it, beſtow a Pale-full of Water on every Tree once a Day, till you ſee the Fruit ſet. I know one that did ſo, and ſeldom failed of a Crop.

When *Fruit-Trees* are old and decaying, lop off ſeveral of the Boughs, and lay *Lime* or *Chalk* to them, and it will cauſe them to ſtrike new Roots, and to thrive, and bear well; it being a renewal of their Age. Mr. *Moor* in his deſcription of *Northamptonſhire page* 487. ſays an old Orchard of Apple trees growing moſſy and bearing but indifferently, the owner, the beginning of *April*, disbarked the Trees from near the bottom, almoſt up to the part of the Tree, where the head divides into Branches; and the Bark grew again in Summer: and by the latter end of the Year, was firme and ſmooth; and had a fine thin rind, and bare every Year after very well; not one of them miſſing. And one in *Eſſex* told me of a Pear-Tree which he barked, about ſix Inches round; and it occaſion'd it to bear every Year tho' it never bore before. In manuring of Trees, do not lay it too near the Body, but at ſome diſtance, that it may ſoak down to the ſmaller Roots to nouriſh them, they being as it were the Mouths that ſuck in the Nouriſhment for the Trees.

If you have *Vines* or *Fruit-Trees* that run upon the Tyles of any Building, or upon a ſloaping Wall, according to Mr. *Fatio*'s Propoſal, you may ſet *Melon* glaſſes on the Fruit, which will much forward its ripening.

But from what I have already mentioned concerning an Experiment made by a Friend of mine, in my former Treatiſe of Husbandry, at *pag.* 527. concerning *Walls*, I cannot but prefer Mr. *Langford*'s Contrivance of building of them in the Form of a Semi-circle, which he propoſes to be about eight Yards in circumference on the inſide, and about ſix in diameter, which he ſays is found by experience to do well; of which opinion, I muſt own my ſelf, eſpecially for *Vines*, becauſe they bear only on the Shoots of the ſame Year; which Shoots are apt to grow upright, which muſt occaſion them to be at a greater diſtance from the

Wall,

Wall, if they grow upon a floaping Wall, than where they run parallel to an upright one ; and befides, thefe circular Walls may be conveniently covered, which will be not only a great Advantage to Vines, in that you may fecurely let the Grapes hang on the Vines to ripen, as long as you pleafe, without any danger from the Frofts ; but you may alfo by the fame means, in the Spring, cover your forward Fruit from being prejudiced by the early Frofts, which fhould upon other Walls have Matts or Canvafs hung before them, from the time of their firft beginning to bloffom, until they are well knit or fet, and for fome time after, if you find occafion.

In ordering of *Wall-Trees*, you may prevent unneceffary Branches, by rubbing or cutting off fuch Buds as come forth, where there not convenient room for them to be laid.

A Gentleman of *Hampfhire* fends me word, that he has obferved but few good Apples to do well in divers Parts of that Country, except the *Golden-Rennet* and *Codlin*, efpecially on a gravel or fandy Soil that have Springs within three or four Foot of the Surface, and that *Oaks* on fuch Land thrive very well.

To which I muft add, that I have obferved the *Golden-Rennet*, in other places, to thrive on Gravel er Sandy Land, the beft of any Fruit-Trees, except *Plumbs* ; and in *Hampfhire* I have met with greater variety of Apples of the *Golden-Rennet* kind, than in any other part of *England* ; and I have obferved, that moft of the largeft forts of Apples do beft on Gravel, and are the leaft fubject to Mofs or Cenker.

The *Red ftreak* is reckoned where it yields the beft Cyder, not to grow fo large as other *Fruit-Trees*, and therefore it may be planted nearer than other Trees.

Chap. X. *A Catalogue of Fruits,* &c.

THE Aromatick or Golden-Ruffeting hath no compare, it being of a Gold-colour'd Coat under a Ruffet Hair, hath fome Warts on it, its Flefh of a yellow colour, its form of a flattiſh round. This Fruit is not ripe till after *Michaelmas*, but keeps over the Winter; and is, without diſpute, the moſt pleaſant taſted Apple that grows, having a moſt delicate Aromatick reliſh, and melting in the Mouth.

The Orange Apple, ſo call'd from its likeneſs in colour and form to an Orange, deſerves the next place, having a fine rough Gold-colour'd Coat, reſembling the Golden Pippen, only fairer, keeps long, and is of a very pleaſant taſte.

The Golden Pippin is ſmaller than the Orange-Apple, elſe much like it in colour, taſte and long keeping, being the beſt of Apples for Cyder, Eating and Baking.

The Ruſſet Pearmain is a very pleaſant Fruit continuing long on the Tree, and in the Conſervatory, partakes both of the Ruſſeting and Pearmain in colour and taſte, the one ſide being generally Ruſſet, and the other ſtreak'd like a Pearmain.

The Pearmain, whereof there are ſeveral ſorts, is ſo excellent an Apple and ſo well known, that no more need be ſaid of it; only the larger ſort keeps not ſo well, neither is the Summer Pearmain ſo good as the Winter; they are all good Cyder-Apples, but no great Bearers.

Pippins are of ſeveral ſorts, and take their Name from the ſmall Spots or Pips that uſually appear on the ſides of them. Some are call'd Stone Pippins from their obdurateneſs; ſome are call'd *Kentiſh* Pippins, becauſe they are a Fruit that agrees well with that Soil; others are call'd *French* Pippins, having their original from *France*; which is the beſt bearer of any of theſe ſorts of Pippins, the *Holland* Pippin from the ſame cauſe, and the Ruſſet Pippin from its Ruſſet hue, with divers others denominated from the ſeveral places of their growth, but ſuch as are diſtinguiſh'd by the Names of Grey and White Pippins, are of equal goodneſs. They are generally a very pleaſant Fruit, and of a good Juice, fit for the Table, Conſervatory or Kitchen, but they are ſlender bearers.   The

The Golden Rennet is a very pleasant and fair Fruit, of a yellow Flesh, and the best of bearers for all sorts of Soil, of which there are two sorts, the large sort and the small ; the smallest keeps the best, and is the best flavour'd, but the other is a mealy Apple if kept after *November*, and neither of them good Cyder-Apples alone ; but Mr. *Worlidge* commends them if mix'd with the *Red streak*. They sell well in the Market, and are good eating Apples during the first part of Winter.

We have in *Essex* an Apple call'd a Snow Apple, which is an extraordinary certain bearer on the light brick Earths, but a very ordinary Apple of its kind ; yet for its constant bearing I cannot but value it, for I had rather have any indifferent Apples, than none at all.

There is also in *Hertfordshire* an Apple much resembling a Gennet-Moil, which they call a *Wiltshire* ; it is both a good bearer and a good Cyder-Apple.

The Leather-coat or Golding Russeting, as some call it, is a very good Winter Fruit, lasts long, and is of a good, firm and yellow Flesh ; and an Extraordinary good bearer, and not so subject to Moss as many other Fruit-trees are.

The *Bartlet Queening*, is a very good Cyder Apple, especially if mixed with the *Golden Rennet*, which will not do well for Cyder without it. I likewise met with in my Neighbourhood a *Yellow-Queening*, which is a very juicy Apple, and one of the richest tasted Apples that ever I met with. I call it a *Queening*, because exactly of the same shape, but for colour is wholly yellow, both inside and out. The *Royal-Apple* is also a good Apple, the *Rose-Apple*, *Cotton Apple*, the *Sage-Apple*, the *Gaunt-Apple*, *Kentish-Codlin*, *Powel-Apple*, *Violet*.

The Green Russeting is a tough and hard Fruit that lasts long.

The Red Russeting is of a lesser size, long lasting, and are all of them of a pleasant Relish.

The

The Sharp Ruffeting is an extraordinary bearer, and a good Cyder, and keeping Apple.

The *John* Apple, or *Deux-Ans*, so called from its continuing two Years before it perisheth, is a good relished sharp Apple the Spring following, when most other Fruit is spent, they are fit for the Cyder Plantations being great Bearers, and though dry Fruit, they yield very good Juice, but must be ground before *January*. There is a Summer *John* Apple, that is very much commended also.

The Marigold Apple, so called from its being marked in even Stripes in the Form of a Marigold ; sometimes the Onion Apple, from the reddish brown Colour resembling a well coloured Onion ; sometimes called the Kate Apple, and sometimes *John*'s Pearmain, from its likeness to a Pearmain, is a very good Fruit, long lasting, and fit for the Table, Conservatory, Kitchen, or Press, yielding a very good Juice ; it bears every other Year, even to Admiration : There is another sort of them called Summer Marigolds.

The Harvey Apple, and the round Russet Harvey, are very pleasant Fruit, and good Cyder Apples, but are no good bearers.

The Queen Apple ; those that are of the Summer kind, are good Cyder Apples mixed with others, being of themselves sweet : The Winter Queening is good for the Table.

The Paradise Apple is a curious Fruit produced by grafting a Pearmain on a Quince.

The Pome-Roy is a Fruit of good taste, a pulpy Substance, but not yielding much Juice.

The Pome-water is an indifferent good lasting Fruit.

The Golden Douset, or Golden Ducket, is much commended.

The *Westbury* Apple, taking its Name from *Westbury* in *Hampshire*, from whence they are much dispersed into the adjacent Parts, is, as Mr. *Worlidge* says, one of the most solid Apples that grows, of a rough Rind and obdurate Flesh, sharp and quick taste, lasts long, and yields a very excellent and plentiful Juice, making Cyder equal to the best of Fruits ; and for the Kitchen, few or none exceed it.

The Gilliflower Apple is of a pleasant relish and long lasting; of a thick Rind, hard Core, striped, and a good Cyder Apple, making an excellent mixture.

The *Margaret* Apple is the best and most early, usually ripe about St. *Margaret*'s Day in *June* ; it is a fair and beautiful Fruit, of a pleasant taste and scent, and deserves a more general Propagation.

The Jenneting is next to be esteemed, as well for its early ripening as its pleasant taste.

The *Devonshire* Quarrington is also a very fine early Apple.

The Summer Pippin is a very pleasant Apple in colour and taste, yielding a delicate Juice.

The Creeper is an Apple so called from the Tree, which grows low, and traileth its Branches near the Ground.

The Ladies Apple is very beautiful, and begins to be good about *December*, and lasts 'till *March* and *April*; it is a great bearer, and never wrinkles with keeping.

The Ladies Thigh is a kind of Russeting in shape and colour, with a very juicy and tender Pulp, a little musked ; its Tree is very long before it bears, but afterwards produces abundance, and is ripe the beginning of *July*.

The Violet Apple is of a whitish ground Colour, a little speckled in those parts that are from the Sun, but striped with a deep Red on the Sunny side ; the Pulp is very fine and delicate, and is to be eaten as soon as gathered, tho' it will continue good 'till *Christmas*.

The Codling, so called from the use it is put unto, makes a good Summer Cyder, and is a good bearer, either in Hedges or Standards.

The Claret Wine Apple is fair, and yields plenty, of a pleasant sharp Juice, from whence it takes its Name; being well ordered, it excells most other Cyders ,especially with a mixture of sweet Apples.

The

The White Wining is a small white Apple, and a good bearer, the Fruit juicy and pleasant, but soon perishing, and the Cyder made thereof small.

The King Apple, tho' not common, yet is by some esteemed an excellent Apple, and preferred before the Jenneting.

The Famagusta is also in the number of the best of early Apples.

The Giant Apple is a large Fruit, well tasted, and the best of any Summer Apple for the Kitchen.

The *Bontra-due*, or Good House-wife, is the largest of Apples, a great bearer, good for the Kitchen. It makes good Summer Cyder.

The Cat's-head, by some called the Go no-further, is a very large Apple, and a good bearer.

The Spicing Apple, of all Apples that are marked Red, is the meanest.

The Gennet Moil is a pleasant and necessary Fruit in the Kitchen, and one of the best Cyder Apples, and a good bearer.

The White Must is a very pleasant Apple, yielding great plenty of Vinous Liquor, bearing this Name in *Herefordshire*, and is thought by some to be the same with the Golden Rennet in *Hampshire*.

The Red Must is also of the same Nature.

The Fox Whelp is esteemed among the choice Cyder Fruits.

The Bromsbury Crab, altho' little better than the common, yet kept on heaps 'till *Christmas*, yields a brisk excellent Cyder, and very strong.

Eleots are Apples much in request in the Cyder Countries for their excellent Liquor, but not known by that Name in several parts of *England*.

The Stocken, or Stoken Apple is likewise in esteem there, altho' not known by that Name in many places.

The Bitter Scale is an Apple much esteemed of in *Devonshire* for the excellent Cyder it yields, without the mixture or assistance of any other.

The Dean's Apple is there well esteemed of for the same reason.

As alſo is the Pleaſantine, perhaps the ſame with our Marigold.

The Pureling. Its Name is not uſual, but in the ſame Parts.

The Underleaf, whoſe Cyder is beſt at two Years; 'tis a very plentiful bearer, hath a Rheniſh Wine flavour, the very beſt of all Cyders of this kind; the Apples ſhould be hoarded a little within Doors; and the longer you would keep your Cyder, the longer you muſt hoard your Fruit.

A long pale Apple called the Coleing, about *Ludlow*, is an extraordinary bearer.

The Arier Apple, a conſtant bearer, making a ſtrong and laſting Cyder; ſome call them *Richards*, ſome Grange Apples; and indeed they make ſo excellent a Drink that they are worthy to be recovered into uſe.

The Olive, well known about *Ludlow*, may, I conceive, be accounted among the Winter Cyder Apples, of which it is the conſtant Report, ſays Mr. *Evelyn*, that a Hogſhead of the Fruit will yield an Hogſhead of Cyder.

Fillets, whereof alſo there are the Summer and the Winter, in very high eſteem for the delicate Vinous Liquor they yield. The Summer Fillet for the preſent, and the Winter Fillet for laſting Cyder.

The Red-ſtreak, of all Cyder Fruit, hath obtained the Preference, being but a kind of Wilding, and tho' kept long, yet is never pleaſing to the Palate; there are ſeveral ſorts of them, the Summer and the Winter, the Yellow and the Red, and the more green Redſtreak: Some ſorts of them have red Veins running thro' the whole Body of the Fruit, which is eſteemed to give the Cyder made thereof the richeſt Tincture if they are kept 'till mellow; the Cyder at firſt is very luſcious, but if ground more early, it is more racy.

The Quince Apple, ſo called from its Colour, is a good Table Fruit as well as Cyder Apple.

The Nonſuch is a long laſting Fruit.

The Angel's Bit is a delicate Apple for taſte, and the

the Tree, or its Name, proper to *Worcestershire*, and those parts.

The Peeling is a lasting Apple, makes very good Cyder, agrees well with this Air, and is a good bearer.

The Oaken-pin, so called from its hardness, is a lasting Fruit, yields excellent Liquor, and is near the nature of the *Westbury* Apple, tho' not in Form.

The Greening is of a colour green, and keeps to a second Year, and is a good Apple.

The Lording is a fair, green, and sharp Apple, a constant bearer, being a hardy Fruit, for the Kitchen only.

Sweet Apples there are of several sorts, and their Names change in every place, so that they are rather known by their Colour and Size than their Names; there is one called the Honey-comb in some places, which is a fair Apple, and mixed with other Fruit makes admirable Cyder; so doth the small Russet sweet Apple, whose Tree is always Cankery.

Pome-appeale is an Apple newly propagated, the Fruit is small and pleasant, and yields no unpleasant Scent; the Tree is a good bearer, and it is supposed that this is that which is called the Lady's Longing.

The Fig-apple is also newly propagated, the Tree yielding no Blossoms as is usual with other Apple Trees, nor hath the Fruit in it any Core or Kernel, in these resembling a Fig, and differing from other Apples, yet is a good Table Fruit, and lasting.

The *Sodom* Apple, or Bloody Pippen, is a Fruit of more than ordinary dark Colour, and is esteemed a good Apple.

The *Muscovy* Apple is a good Winter Fruit, and a great Curiosity, for that it is transparent.

Belle and Bon are of two sorts, the Summer and Winter; it is a fair Apple, and a good bearer, but the Fruit not long lasting; it makes indifferent Cyder; the Winter Belle Bon is much to be preferred to the Summer in every respect.

The Pear Apple is a curious pleasant Apple of a rough Coat, but is no good bearer.

U 3   There

There are also the Apples called Esquire *Vernon's* Apple, the Grutchling, the Pear Russet, the *Stoak* Apple, the *Suffolk* Apple, which are much commended for the Table and Kitchen.

The Pell-mell Apple, the Thrift Apple, and the Winter Glory, are very good lasting Apples.

Crabs, when kept 'till they are mellow, may be reckoned amongst Apples, and being ground with other mellow Fruit, do much inrich the Cyder, and is the best refiner of foul Cyder.

The Castard Parsley Apple, the *William*, the Cardinal, the Short-start, the Winter Red, the Chesnut Apple, and the Great Belly, are in many places Apples of esteem, but being not acquainted with them I can only name them. Many more there are both *French* and *English*, which either are not made familiar to us, or else are peculiar only to some Places, or their Names changed in most Counties, or else are of small account, which to enumerate would be tedious and useless.

Chap. XI. *Of Pears.*

THE next in esteem are Pears, so called from their Pyramidical Form, whereof there are so great variety, that the Kitchen and Table may be furnished throughout the Year with different Species.

The early Susan is the first ripe, being a small round Pear, little bigger than a large Cherry, the Colour is green, and the Taste pleasant.

The *Margaret*, the *Maudlin*, the Cluster Pear, the Lenthal Primet, the Sugar, the *Madera*, the Green Royal, *July* Pear, St. *Laurence*, Green Chesil, and many other early Pears are in esteem for the Tables in *July*; but after them you have

The *Windsor*, the Green-field, the Summer Burgamot, the Orange, the Sovereign, several sorts of *Katharines*, whereof the Red *Katharine* is the best, the Denny Pear, *Prussia* Pear, Summer Popperin, Lording Pear, Summer *Bon Chrestien*, the Orange Burgamot,

Hampden's

Hampden's Burgamot, Bezy de Hery; the Violet Pear, the Painted Pear, so called from its delicate striped Colours; the Rosewater Pear, the Shortneck, so called from the shortness of its Form and Tail; the Binfield, or Dove Pear, the great Muck Pear, the great Russet of *Remes*, Amadotte; the Rouselet *Norwich* Pear, the Pomegranate Pear, so called from its shape, and the *Edward* Pear very pleasant; the Meola ala Busk, Crown Pear, St. *Michael*'s Pear, *Carlisle* Pear, Roshea, esteemed an extraordinary Pear, King *Katharine*, Rouselet Petit, Rouselet Hastife, Musk Blanquet, Dove, Musk Burgamot, Queen Pear, White *Robert*, and the desirable Pear, are all very good Table Fruit for their Season before or at *Michaelmas*.

The *Beure du Roy* is esteemed for the Table the best of all Summer Pears; 'tis a fair brown Pear and excellent in its Season, melting in the Mouth, and thence called the Butter Pear, and bears well against a Wall. The green *Beure* Pear is more green and larger than the former.

The *Lewis* Pear, or by some the Maiden-heart, is a very good bearer, and the best of all Pears to dry.

The Bloody Pear is a good Pear, taking its Name from the Red Juice it hath within its Skin, and is a very great Curiosity.

The *English* Warden, the *French* Warden, and the great *Spanish* Warden, the White Warden, the Stone Pear, the *Arundel* Pear, the Bishop's Pear, the Caw Pear, Winter Musk Cashurine, the Lady *Hatton*'s Pear, the Quince Pear, the Davis Pear, Malborne Pear, the Red *Roman* Warden, the Green Warden and Winter *Norwich* are excellent baking Pears.

The great black Pear of *Worcester* or *Perkinson*'s Warden bears well against a Wall, they usually weighing twenty Ounces, or more, and being twice baked with Sugar exceed most Fruits.

The Diego Pear, Monsieur *John*, the Gilliflower Pear, Pear Royal, Bowden Musk, *French* Violet Mogul Pear, Virgo; Lait, Sovereign Pear, Okenbury Pear, the white

Worcester

Worcester Roufelet-durine, *Montpelier* Imperial Pear, Pear *de Lyons,* a rare Winter Pear for the Table; the Burgamot Bougee, Rowling Pear, Balfam Pear, Bluster Pear, Emperor's Pear, Queen-hedge Pear, Frith Pear, Brunfwick Pear, *Bing's* Pear, Winter Poppering, Thorn Pear; the Portail, the Nonfuch Dioniere, Winter *Katharine,* Clove Pear, Lambert Pear, Ruffet Pear, Saffron Pear, the *Petworth* Pear, or Winter *Windfor,* Winter Burgamot, Pound Pear, and Hundred Pound Pear, Long Burgamot, Burncat, Lady Pear, Ice Pear, Dead-man's Pear, Bell Pear, the Squab Pear, Spindle Pear, Dogoniere, Virgin, Gafcoign Burgamot, Scarlet Pear, and Stopple Pear, all are very good Winter Pears, and keep throughout the old Year.

Pears that ufually keep until the fucceeding Spring are the *Bon Chreftien,* the beft of Winter Pears; the great Surrein, or Serene, Little Dagobert, the Double Bloffom Pear, the longeft liver of all, and takes very well in the Spring; the Oak Pear, the great Kairville, the little Black Pear of *Worcefter.*

Pears that are efteemed for their Vinous Juice in making of Perry in *Worcefterfhire,* and thofe adjacent Parts, are the Red and Green Squafh Pears, the *John* Pear, the green Harpary, the Drake Pear, the *Mary* Pear, the Lullam Pear; but above the reft are efteemed the Bosbury, and the Bareland Pears, the White and Red Horfe Pears, and above all, that which is moft commended is the *Turgovian* Pear mentioned by Mr. *Evelyn.*

When I lived in *Hertfordfhire,* a poor Man brought me three or four Bufhels of fmall Pears, which were very fmall, hardly fo big as the fmalleft Crabs, having fomething of a muskifh Flavour, though with it fo rough a tafte that the Hogs would hardly eat them; which made me think them a good Perry Pear, and accordingly I made Perry of them, which was fo rough the firft Year that no Body would drink it: But I found that as the roughnefs wore off, the fine Flavour increafed, fo that at four Years of Age it had the co-
    lour

lour of Canary and was as ſtrong, and had as fine a flavour, being valued by all that taſted it before it; but the Man that brought them me dying before I knew the excellency of my Liquor, I could not be ſo certain of the Pear, as I might if I had had his Directions, there being three or four Trees cut down in the Field where he gathered them. Since, I got ſome Grafts of a Pear juſt like it for ſhape and taſte, which I believe to be the ſame; but the Trees I have raiſed being not yet big enough to furniſh me with Pears to make the trial of their Juice, I cannot be poſitive of it 'till they do, which I mention only to adviſe others to more caution when they meet with any rarity of this kind.

Chap. XII. *Of Cherries.*

IN the next place the Cherry is admitted to be a Fruit of general Uſe, eſpecially for the Palate, and for the Conſervatory. They are ripe on the Trees but three Summer Months, *May*, *June*, *July*, afterwards to be had only in the Conſervatory.

In *May* the Cherries then ripe are uſually ſo called from the Name of the Month, the Duke and Archduke, againſt a good Wall, are moſt Years ripe before the end of this Month.

In *June* are ripe the White, Red, Black, and Bleeding Hearts, Lukeward, one of the beſt of Cherries, the Early *Flanders*, the Cloſter Cherry bearing three, four, or five uſually on a Stalk, the White *Spaniſh* Cherry, the Amber Cherry, the Black *Orleans*, the White *Orleans*, Nonſuch, the *Spaniſh* Black and the *Naples*.

In *July* uſually ſucceed the late *Flanders*, commonly called *Engliſh* Cherries, Carnations, a delicate Fruit for the Table or Conſervatory; Morella, or the Great Bearer, being a black Cherry fit for the Conſervatory before it be through ripe, but 'tis bitter eaten raw, only it is to be eſteemed being the laſt Cherry that hangs on the Tree, the *Morocco* Cherry, Great Amber,

ber, the Egriot, Bigarreaux, the Prince-Royal, the *Portugal* Cherry, the King's Cherry, the Crown Cherry and the Biquar, both ill Bearers, the Great Purple Cherry one of the best and latest Cherries, and a good Bearer, the Ounce Cherry, so called from its fairness, the Dwarf Cherry, so called from the smallness of its Twigs and Fruit; there is also the common Black Cherry much in esteem for its Physical Properties.

Chap. XIII. *Of Plumbs.*

THere is great variety of Plumbs, and they are also appropriated to several Uses; they continue longer on the Trees than Cherries, and are by some esteemed a more pleasing, but not so wholsome a Fruit.

The first ripe are the Red, Blue, and Amber Primordian Plumb, the Violet, Red, Blue, and Amber, the Matchless, the black Damascen, the *Morocco*, the *Barbary*, the *Myrobalan*, the Apricock Plumb, a delicate Plumb that parts clean from the Stone, the Cinnamon Plumb, the King's Plumb, the *Spanish*, the Lady *Elizabeth*'s Plumb, the great Mogul, and the Tawny Plumb.

After them are the White, Red, and Black Pear Plumbs, the two former little worth, but the Black a pleasant Fruit, the Green Osterly Plumb, the Mossel Plumb, one of the best of Plombs, the *Catalonia* Plumb, much like the former, the White Prunella, the Black Prunella, the Bonum Magnum a fair yellowish green Plumb, excellent for the Kitchen and Conservatory, the Wheaten Plumb, the *Laurence* Plumb an ill tasted Fruit, the Bole Plumb, the Cheston Plumb, the Queen Mother Plumb, ' one of the best sort, the Diapered Plumb, the Marbled Plumb, and the Blue Marbled, the Damasco Plumb, the *Foderingham* Plumb, the Blue and Green Podrigon, and the White, not so good a Fruit, the Verdoch good only to preserve, the Peach Plumb, the Imperial Plumb one of the largest of Plumbs, the Giant Plumb, the Denny Plumb, the *Turkey* Plumb, the

the Red, White, and Green Peafood Plumbs, the White, Yellow, and Red Date Plumbs, the Nutmeg Plumb, the Great *Anthony*, the *June* Plumb, the Prince Plumb, the laſt ripe and good for ſeveral Uſes. Many other ſorts of Plumbs there are whoſe Names are uncertain, and are therefore here omitted.

There are ſeveral other ſorts of Plumbs, as the Friars Plumb, Becket Plumb, Chryſtal Plumb, White Muffel, White Prunella, *French* White Nutmeg Catholick Plumb, *Turkey* Plumb, Amber Plumb, and the Graſs Plumb, all of them curious and well taſted Fruits.

There are two ſorts of *Damaſcens*, the Black which is the moſt uſeful and beſt of all Plumbs, and the White which is not ſo good as the Red: Theſe are natural to our *Engliſh* Soil, as are the White and Black Bullace, whereof the White are pleaſant in *October* and *November*, and the Black neceſſary for the Kitchen in *December*, they uſually hanging on the Trees 'till *Chriſtmas*.

There is alſo a *Cornelian* Plumb or Cherry, which may be increaſed by Layers, and will ſometime grow of Slips or Branches; alſo of the Stones only, they will lie ſometimes a Year in the Ground before they come up.

Chap. XIV. Of *Apricots, Peaches, Malacotounes* and *Nectarines.*

THE *Apricots*, ſo called from *Apricus*, delighting in the Sun, is a kind of Plumb, but far exceeding any of the former in every reſpect; whereof

The *Algier* Apricot is the earlieſt ripe; it is a ſmall round and yellow Fruit, ripe in *June.*

The Maſculine Apricot is a better and earlier Fruit than the former, but not ſo good a bearer.

The long White and Orange Apricot differ from the common Apricot, as their Names tell you; there is alſo the *Turkey* Apricot.

The Green *Roman* Apricot is the largeſt of all the kinds, and therefore beſt for the Kitchen and Conſervatory.

Gum

Gum is hurtful to Apricots and Peaches, and should be taken off to the quick, and some Cow-dung clapped on the Wound, wraping of it round with some Linen, which tie on with Pack-thread.

Chap. XV. *Of Peaches.*

PEaches, from the *French* Name *Pesche*, are of longer continuance than Apricots, and of a more rich, noble Guſt and Flavour.

The moſt early are the Nutmeg, both White and Red, the *Troy* Peach, next the *Savoy* Peach, *Iſabella, Perſian,* the White Monſieur, *Newington* Bellice Peach to be preferred to the former, the Queen Peach, the *Magdalen* Peach, and the Double Bloſſom Peach.

After theſe come the Rambouillet, the Musk Peach, and the Violet Musk, both uſually eſteemed the beſt of Peaches, the Crown Peach, the *Roman* Peach, Man Peach, Quince Peach, Grand Carnation, *Portugal* Peach, *Bourdeaux* Peach, late *Newington* Des Pot being ſpotted, *Verona, Smyrna, Pavia* Peach, and the *Coleraine* Peach; one of the lateſt is the Bloody Monſieur, an excellent Peach, very Red within and without.

The *Modena, Orleans,* Red Peach, *Morello* Peach, *Navarra* and *Alberges* are very good Fruit, and come clean from the Stone.

There are ſeveral other ſorts of Peaches, as the *Arundel,* the Admirable, the *Sion* Peach, the *Uvedale* Peach, the Superintendant, the *Eaton* Peach, the *Laurence* Peach, the *Mountaban,* the *Perſick,* the Minion, the Perprice, the Supreme Peach, and the *Arabian* Peach, all of them very curious Fruit. But the Ricket Peach hath lately gained the Reputation of being the beſt of Peaches.

Of Malacotounes, as much as to ſay, Apples with Cotton on them; there are two or three ſorts, but being late ripe, and old Fruit, they are not much valued.

Nectarines, of the ſavour and taſte of Nectar, are very pleaſant Fruit, whereof the Red *Roman* is the
faireſt,

faireſt, and by moſt eſteemed the beſt and moſt delicate Fruit for its guſt, that this Iſland yields : By ſome
the Muroy is preferred, by others the Tawny, neither
of them ſo large as the Red *Roman.*

The Red or Scarlet Nectarine is by many much e-
ſteemed, becauſe it leaves the Stone.

Beſides all which, there are the Great Green, the
Little Green, the Cluſter, the Yellow, the White, the
Paper White, the Painted, the Ruſſet, the *Genoa*, the
Aegol, the *Perſian*, and the Orbine Nectarines that are
very good Fruit.

Peaches thrive and bear beſt in a moiſt Soil, and
therefore they ſhould be well watered if planted on
dry Land. Stones of Peaches will produce Trees, that
will bear Peaches ſometimes better than the Peaches
out of which the Stones were taken, by which means
the Gardiners by ſetting many Stones have raiſed
new Sets, but this cannot be ſo well practiſed by private Perſons, becauſe Peach Trees ſo raiſed, will be
longer before they bear than thoſe which are inoculated ; and becauſe, though ſome prove better, yet
many prove worſe, and ſo there is a hazard run in not
filling of the Walls with good Fruit.

Chap. XVI. *Of Quinces.*

THE *Portugal* Apple Quince is eſteemed the beſt ;
it is a large yellow Fruit, tender, pleaſant, and
ſoon boiled.

The *Portugal* Pear Quince is much like the former,
except in its form.

The *Barbary* Quince is leſſer than the other, as is
the *Engliſh* Quince, which is a harſh Fruit, and covered
with a Down or Cotton.

The Lyon's Quince is a large Yellow, and the *Brunſ-
wick* Quince a large White, both very good ; but all
inferior to the two firſt ſorts.

Chap.

Chap. XVII. *Of Figs.*

FIGS are highly esteemed by some, whereof the Great Blue Fig is most accounted of ; next unto it the Dwarf Blue Fig, being much less in Tree and Fruit, but better tasted, and sooner ripe.

Chap. XVIII. *Of the Cornel Tree.*

THE Cornel Tree beareth the Fruit commonly called the Cornelian Cherry, as well from the Name of the Tree as the Cornelian Stone, the Colour whereof it somewhat represents. This Fruit is good in the Kitchen and Conservatory.

Chap. XIX. *Of the Pruning of Fruit Trees.*

IT conduceth very much to the proof and growth of a Tree to be well pruned from its unnecessary and injurious Branches, and also to the making of it Fruitful.

If a Tree is to be transplanted, and you are obliged to lessen the Roots by taking of it up, you must take care to lessen the Head, that there may be a proportion between the one and the other ; because the Head depends upon the Roots for its Nourishment.

The best time to prune Trees is in fair Weather, and in the decrease of the Moon.

You must not prune a Graft the first Year, though it shoot never so strong.

But a Peach, the more it runs to Wood, and the stronger Shoot it makes, the better it will bear.

Wall Trees are to be pruned in Summer, and in Winter; the Summer pruning is to be about *June* or *July*, to take off the superfluous Sprigs or Shoots of the same Years growth from Vines, Apricots, and other Trees that put forth large Shoots that impede the Fruit from its due Maturity, and contract much of the Sap of the Tree to themselves, and thereby rob the other. In

In Winter, as soon as the Leaves are off the Trees, you may prune and cut away the residue of the Branches, and place those that are fit to be left in order: This Work may be continued throughout the Winter, except in great Frost; but in *February* or *March* is the best time to prune Trees, only you must observe to prune the most luxuriant and vigorous last, and to cut your Boughs close to the Body, and not leave them any length from the Tree, because by that means they become hollow, and serve only to convey water to the Body to rot it, and do not let your Lop grow large upon your Trees, for it makes the Scars the larger, and causes the Trees to be unthrifty, or die the sooner, where you cut them quite off; only Peaches and Nectarines are not to be cut till they begin to Bud, and observe to cut away superfluous Branches, or such as cross one another, or grow too thick, or that offend any other Tree or Place, or that are broken, bruised or decayed, and all the *August* Shoots, where-ever you find them, unless the place be naked, and that you suspect the next old Branch will not suffice to cover it, and Branches that shade the Fruit too much.

In pruning of Trees, especially Wall Trees, be sure to leave the small Twigs that are short and knotted, and that blossomed the succeeding Year, for you may observe that most Apricots, Peaches, Plumbs, Cherries, &c. grow on these Sprigs, being usually of two Years growth; they are therefore to be carefully nourished, and not cut off, as is usually done to beautify the Tree.

Apples and *Pears,* which bear Fruit also on the Branches that are of two Years growth, and 'tis necessary to be often taking off of some of the old Wood of Fruit Trees, that you may have a Succession of Branches to bear; and when your Trees are young, let them not fill the Wall too thick, because it will hinder their bearing, and oblige you to cut them too much when they grow old.

When

When an old Tree puts forth stronger Branches at the Bottom than at the Top, and the Top is unthrifty, cut it off; and bring your Tree into its Form from the lower Branches.

Every Bud which hath but a single Leaf produces only Wood; but that of Fruit hath many Leaves, and the more it hath, the sooner it will bear, and the greater will be its Fruit.

The Fruit Buds which grow on the Body of the Tree, produce fairer Fruit than such as break out of the collateral Twigs, and tops of Branches.

Rub off all the Buds which sprout out either before or behind your Wall Trees: And,

If you design to have your Tree soon furnished on both sides, hinder it from shooting in the middle; and note, that the more you prune a Tree, the more it will shoot.

If any Boughs of Fruit Trees bend downwards with the weight of their Fruit, the next Spring cut off some of the superfluous Twigs, and let not Fruit Trees grow high, because it takes too much of the Sap from the Fruit, and makes it troublesome and dangerous to gather; therefore make them spread as much as you can.

In pruning of Fruit Trees, do not thin the Boughs next the Body, except they cross or gaul one another. But thin them most at the outmost Branches, or where the Branches are the thickest, except you meet with a Branch that has a bearing Bud at the end; that be sure to spare.

Vines, Goofeberries, Currants, &c. bear Fruit for the most part on the Branches they put forth the same Year; so that in pruning of them, you may cut off much of the Shoots of the foregoing Year, and of the old Wood, and there will be more Sap to put forth fresh ones the next Year, provided you leave plenty of Buds for them to put forth at; and with this caution, that such as grow luxurious in Wood

are

are not apt to bear, and the more you cut off, the more they will run to Wood.

Stone Fruit Trees generally bear on the Branches of the foregoing Year; therefore in pruning of them leave a sufficient Number of such Branches.

Make as few Wounds in a Tree as you can, and rather extenuate a deformed Branch, than haggle it in several places; on unthrifty Trees cut your Boughs shorter, and leave fewer on them than on thrifty ones.

In Wall Fruit cut off all gross Shoots, how fair soever they seem to the Eye, that will not without much bending comply well with the Wall: For if any Branch happen to be wreathed or bruised in the bending or turning (which you may not easily perceive) tho' it doth grow and prosper for the present, yet it will decay in time, and the Sap or Gum will be spewing out of it, which is the cause of the decay of many a good Tree.

In pruning of Trees or Vines leave some new Branches every Year, and take away (if too many) some of the old, which much helpeth the Tree, and increaseth its Fruit.

When you cut your Vine, leave two knots at the next Interval, for usually the two Buds yield a Bunch of Grapes, the not taking care of which often makes Vines unfruitful.

If you cut off any Boughs or Branches, cut them sloping, so as the Rain and Wet may fall off from them, and near to a Bud, that they may the sooner heal without leaving of any Stubs.

It is good also where your Tree is too full of Fruit to disburthen it of some of them, and the rest of the Fruit will be the fairer.

The great thing to be taken care of in pruning and nailing of Trees, is to spread it like a Fan, that it may handsomely cover the Wall. See *pag.* 394.

Chap. XX. *Of some other necessary Observations about Fruit-Trees.*

STrong or hot Dung is not good for Fruit-Trees till it is throughly rotten and cold, but on rich warm Land, Mud or Soil that lies in Streets or Highways or any uncultivated Earth where it may be had, is best, especially for Apple-Trees.

Many applying of Soil and Manure to their Trees, commonly lay it near to the Stems, whereas they should lay it at a proportionable distance to the spreading of the Roots, according to the Age and long standing of the Tree.

If you have an Orchard or other Plantation that is old, and you have a mind to extirpate it upon the account of the decay of the Trees, either set out fresh Ground, or dress and dig the holes a Year before you design to plant them, letting of them lie open to take the Air, that the Sun and Frost may refresh the Earth, and do not plant your Trees in the places where your old ones stood, lest the old putrid Roots corrupt and spoil the young ones.

Winter Fruit, where there is Sun enough to ripen them, are more durable and lasting that grow upon stiff Land, and commonly the best flavour'd: But Trees that grow upon rich Land are the most thriving, and bear the largest Fruit, tho' not of so good a relish.

However, for them that live in the Northen parts of *England*, I would advise them to plant chiefly Summer-Fruit, because the other seldom ripens kindly: Only this may be consider'd, that where Plantations are upon a gravelly, sandy, rocky or Lime-stone Soil, there is at least two degrees difference between such a Soil in the North, and a cold Clay in the South. Besides, the declivity of a Hill of a Southern Aspect, being well shelter'd, gives a great advantage to the ripening of Fruit. All which things are necessary for a Planter to consider, that he may accordingly suit his Plantation and Situation to one another.

Where

Where Fruit-Trees are old, it is good to Prune or Lop them well, and so Manure them often with Dung, rich Earth, or, which is best, with Lime or Chalk, where it is to be had. Sir *Hugh Platt* advises the taking of *two* Quarts of Ox or Horses Blood, and temper it with Pidgeons-dung, till it make it into a soft Paste, which he says is a most excellent thing to apply to the Roots of old Trees, the Roots being first open'd, and laid bare a few Days; this will recover a Tree or a Vine almost dead, and must be laid to the Tree about the midst of *February*, and to a Vine about the beginning of *March*.

I shall conclude this part of Husbandry relating to Fruit-Trees, with Recommendations of the Vine, the Juice of which being so much desir'd, and considering the Advantages that it brings to those Climates and Countries that it is natural to; I could not omit it without making some Essay towards the Propagation of so useful and beneficial a Commodity, especially since it is plain that Vineyards have formerly been in *England*, and that they are now in many Places of the same Climate with us, where they thrive to the great Advantage of the Owner; and therefore I cannot but think the want of *English* Wine to proceed only from Negligence, and our easie procuring of it by means of our Navigation, which tho' it may seem to be an increasing of our Trade, yet it was procur'd upon a very uneven Balance while we had it from *France*. However, let any Commodity be procur'd upon the best Terms of Trade that can be propos'd, it is much short of the Advantage that any thing of a Nations own Product will amount unto: And therefore I could wish that a greater Diligence were us'd for the Promotion of it, especially in the South parts of *England*, which I should think the Essay of the Vineyards of that Worthy Gentleman Sir *William Bassit*'s, near the *Bath*, should incourage; since I have drank Wine made of his Grapes (as I have been inform'd) that I think was as good as any of the Wines that I have drank, either

in

in *Paris* or *Campaign*. What Art was us'd to it I could not learn, but it is what I think is worth inquiring after; and tho', I suppose, I may not propose the same Method, yet when I come to treat of *English* Liquors, I hope I shall be able to do somewhat towards the Improvement of it, and therefore I shall at present confine my self only to what relates to the Propagation and Culture of the Vine.

Chap. XXI. *Of Vines.*

THE Vines most proper for our *English* Climate I think are, First, The small black Grape, by some call'd the Currant, or Cluster-Grape, which I reckon the forwardest of the black sort. Secondly, The White Muscadine, the Parsley-Grape, and the Muscadella, which is a White Grape, not so big as the Muscadine, tho' as soon ripe; and the White and Red Frontiniaque, if planted in a very warm Place.

The best Soil for Vines, is the hottest Gravel, Sand or rocky Ground, provided they be kept well water'd and shaded at first planting; and if the aforemention'd Soils run much to Brambles, it is a promising sign of the Vines thriving; but whatever the Soil be, it ought to be fresh, and not to have been plow'd up of a long time. The Soil will much forward their ripening, as I observ'd before.

The next Advantage to be given to Vines in these cold Climates, is that of a warm situation and good shelter, which the Declivity of an Hill lying to the South will best afford, especially if well shelter'd from the North, and incompas'd with a good Brick-wall, because Hills are not so subject to the Morning Fogs, nor infectious Mists, as low Grounds are: Besides, flat Land does not so soon enjoy the benefit of the rising Sun; nor doth it stay so long upon them in the Evening; for since the Vine doth above all things affect a dry Soil, especially after the Fruit begins to be form'd, and approach to its Maturity, there is nothing more

noxious

noxious to it than at that Seafon to be infected with the cold heavy Damps of thefe Fogs. It is in that as much as in any other thing, wherein our more Southern Climates have the Advantage of us.

Vines may be increas'd by Layers, which may be laid any time in Winter before *January*, and will of-ten grow of Cuttings only ftuck in the Ground in a moift Place, and well water'd in Summer, if it prove a dry time, or of Suckers.

For to plant a Vineyard, in *July*, when the Earth is very dry and combuftible, plow up the Swarth and burn or denfhire it, as is before directed about plow-ed Land. In *January* following fpread the Afhes.

The Ground being thus prepar'd, make your Trenches crofs the Hill from Eaft to Weft, becaufe the Vines ftanding thus in Ranks, the rifing and fetting of the Sun will by this means pafs thro' the Intervals, which it would not do if they were planted in any o-ther Pofition; nor yet would the Sun be able fo well to dart its Beams upon the Plants during the whole courfe of the Day.

To plant the fetts, ftrain a Line and dig a Trench about a Foot deep, and fet your Plants in it about three Foot diftance every way one from another, trim off the fuperfluous Roots of your Setts, and leave not above three or four Eyes or Buds upon that which is above the Ground, and plant them about half a Foot deep, fetting of them floaping as they commonly fet Quick, fo as that they may point up the Hill: Which being done, take long Dung, or Straw, and lay on the Trenches of a reafonable thicknefs to cover the Earth, and to preferve the Roots from the dry pier-cing Winds which would otherways much prejudice them, and from the burning heat in Summer. Keep them well how'd and clean from Weeds, and if need be water them. The beft time to plant them is in *January*.

The firft pruning of the new fet Vine, ought not to be till *January* after its planting, and then you fhould cut off all the Shoots as near as you can, sparing only one of the most thriving ones, on which you should leave only two or three Buds, and so let it rest till *May*, the second Year after planting ; and then be sure to clear the Roots of all Suckers which do but exhaust and rob your Setts, for the small Branches of Vines produce no Fruit, and leave no Branches but what break out of the Buds you left before, continually taking care to suppress the Weeds. The same Method is to be taken the third Year, by cutting off all the Shoots in *January*, sparing only one or two of the most thriving ; which being done, dig all your Vineyard, and lay it very level, taking great care that in this Work you do not cut or wound any of the main Roots with your Spade: As for the younger Roots, it is not so material, for they will grow but the thicker, and this Year you may enjoy some of the Fruit of your Labour, which if answerable to your Expectation, will put you upon providing of Props for them of about four Foot long, which must be placed on the North-side of your Plant. In *May* rub off such Buds as you think will produce superfluous Branches. When the Grapes are about the bigness of birding Shot, break off the Branches with your Hand at the second Joynt above the Fruit, and tie the rest to the Prop. The best way is to break, and not cut your Vine, because Wounds made with any sharp Instruments are not apt to heal, but cause the Vines to bleed.

Fourth Year.   The following Year after its bearing you will be likely to have three or four Shoots to every Plant, and therefore in *December* cut off all the Branches except one of the strongest and most thriving, which leave for a Standard about four Foot high, cutting of the rest very close to the Body of the Mother-Plant, which tie to your Prop till it is big enough to make a Standard of it self: And then you must suffer no Shoot to break out but such as sprout at the top about four Foot from the Ground, all which Sprouts the *French* Prune off every Year, and trust only to the new Sprouts

which

which are the only bearing Shoots. But others propose to leave two or three Branches, the one fuccessively after the other, and so they always cut off the oldeft every Year, and Nurse up the other young ones; but the number of the Branches fhould be proportionable to the Thriftinefs of the Vine.

In *August*, when the Fruit begins to ripen, break off fuch Shoots as you find too thick; but this Work you muft do with Difcretion, and only fo as to let in the Sun for the ripening of the over-fhadow'd Clufters, but not to leave them too bare, left you expofe them too much to the fcorching Heat by Day, and the moift Dews by Night. If you find a Vine to bleed, rub fome Afhes upon it; and if that will not do, fome commend the fearing of it with a hot Iron.

When thro' often ftirring of it you find your Vineyard poor (which the weaknefs of the Crop will foon difcover) prune your Vine, as is before directed, and fpread good rotten Dung mix'd with Lime, over the whole Ground, letting of it lie a whole Winter to wafh into the Ground, mixing of about ten Bufhels of Lime with a Load of Dung, and if fome Afhes or Soot be likewife fpread on it, it will do well, which Manure turn in about *February* with a flight digging, but not too deep, which fhould be perform'd in a dry Seafon and not in wet, left it occafion the Ground to bind too much, and caufe the Weeds to grow. But to forward Grapes ripening, and to make them fruitful, the Blood of Beafts mix'd with Lime or Soot is very good to lay the Roots of the Vines in *December* and in *July*; and if the Seafon is very dry, the watering of Vines in *August* is a great Advantage.

But in our cold Climate where we are oblig'd to plant them againft a Wall, or other Shelter, Vines fhould be prun'd only of thofe Branches that are unthrifty, which are flat and grow dry in Winter; fo that you perceive no Sap in them when you cut them, for the plenty of Grapes, which they cannot bear, extracts fo much of the Nourifhment of the Vine, that

it

it will rather decrease every Year than grow too luxuriant. And besides, the more Wood you have, the more Fruit you may expect, because 'tis only the present Years Shoot that bears; and therefore take away as much of the old Wood as you can, that has only a few good Branches, and bend them downwards as low as you can well do, and from them will grow young Roots; but if they run too much to Wood, cut off the worst of the Branches, rather than shorten them, because it causes them to be rough, and not to shoot out young Branches. If an old Vine bear not well, lay down a Layer of some of the strongest branches of the foregoing Year that grows low, and from that Layer nurse up a young Vine, and cut the old Vine away as the new one spreads upon the Wall.

Gather your Grapes in a dry Day, when they are very plump and transparent, which is when the Seeds or Stones are black and clear, not viscous or clammy, when the Stalks begin to shrivel at the part next the Branch, which is a sign it hath done feeding; only you must take care if Rain come, and Frost immediately follow, to gather them as soon as you can.

It is best to cut and not to pull the Grapes from the Vine, and to put them in Baskets, out of which, empty them gently, and lay them on heaps on a Floor, to sweat for four or five Days, or a Weeks time, which will ripen them much.

If you would make Claret, let it remain with the Murc or Husks till the Tincture be to your liking, but the White Wine may be press'd out immediately.

When the White Wine is turned, some propose to stop it up immediately, and say that it will not hurt the Cask, and leave half a Foot or more void; and for Claret leave something more, which they replenish at ten days end (when the fury of working is over) with some proper Wine that will not provoke it to work again. This must be frequently repeated, for new Wine will spend and waste somewhat till it is perfect.

This

This is the manner of *Languedoc*, and the Southern parts of *France*, and about *Paris* they let it abide with the Marc in the Must two Days and Nights for White Wines, and at least a Week for Claret ; but then they observe to let it be well cover'd.

In some parts of *France*, they Tun it when it hath wrought in the Kelers, filling of it up (as before is described) with what is squeez'd from the Husk, which some think very practicable with us.

Whilst the working and filling of it up continues, keep it as warm as you can, by closing up any Northern Windows, if you have any in your Cellar, left it sour the Liquor, and about the expiration of *March*, stop your Vessel for good and all. Some about this time roll their Cask about the Cellar to mix it with the Lees, and after a few days Re-settlement, they rack it off with great Improvement.

Put into your Vessel the plaining, or chips of green Beech, the Rind being carefully peel'd off; but first boil them in clean Water about an Hours space to extract their rankness, and then dry them in the Sun, or an Oven: Less than a Bushel of Chips will be sufficient to fine a whole Tun of Wine, and it will set your Wine in a gentle working, and purifie it in twenty four Hours, giving of it a good and agreeable Flavour. *Wine to fine.*

These Chips may be wash'd again, and will serve the better upon the like Occasion, even till they are almost consum'd. Let your Chips be plained off as long and large as you can get them, and put them in at the Bung.

Some sweeten their Wines (to prevent harshness) with Raisins of the Sun trodden into the Fat being a little plump'd before, or by boiling one half of the Must or Liquor in a Vessel for an hour, scumming of it, and tunning of it up hot with the other.

But the best Method that I have met with to make *English* Wine, is after the Grapes are pick'd from the Stalks to press them, and to let the Juice stand twenty four hours in the Fat, draw it off from the gross Lees, and put it up into a Cask, and to every Gallon *English Wine.*

of

of Juice add a Pint or Quart of strong Red or White Port, according as you desire it in Strength. Let it work together, and when it hath done, Bung it up close, and let it stand 'till *January*, at which time in dry Weather Bottle it ; this way I have made as good Wine, as any *French* Wine without any Adulteration, which consisting of four parts of our own Product, and but one of Foreign, must be of advantage for the promotion of our own Grapes.

Chap. XXII. *Of gathering of Fruit.*

YOUR Trees having attained to their desired end of bearing Fruit, it will be necessary to consider the Methods to be used in gathering, transporting and keeping of it.

Gathering of Fruit. As to the gathering of Fruit, care must be taken to do it without bruising, especially of such as you design to keep, and that you do it when they are arrived at their due Maturity, at which time they are not only best for eating, but keeping too. Fruit ripens sooner or later, according as the sort is, and the Season of the Year falls out, or that they are situated and sheltered, and that the Soil is either hot or cold. But the best time for the gathering of Winter Fruit, is about *Michaelmas* after the first Autumn Rains come, when the Tree, being sobbed and wet, swells the Wood, and loosens the Fruit : Or when the Frosts advertise you that 'tis time to lay them up, beginning to gather the softest Fruit first, but mind never to gather Fruit in wet Weather.

Transporting of Fruit. For the Transportation of Fruit, or the carrying of it to Market, &c. Apricots, Peaches, Figs, Strawberries, Cherries, Rasberries, &c. require Water-carriage, or to be carried on Men's Backs ; but for Peaches or Apricots, they should be laid upon that part that the Stalks grow out of without touching of one another, and to be laid upon a Bed of Moss, Fern, or Leaves, or to be wrapped up in Vine Leaves. And

in

in case several Beds be laid one upon another, a good quantity of Moss ought to be laid between them.

Figs are very tender, and therefore each Fig should be wrapped in a Leaf, and small Partitions made with Splinters, like the bottom of Sieves, to part each Layer in the Basket, that so they may not lie one upon another.

Plumbs may be put in a Basket without any other Ceremony, than the laying of Leaves at the bottom and top.

Strawberries and Rasberries are commonly put into small Baskets made on purpose for them, and the Leaves laid at top and bottom, and stuffed by the sides.

Apples and Pears are commonly packed in Baskets, with a good quantity of Straw at the bottom and top.

As to the preserving of Fruit, if it is Summer Fruit (especially Peaches) they must be laid in a dry place on Shelves with the Windows always open, and upon dry Moss, or other soft things that have no ill-scent or savour; for Peaches like Melons eat better for being gathered a day or two before they are eaten. All Fruits must be visited daily, and the rotten ones pick'd out, lest they should infect the other. Pears may be placed with their Eye downward, but beware of laying of them, or Apples upon Hay, Wheat or Rye Straw, which will give the first an ill-flavour, and leave the other none; the best Straw is that of Oats, but Fern or Blankets is much better.

The best way to keep Grapes is to hang them up in the Air fastned to a Packthread; but if any are desirous to preserve them 'till towards Spring, they must be gathered before they are perfectly ripe, and care must be taken constantly to pick out those that are rotten. Some say, the best way to keep them, is to hang them up in a Barrel, which must be headed up so close that no Air may come at them. Some lay them in a Cask in Oat-chaff.

But as Apples and Pears are of long duration, it will be necessary for those that are curious in keeping of
them,

them, to have a Conservatory or Store-house made
after this manner: Choose some place in your House
the most convenient for this purpose, which should
have the Windows and Overtures narrow, to prevent
the Extremity both of Heat and Cold. These should
always be kept shut, except in very fine Weather.

About the Room should be Shelves made one above
another, and the middle be left to lay Fruit in on
Heaps, such as are the most common, or that you design for Cyder; but if your Room be narrow, then only Shelves on one side, and the two ends will be enough.

Let your Shelves be laid upon Brackets, being about
two Foot wide, and edged with a small Lath to keep
the Fruit from rowling off of them, placing of them
about a Foot asunder.

And as you gather your Fruit, separate the fairest
and biggest from the middling, and such as are fallen
off of themselves, or that were thrown down in gathering: And putting each sort into Baskets, as fast
as you gather them, carry them into your Store-
house, and range them upon your Shelves, so as that
they may not touch one another, laying of Fern under
them, and having of a good quantity more of Fern by
you, cover them well up with it, and in case of Frost
you may lay Blankets and other things to secure them;
but in very severe Frost, some commend a wet Sheet
to lay over them, as the best thing to preserve them.
Be sure your Fern is very dry, let it be cut in Summer
while the Sap is in it, and that it have contracted no
ill favour or mustiness.

Where you keep your Fruit, 'tis a good way to lay
each sort by themselves, especially those which are
least lasting, and the most durable by themselves.

All Fruit at a Thaw will give and be moist, at which
time let them lie without touching, except those you
take for present Use; and so likewise during great
Rains only as in Frosts, 'tis best to keep them as close
as you can, so in wet Weather 'tis best to let in all
the Air, especially about the middle of the Day.

And

And every other Day look carefully to the Apples and Pears, and take out all that are specked or rotten, least they infect the others.

As for the time of Fruits being in Season and their lasting, I shall have occasion to mention a great number of them in the Kalendar.

BOOK XVII.

Chap. I. *Of English Liquors.*

 Aving given an Account of the way of ordering, managing and improving of Corn, Fruits, and Flowers; I shall in the next place endeavour the Improvement of *English* Liquors, which is a part of Husbandry that I think is too much neglected; and therefore I shall give the best help towards it that I can, and begin with Beer, as the most common Liquor, and what for the want of good Management, is generally the most spoil'd, of any Liquor we make.

Chap. II. *Of Beer and Ale.*

IN the brewing of Beer, two things must particularly be taken care of; First, Good Malt, which I have already given an Account how to make. And, Secondly, Good Water that is soft, and will bear Soap, for harsh Water makes not only unpleasant Beer or Ale, but likewise requires much more Malt than soft, and that in proportion to the harshness or softness of it; and Lastly, Being provided with good Hops, First,

First, Heat a Hogshead of Water, and cover it with Bran; when it is scalding hot, put one third part of it into the Mashing-tub, and there let it stand 'till the Steam is so far gone that you may see your Face in the Liquor, then stir in four Bushels of Malt, and let the remainder of the Water in the Copper boil a little, then draw out the Fire, that the heat of the Water may be qualify'd before you put it to the Malt, and when it is of a due heat, add it to the other part that was put into the Mashing-tub before, and stir it well again, putting up two or three Shovels full of hot Wood-coals upon it, to take off any ill Taint of the Malt: Then let it stand two Hours, in that time heat a Hogshead more of Water; and when your first Wort is drawn off, put part of it upon the Grains, and stir in three Bushels of fresh Malt; if you intend to make Ale at the same time, then add the rest of the Water and stir as before; after which put your first Wort into the Copper again, make it scalding hot, and put part of it into a second Mashing-tub, and when the Steam is gone, stir in three Bushels of fresh Malt, then put up the rest of the Wort, and stir it well, as before, letting of it stand two Hours, and put another Hogshead of Water into your Copper, and when what was put in the first Mashing-tub has stood there two Hours, draw it off, as also that Liquor in the second Mashing-tub, and take the Grains out of the second Mashing-tub, and put them into the first, and put the Water that was scalded in the Copper to it, which let stand in the Mashing-tub an Hour and in half at most; and while that is standing, get ready another Copper of Water (the Copper containing about a Hogshead) which put upon the Grains, and let it stand as before; only note, that in all the Mashings (when you think that the Liquor hath stood long enough upon the Malt) before you let it run out, you draw out some of the Liquor first, and see if it run clear; if it doth, draw it off; if not, fling it up again, and let it stand 'till it doth. Then take the first

Wort

Wort and boil it with two Pound of Hops, two Hours,
or 'till you find it look curdly; after which boil the
second Wort for Ale an Hour and an half, with three
quarters of a Pound of Hops, and the Hops that were
boiled in the first and second Wort, boil in the remaining Liquor an Hour and an half, which quantity will
make a Barrel of Strong-Beer, and a Barrel and a half
of Ale, and one Hogshead and a half of Small-Beer.
This is the best way of Brewing your *March* and *October* Beer.

But for the Brewing of Small-Beer, or common
Ale, take something above the quantity of a Barrel
of Water scalding hot, which put into your Mashing-
tub alone; let it cool 'till you can see your Face in it,
and put to it four Bushels of Malt, pouring of it in
by degrees, and stirring of it well: Let it stand on
the Malt two Hours (observing the same Method as
before proposed for Strong-Beer) then draw it off,
and let it boil an Hour and an half in Summer, or an
Hour in Winter; and when it is boiled enough, it
will look curdled. Of this first Wort you may make a
Barrel of Ale: After this is boiled, scald about a Barrel of Water more, and put it upon your Malt, letting
it stand an Hour and an half: This draw off, and put
the same quantity of hot Water on again, observing
the same Rules, as before directed, of this you may
make an Hogshead of Small-Beer. When you put it
together to Work, take care that it is not too hot,
and when you put Yeast to it, put it to a small quantity at first, and add more and more to it by degrees,
and when it hath work'd twenty-four Hours in the
Tub, Tun it up. But if you brew Small-Beer alone,
two Bushels of Malt and a Pound and a half of Hops
will make a Hogshead of good Small-Beer; or eight
Bushels of Malt will make a Barrel of Ale, and three
Hogsheads of Small-Beer.

These Proportions of Brewing are for a small Fami-
ly, which I chuse to Instance in, because others may
easily proportion it to larger Quantities as they please.

To

To what hath been already mentioned, I shall add the manner of brewing of Ale and Beer, published by Sir *Jonas Moor* in a small Treatise of his; which as it contains a great many particulars, and is recommended from his own Experience, may be of use to the Publick, which take in his own Words.

In the brewing of Ale and Beer, after you have made a discreet choice of your Materials, you must first consider what sort of Drink you design to Brew, and accordingly proportion your quantities. If you design your first Wort for Strong-Ale, or *March* or *October* Beer, you must proportion five Gallons of Drink to every Bushel of Malt (that is to say avoiding Fractions) eleven Bushels of Malt to an Hogshead of Ale or Beer. But it must be remembred, that in so great a Disproportion of Malt Drink as eight to five, almost a third of your Liquor in the first Wort will be absorbed by the Malt never to be returned, and an allowance is to be made of about a sixth part to be evaporated in boiling; so that if you expect to clear a Hogshead of Drink, that is, fifty four Gallons, from your first Wort, you must put into your Mash-tub near ninety Gallons of Liquor. But for your second or third Worts, the Malt being wet before, you need put up no more Liquor than you intend to make Drink, except an Allowance of about a tenth Part for waste, that not boiling so long as your first Wort: And you may of your second Wort make one Hogshead of good middle Beer or Ale as strong as the common Ale-house Drink in *London*; and your third Wort will make one Hogshead of good Small-Beer.

I propose, in this Case, the drawing off three Worts, because of the great quantity of Malt to a smaller of Liquor; otherwise in ordinary Brewings, where you design not very strong Drink, six or seven Bushels of Malt will make one Hogshead of good Strong, and another of Small-Beer. And in such Cases, two Moakses will as well take out the strength of your Malt, as three in the other.

The

The proportion of Hops may be half a Pound to an Hogſhead of Strong-Ale, one Pound to an Hogſhed of ordinary Strong-Beer to be ſoon drunk out, and two Pounds to an Hogſhead of *March* or *October* Beer. And for the after Worts which are not to be kept long, what comes from the firſt Wort will ſerve well enough to boil again with them.

If you put into your firſt Wort a greater proportion of Hops, and boil them all the while your Wort boils, you will make it too bitter. But I conceive it adviſable to double the proportion by taking out the firſt Parcel, when your Wort has boiled half the time you deſign it ; and then adding the ſame quantity of freſh Hops, to continue boiling till you take your Wort out of the Copper. This will ſomewhat encreaſe your Charge, but that will be very inconſiderable, if you furniſh your ſelf in a cheap Year of Hops.

Hitherto of the Qualities and Proportions of your Materials; now concerning the manner of putting them together.

After you have put your Liquor into your Copper, ſtrew an handful, or two or three handfuls of Bran or Meal upon it, not ſo much to ſtrengthen your Liquor, as to make it heat quickly, for ſimple Water alone will be long e'er it boil. But you muſt take your Liquor out of the Copper, when it begins to ſimmer, and not ſuffer it to boil; for tho' it were granted that the boiling did no harm to your Liquor, by evaporating the natural Spirit of the Water; yet it is a needleſs expence of Fuel and Time, firſt to make it too hot, and after to ſtay till it is cooler again. For you muſt by no means mix your Malt with boiling hot Liquor, which will make Malt clot and cake together, and the moſt flowery parts of it run whitiſh, glewy and ſizie, like Sadlers Paſte, ſo that it will never mix kindly, nor give out its ſtrength equally to the Liquor.

I had not dwelt ſo long on this Head, but that I know many put their Malt firſt in the Maſh-fat, and then pour in their Liquor for the firſt Wort, which is indeed neceſſary in the ſecond and third Worts.

The contrary Practice of putting in your Liquor firſt hath theſe Advantages;

Firſt, You can never otherwiſe gueſs when your Liquor is juſt cool enough to be mingled with your Malt. But in this caſe you have a certain Criterion and Rule to judge, that is, you muſt let your Liquor remain in your Maſh fat, till the Vapours from it be ſo far ſpent, that you can ſee your Face in the Liquor: And then pouring your Malt upon it, you have this further Advantage, that you keep your Liquor longer hot, and it ſinks gradually, diſtributing its ſtrength to your Liquor equally without matting; and if it does not deſcend faſt enough of it ſelf, you muſt preſs it down with your Hands or Rudder, with which you uſe to ſtir your Malt or Moaks. This muſt be done by degrees, always remembering that you ſhake your Sacks before you remove them, over the ſide of your Maſh-fat, to get out the Flour of your Malt which ſticks to them; and after all your Malt is ſettled and your Liquor appears above it, you muſt put up in your Maſh-fat as much more hot Water out of your Copper, as will make in all ninety Gallons for one Hogſhead: Then ſtir it almoſt without ceaſing, till it has been in the Maſh-fat about two Hours from the firſt putting up your Malt, in which your Servants may help and relieve one another.

After this pull out your Rudder, and putting a little dry Malt at top, cover it cloſe, and let it ſtand half an Hour undiſturbed, that it may run off clear, and the Malt being ſunk to the bottom, the Liquor at top will run thro' it again, and bring away the ſtrength of it. After this, you muſt lift up your Tap-ſtaff, and let out about a Gallon, not into your Tub underneath or underback, which is two receive your Wort, but into your long Hand-jet, and put it back again, ſtopping your Tap-hole: This do two or three times, till you find it runs clear, which it will not do at firſt, tho' your Tap-hole be never ſo well adjuſted. Thro'-

Throughout the whole courfe of your Brewing, you muft be very careful to do all you can to promote the finenefs and clearnefs of your Drink.

In the North of *England,* where much the beft Malt-drink is made, they are fo careful of making their Drink fine, that they let their firft Wort ftand in their Receivers till it is very clear, all the grofs Parts being funk to the bottom; this they continue to do about three Hours in Summer, and ten or twelve Hours in the Winter, as occafion requires, which they call Blink-ing: After which, leaving the Sediment behind, they only lade out the clear Wort into the Copper. This Cuftom is peculiar to the North, and wholly unpra-ctis'd in others parts.

When all is run out into your Receiver or Under-back, Lade or Pump out your fecond Liquor, ordered fo as to be then juft ready to boil, on your Moaks, and putting your firft Wort in your Copper again, let it boil reafonably faft (which boiling, the Hops, put on it will much accelerate) for about one Hour and an half for *March* or *October-Beer* to keep long, and one Hour for Strong-Ale to be drank new. I know that a longer boiling is generally advifed; but that I fhall anfwer when I come to fhew the Reafons, why common Brewers feldom or never make good Malt-drink. I advife the Wort rather to be boiled reafonably faft for the time, than to ftand fo long to Simmer, becaufe common Experience fhews it waftes lefs, and ferments better after fo long boiling than fimmering.

Your firft Wort being thus boiled, muft be pump'd or laded off into one or more Coolers, or Cool-backs, in which leave the Sullage behind; and let it run off fine. The more Coolers, and the thinner it ftands, the fooner it cools (efpecially in hot Weather) the better; let it run from your Cool-back into your Tun very cool, and fet it not there to Work in Summer till it is cool as Water. In Winter it muft be near Blood warm at leaft; the Bowl in which you put your Yeaft to fet the reft on Working, muft have a mixture of Wort hot enough to make it all ferment: When you find it begins to Work up thick to a Yeast, mix it again with your Hand-jet: And when it has wrought it self a second to a Yeast, if you defigned it for Ale, and fpeedy drinking, and hopped it accordingly, beat in the Yeast every five Hours for two Days together in the Summer time, or more, according as the Weather is, and for three or four Days in the Winter, covering your Fat clofe, that it fall not in your Working-Tun.

When your Yeast begins to Work fad, and upon the turning of the Concave of your Bowl downwards fticks faft to the infide, then skimming off the Yeast, firft cleanfe the reft into your Veffel, leaving all your Dregs in the bottom of your Tun, and putting only the clear up. After it has a little fermented in your Veffel, you will find it in a few Days fine and fit for your drinking, tho' according to the quantity of your Hops, you may proportion if for longer keeping.

If you Brew in *March* or *October*, and have hopped it for long keeping, you muft then upon its fecond Working to a Yeast (after once beating in) cleanfe it into your Veffel with the Yeast in it, filling it ftill as it Works over, and leaving when you ftop it up a good thick head of Yeast to keep it.

In brewing *March* and *October*-Beer, it is advifable to have large Veffels, bound with Iron Hoops containing two, three or four Hogfheads, according to the Quantity you intend to make, putting all into one Veffel; this fort of Drink, keeping, digefting and mellowing beft in the largeft Quantities.

Your Veffels muft be Iron-hoop'd, elfe your *March*-Beer will be in danger to be loft or fpoil'd; leaving your Vent-peg always open palls it; if it happens to be faften'd but fix Hours together in the Summer, a fudden Thunder or ftormy Night may happen next Morning to prefent you in your Cellar an empty Veffel, and a cover'd Floor.

It is pretended, that *March* is the beft Month for brewing, and the Water then better than in *October*;

but

but I always found that the *October*-Beer, having so many cold Months to digest in, proves the better Drink by much, and requires not such watching and tending as the *March* Beer doth in opening and stopping the Hole on every Change of the Weather.

Many Country Gentlemen talk of and magnifie their stale Beer of Five, Ten or more Years old; it is true, more Malt and Hops than I propose will keep Drink longer than I use to do; but to small purpose, for that it will not exceed mine in any thing desirable, except such an extraordinary strength as few Men care for: I always broach mine at about nine Months end, and my *March*-Beer at *Christmas*, and my *October*-Beer at *Midsummer*, at which time it is generally the best; but will keep very well in Bottles a Year or two more. Stop your Vessel close with Cork not Clay, and have near the Bung-hole a little Vent-hole stopp'd with a Spile, which never allow to be pull'd out till you bottle or draw off a great quantity together; by which means it is kept so close stopp'd, that it flushes violently out of the Cock for about a Quart, and then stops on a sudden, and porles and smiles in a Glass like any bottled Beer, tho' in the Winter time. But if once you pull out the Vent-pest to draw a Quantity at once, it will sensibly lose this briskness, and be some time before it recovers it.

I propose no Directions for the second and third Worts; he that can manage the first well, can never fail in the rest. Your third Wort being pour'd on hot Goods, may be only cold Water.

But which is the best Method to be us'd, I must refer to Experience.

Chap. III. *Of* Nottingham-*Ale.*

THE chief thing that they observe in making of it, is, only when it is working, to let it stand in a Tub four or five days before they put it into the Cask, stirring of it twice a day, and beating down the Head or Yeast into it; this gives it the sweet Aleish taste.

If Ale or Beer do not fine well, put into a Hogfhead two or three Bottles of old ftale Beer or Ale, and it will much help it.

If Beer be flat or fowre, put into a hogfhead a pint of horfe bean flower, a pint of wheat flower, mix it with yeaft to the confiftency of a Pudding, and it will recover it.

Take Grains, and lay a layer of them and of your boiled Hops, upon them, and fo on, and they will keep good a long time ; and if they are given to a hide-bound Horfe, they will ftrangely recover him.

Mr. *Martin* in his difcription of St. *Kelda* fays, that the inhabitants when they brew Ale, take the juice of Nettle-Roots, which they mix with a little Barly meal dough : Thefe Sowens (that is Flummery) being blended together, produce good Yeaft.

Chap. IV.   *Of Cyder.*

NEXT unto Beer, Cyder is of the moft common ufe, of which excellent liquor there are feveral ways of making it according to the Skill of the Operator, and the Palates of thofe that are to drink it, fome efteeming one fort of Cyder beft, and fome another, according to the Fruits it is made of, and the Methods us'd by them that make it.

Now Cyder-Fruit may be reduc'd to two forts or kinds, either the wild, harfh and common Apple growing in great plenty in *Hertford, Worcefter* and *Gloucefterfhires,* and in feveral other adjacent places in the Fields and Hedge-rows, and planted in feveral other Places of *England* for Cyder only, which are not at all tempting to the Palate of a thievifh Neighbour, not requiting the Charge and Trouble of the more referv'd Inclofures.

Or the more curious Table-Fruit, as the Golden Pippin, the *Kentifh* Pippin, and Pearmain, &c. which are by many preferr'd, having in them a more cordial and pleafant Juice than other Apples.

For

For the former, the beſt ſorts for Cyder are found to be the Red-ſtreak, the White Muſt, the Green Muſt, the Gennet Moil, *Eliut*'s Stocken Apple, Summer-Fillet, Winter Fillet, Broomsbury Crab, the Olive Under-leaf Apple, and the Fox Whelp; the Cyder of which comes not to be good till 'tis three or four Years old.

The greater part of them being meerly Savage and ſo harſh that hardly Swine will eat them, yet yielding a moſt plentiful, ſmart and vinous Liquor, comparable, if not exceeding the beſt *French* Wine; and for the advantage of planting of them they claim the Prefe-rence before Pippins, or any other of our Garden-Fruit.

The other ſorts of Fruits for the making of Cyder are, as I ſaid before, the Golden Pippin, Kentiſh Pippin, Pearmain, Gillyflower, Kirton Pippin, Mother Pip-pin, *&c.*

The beſt ſorts of Cyder-Fruit are far more ſuccu-lent, and the Liquor more eaſily divides from the Pulp of the Apple than in the beſt Table-Fruits.

Some obſerve the more red any Apple is the better it is for Cyder, and the paler the worſe, and that no ſweet Apple that hath a rough Rind is bad for Cyder; but the more inclinable to yellow the fleſhy part of an Apple is, the better colour'd the Cyder will be.

Apples of a bitter taſte will ſpoil your Cyder, but *Gather-* the Juice of them, and of Crabs, will make as good Spi- *ing of* rits as the beſt Apples when fermented; for neither *Apples.* the ſowre nor the bitter taſte ariſes with the Spirit.

Let your Apples that you make Cyder of be tho-rough ripe, and be carefully gather'd without Bruiſes in dry Weather; it very much conduces to the good-neſs and laſting of the Cyder, to let them lie a Week or two on Heaps; the harſher and more ſolid the Fruit is, the longer they may lie, and the more mellow and pulpy the leſs time, which makes them ſweat forth their Aqueous Humidity, and digeſteth and meliorates the remaining Juice, but they will yield more from the Tree, than ſo kept.

Y 4

Such

Such as are windfalls, bruis'd, or any ways injur'd, or unripe Fruit, divide from the sound and ripe.

For it is better to make two sorts of Cyder, the one good and the other bad, than for all to be bad ; the sooner such Fruit is press'd the better ; and from your Apples take away all Stalks, Leaves and rotten Apples ; because Stalks and Leaves give an ill taste to the Cyder, and rotten Apples make it deadish.

Let such Apples as fall before they are ripe be kept till the time of the full maturity of the other Fruit, or else the Cyder will not be worth drinking.

About twenty or twenty two Bushels of good Cyder-Apples, just gather'd from the Tree, will make an Hogshead of Cyder ; after they have lain a while in heaps to mellow, about twenty five or thirty Bushels will make an Hogshead.

They that have great Quantities usually grind their Apples with a Horse-mill, such as the Tanners grind Bark with, but the new invented Engine described in Mr. *Worlidge's Vinetum Britannicum*, is a very good Mill, and will grind a great quantity.

After your Apples are ground they should be made up in Straw, or in an Hair-Bag, and so commited to the Press ; of which there are several sorts, but the Screw-Press is the best.

But as there are several ways of making of Cyder, as well as several sorts of Fruit to make it of, and that some esteem one sort of Cyder, and some another, according to the manner of its making, and the Fruit is made of, as I said before, I shall endeavour to give you several of the Methods I have met with, and leave them to your Experience. But,

I think the chief way of improving of this Liquor would be a particular Management of it according to the Species of Apples it is made of (especially what is made of the chief Cyder-Apples.)

Chap.

Chap. V. *Several Ways and Methods of making Cyder.*

AS firſt that of Mr. *Worlidge*, who propoſes, that when your Cyder is preſs'd out it ſhould ſtand a Day or two, or more, in an open Tun, or cover'd only with a Cloath or Boards to keep it from Duſt, or in a Hogſhead or other Veſſel not quite full, with an open Bung, till the more groſs Parts ſubſide, and then to draw it into Pails, and fill it up into the Veſſels you intend to keep it longer in, leaving about an eighth part empty. Set theſe Veſſels in your coldeſt Cellars or Repoſitories with the Bung open, or cover'd only with a looſe Cover, that there may be a free Perſpiration of the Volatile Spirit of your Muſt, which would otherwiſe force its way, and that your Muſt may be cool and not kept warm, left it ferment too much.

Thus ſtanding open, the better it will by degrees let fall its groſſer Parts, and in time become clear without the loſs of any of its true and durable Spirits. For coldneſs is here the cauſe of its purifying, warmth occaſioning the ſolution and detention of thoſe Particles that ſpoil the Colour and Taſte of Cyder, and which would otherwiſe precipitate.

As for the time of its ſtanding open in the Veſſel, it varies according to the nature of the Fruit ; if the Fruit were mellow or ſweet, the more of the groſs Particles will be preſs'd out with the Liquor, and ſo the longer time will be requir'd for their Precipitation; But if the Fruit were hard or ſharp, the thinner doth the Liquor iſſue out of the Preſs, and the ſooner will your Cyder become fine : And you muſt be ſure to obſerve, that as ſoon as this Cyder of hard Apples is fine, you muſt draw it off from its precipitated Lees, left it become acid, or acquire ſome ill taſte from them.

This ſtanding open of the Veſſel cauſeth an expence of that Wild or Volatile Spirit, which being pent in, would beget a continual Fermentation, much prejudicing

cing the Cyder ; and in cafe it doth not otherwife work
its way out, would in time break the Veffel that de-
tains it.

The principal Caufe that there hath been fo much
bad Cyder made in moft parts of *England*, was the too
early ftopping of it up : It being ufually prefcribed,
and as ufually practis'd, that as foon as Cyder is preft,
ftrain'd and fermented, they ftop it clofe with a very
great Confidence ; that unlefs it be clofe ftopped it
will decay and become of no ufe ; fo that when thefe
Cyderifts have taken care for the beft Fruit, and ordered them after the beft manner they could, yet hath
their Cyder generally proved pale, fharp and ill tafted,
&c. and all from the too early ftopping of it. For the
Slopping of Cyder clofe before it be fine, or with its
Fæces in it (although precipitated) begets reiterated
Fermentations, which Fermentations very much impo-
verifh the Liquor by precipitating thofe Particles,
which enrich it with Tincture and Guft.

Whilft its grofs *Fæces*, or any fettling remain in the
bottom, every change of Weather caufesfome Moti on
therein, which is ufually term'd Fermention ; this
doth fo attenuate this Liquor, that it eafily letteth or
fuffereth thofe Particles to fubfide, and leaveth the Cyder thin, jejune, acid, and ill tafted. It is thin and
jejune, becaufe it hath loft its Subftance; acid, becaufe
it hath loft its Sweetnefs ; thofe Particles being the
Saccharine Subftance, or part of the Apple, and of ill
Savour and Guft, becaufe thofe Particles when precipitated, being mix'd with the more grofs, do putrifie
and heat, infecting the whole Mafs in the Veffel : All
which effects are apparently obvious in Cyder made
after the vulgar Method. Thefe *Fæces* are the caufe,
that the Corks fly out of the Bottle, or break the Bottles, or at leaft at the opening of them make the Cyder fly, and mixing with it make the refidue unpleafant.

Thefe things being generally taken notice of, have
fet many Heads at work to provide Remedies : Some
have made ufe of many ways to ferment it and make it
clear

clear by reiterated Fermentations ; others by Additions, as Ifing-glafs, &c. have enforced a Precipitation,
and when they have fo done, findnig it to be thin, pale
and acid, have by Moloffes, Treacle, or courfe Sugar
given it Body, Colour and Guft. What delight or
pleafure there can be in drinking fuch Compounds, or;
how much this muft conduce to Health and long Life,
I leave every unprejudiced and ingenious Man to Judge.

After your Cyder hath ftood open fome reafonable
time, till it is become indifferently fine, which it
may be in three, four or five Weeks, then will it be
convenient to draw it into Bottles, if you have a fufficient Stock, or into other Casks, that it may there become more fine; for after it is feparated from its grofs
Faeces it will more eafily remit the remaining Particles
or flying Lee, than it would have done whilft the
groffer parts remained, renewing its Fermentation on
on every change of Air, or other accidental Occafion.

Its Finenefs will fometimes plainly appear if you
move the Scum afide with a Spoon, or the like ; but to
be more exact, you may take a Glafs Pipe of a Foot or
more in length, open at both ends, ftop the upper end
of the Pipe with your Thumb, and let the other end
down into the Cyder as deep as you think fit, then open
the upper end, by removing your Thumb, and the
Cyder will rife in the Pipe; then ftop the upper end
again with your Thumb, and take out the Pipe and
hold it over a Drinking-glafs, remove your Thumb,
and you may there difcern the ftate and finenefs of
your Cyder.

If your ftock of Cyder be not over-great, or that
you are willing to preferve your choiceft forts of Cyder, the beft way is to have large Glafs-Bottles of one
or two Gallons apiece, more or lefs, enough to receive
the fame, into which draw off or rack your Cyder, and
let the Bottles ftand open, or but barely covered, in
your cooleft Repofitory for a Month, or more, till
you obferve your Cyder, by your interpofing it between a Candle and your Eye, to be very tranfparent ;
which

which then may be call'd Superfine, the remaining Particles, or flying Lee, being precipitated and settled in the bottom of the Glass-Bottle.

If the Quantity of your choicest Cyder be too great for your Bottles, you may instead of them make use of Stone-Bottles, or Jarrs, or Stounds of *Flanders* Earth, or glaz'd Earthen Vessels, the larger the better; which may be plac'd in Rows in your Repositories, Cellars or Vaults, and cover'd with Boards or the like, to preserve your Cyder from Dust, &c. but not from the Air; but by reason that you cannot so easily discern the fineness of your Cyder in these as in the transparent Vessels, you may now make use of your Glass-Pipe before-mention'd.

The reason why Glass-Bottles, or other glazed or stone Vessels are more fit for this second fining than those of Wood, is, for that the coolness of the Vessel very much contributes to the Precipitation of those remaining Particles that would otherwise debase this Liquor.

But if your quantity of Cyder be so great, that these Vessels cannot receive it, then may you rack it into other Vessels made very clean, dry and sweet, and suffer'd to stand slightly cover'd till it be very fine before you stop it up: If you find that your Cyder doth not fine in wooden Vessels so soon as you desire, for want of that coolness that is in glazed Vessels, you may take Flints or Pebble Stones clean and dry, and put them into your Cask of Cyder, this is said (and with great probability) to contribute much towards the nimble Precipitation of the *Faeces*; the like effect hath the applying of a Bag of Salt to the outside of the under part of the Vessel.

When your Cyder has attained its utmost degree of fineness, which after this way of ordering it will do if you have but patience to let it stand open long enough (altho' some will fine in half the time that other requires) then take your Glass, Syphon or Crane, and draw it off from its last *Faeces* into smaller Bottles,

wherein

wherein you intend to keep it for your use. Thus being drawn off, and thoroughly depurated, you may close cork all your Bottles, and place them in your cool Conservatory, where, after a few Weeks standing, your Cyder will acquire a fine briskness, and mantle in the Glass, without any manner of Feculency, and retain its first Sweetness, and change from a pale to a lively Canary, or *Malaga* Colour; but if you have occasion to accelerate its Maturity, place so many of your Bottles as you think you may have sudden occasion for, in some place warmer than your usual Conservatory, and it will soon answer your Expectation:

Sometimes it will happen, that the next Summer after it is become so pure, some Rags or flying Feculencies may appear in your Bottles, which are occasion'd by the warmth of the Season begetting another Fermentation from that fatness of the Body of the Cyder made of a sweeter sort of Fruit, which are not apt to appear in the thinner Cyder; but in some short time these will subside, and you may draw off the Fine from the *Fæces* with your Syphon, without any great prejudice to your Cyder. These later Fermentations in great Quantities of Cyder often spoil it for want of a timely prevention, which cannot be so well done in Vessels of Wood as those of Glass, where you may easily perceive the various Changes that may happen in these Liquors. Thus far you have Mr. *Worlidge*'s Opinion of this Liquor.

Mr. *Langford* proposes in the making of Cyder, to take the Liquor, as soon as press'd, and strain it through a Sive, and so to tunn it up into a Cask, which should want about two Gallons of being full; which stop up only with a loose stopper for two or three days; and then stop it up close with Clay, and put a peg into the vent hole loose, which for a Weeks time or more, you may once a day draw to give it a little vent, then stop it up close and let it stand till you think it clear; and pearce it to see how it fines; the Summer Fruit after a Month, the Gennet Moyl after

the

the firſt Froſt, and the Red ſtreak or other winter Fruit not till *January*

Mr. Cook's *way of making of Cyder is after this manner.*

Let your Fruit hang till through ripe, which is beſt known by the brownneſs of the Kernels, or their rattling in the Apple, or the Apples falling much in ſtill Weather; for if the Fruit be green, your Cyder will be ſowre. Gather your Apples dry, and rejeꞔ ſuch as are bruiſed, becauſe they will rot and ſpoil the taſte of the Cyder.

If you gather not by hand, which is tedious, lay a Truſs of Straw beneath the Tree, and over that a Blanket, diſcreetly ſhaking them down, not too many at a time, but often, carrying them where they are to ſweat, which ſhould be on dry boarded Floors; by no means on Earth, unleſs ſtore of ſweet Straw lie under them. In about ten or fourteen Days they will have done ſweating, then grind or beat them, keeping the Fruit ſeveral, in caſe you have enough to fill a Veſſel of one kind; if not, put ſuch together as are near ripe, for its more uniformly fermenting: Winter-Fruit may lie three Weeks or a Month e'er you grind them; the greener they are when gather'd, let them lie the longer.

Being ground let them continue twenty four hours before preſſing, it will give it the more Amber bright colour, hinder its over-fermenting, and if the Fruit were very mellow, add to each twenty Buſhels of Stampings, ſix Gallons of pure Water pour'd on them ſo ſoon as beaten; the ſofter and mellower the more Water to reſtrain its over-working, and tho' the Cyder be weaker it will prove the pleaſanter: For over-ripe and mellow Fruit let go ſo much of the looſe and fleſhy Subſtance thro' the Percolation, that with difficulty you will ſeparate the Lee from the Liquor before it ferment, and then away goes the briſk and pleaſant Spirits, and leave a vapid or ſour Drink contraꞔed from the remnant groſs Lees; the Cyder made of ſuch Fruit, had need be ſettling twenty four

hours

hours in a large Fat or Veffel, that the *Fæces* may fettle before you tun it up, and then draw it off, leaving as much of the thick Lee behind as you can (which yet you may put among your Preffings for a Water Cyder): If you conceive your Cyder ftill fo turbid that it will work much, then draw it into another Veffel by a Tap two or three Inches from the bottom, and fo let it fettle fo long as you think it is near ready to work; for if it work in your Tubs, but little of the grofs Lees will you be able to get from it. *Note,* That you muft cover it all the time it is in your Tubs, and the finer you put it up in your Veffels, the lefs it will ferment, and the better it will drink; but in cafe you chill the Cyder (as it often happens in cold Winter Weather) fo as it doth not work when put into Casks, caft into it a Pint of the Juice of Ale-hoof, with half the quantity of Ifing-glafs to refine it, which tho' it do not fuddenly, at the Spring it will.

Thefe Directions obferv'd, barrel it up, and when it ceafes working, Bung it clofe, and referve it fo till 'tis fit to Bottle, that is, when fine, fince till then it will endanger their burfting; and if you would have it very brisk and cutting (which moft affect) put a little lump of Loaf Sugar into every Bottle.

Or you may obferve the following Method, which is, That after your Cyder is prefs'd, to ftrain it, and put it into a Tub or Fat with a Tap to it, which cover clofe with Sacks or Cloaths, by which means fome of the Spirits will have liberty to evaporate; whereas, if you put it too foon into a Cask, it will reverberate the Spirits too foon into the Liquor, and caufe a Fermentation before any of the grofs Lees are feparated from it; for the great thing to be taken care of in making of Cyder, is, only to let fo much of the Spirit evaporate as may prevent its fermenting before the grofs Lees are feparated from it, and yet to keep Spirits enough to caufe a Fermention when you would have it; for if it ferment too much it will lofe its Sweetnefs, and become harfh and fmall; and if it ferment not at

all

all it will become dead and sowre, and therefore let it stand twenty four hours or more in the Fat, according as you find it inclin'd to work, so let it stand longer or a shorter time; and when you draw it off leave as much of the gross Lees as you can behind, for Lees of Cyder are apt to put it into a new ferment upon all Changes of Weather. After it hath stood its time in the Fat, put it into the Cask, which fill almost full; but if you find it begin to work much, rack it off again, and take out the gross Lees; and if you find it still upon a fret, repeat the same Operation till you can settle it; for it is a very ticklish Liquor, and very subject to ferment, especially if the gross Lees are not timely separated from it, and therefore if unsettled or moist Weather happen at the time of its working, it will be so much the more difficult to manage, and will require the more care to be taken of it.

When it hath done working, stop it up, only leaving a small vent hole at your first stopping of it up, at which you may sometimes try if it want vent, lest it break your Cask.

Only I think it necessary to premise, that the suiting of the Fruit to the Soil is a great Advantage to the making of Cyder, it being certain, that in many Places, even in the same Country, there is much better Cyder made in one Place than another, tho' both are made the same way, and of the same sort of Fruit; and if particular Remarks were made of the nature of such Soils, and what the natural Production of them is, according to the nature of the several Soils already treated of, I believe it would be of advantage to Planting, and the Improvement of Fruit and Cyder.

Mr. *Worlidge* commends very much brackish Lands near the Sea-side, as excellent for Fruits, and for Winter lasting Fruit, the strong stiff Lands are much the best of any.

But if your Fruit be unripe, or your Cyder small, and that you have a mind to strengthen it, especially if you live in the North-country, you may improve it by the following Receipts. Take

Take Pippins, Pearmains, &c. and to every Gallon *Raifin* of Juice put two pounds of Raisins, which shred small, *Cyder.* cover the Fat, and let them stand two or three days; draw off the Liquor by a Tap, press out the Raisins, and put both Liquors into a Cask that they may ferment, and after a fortnight rack them off. Do not fill the Cask you draw it into, but leave some room for it to ferment in; after which stop it close, only leave a Fosset-hole open or loosely stopp'd and when it hath done working, fill up the Vessel, and when fine, bottle it: Or you may do it another way.

Take your Apples when they relish best, not too green nor too mellow: They who have large Plantations may shake their Trees a little, and gather those that fall off easily, and press them the same day. Fill not your Cask above three quarters full, and let it stand till it grow clear, which is commonly within eight or ten days, and then draw off only the clear, and fill up a clean Cask almost to the top, giving it vent thrice a day, lest it should burst the Vessel, and so continue to do for a Week.

Then for every ten Gallons of Cyder take one pound of Raisins of the Sun, and put them into Brandy for a day or two, and then take only the Raisins, and put them into the Cyder, letting it stand three or four days more: Lastly, stop the Cask very close, but bottle it not till *March*, except it be of Codlings, which will not keep so long.

Another Improvement of Cyder, is, what they call Royal-Cyder, mention'd by Sir *Jonas Moor*, which is done by adding of the Spirits to it, which corrects the Windiness and Crudities of the Cyder, makes it very agreeable to the Stomach, and gives it the strength of Wine, by adding the goodness of two Hogsheads into one: To do which, put one Hogshead of Cyder into a Still, and draw off all the Spirits; after which distil the said Spirits a second time, and put the same into your other Hogshead, and fill it up. Stir it about well, and keep it close stopp'd, except one day in ten or twenty, let it stand open five or six hours, and within a quarter of a Year, this Cyder will be as strong or stronger than the best French Wine.

But if you will have it drink like Canary, you must add more of the Spirits, and as much Sugar or Sweets (the making whereof is hereafter shewed) as will best please your Palate. And as the proportion of one Pint of good Spirits to a Gallon will make it as strong as French Wine, so one Pint and half will make it as strong as Spanish Wine. And by this means, in the like manner, Perry, the Juice of Cherries, Mulberries, Currants, and Gooseberries, may, by adding thereto their proper Spirits, or any other convenient Spirits, be made as strong as Wine.

I mention other Spirits, because Brandy, Spirits of Wine and of Grain, tho' they will do well, yet they are not so natural and good as what is made of the same sort of Fruit: And the Spirits made of Ale and Beer are the worst of any, unless the Ale or Beer be mix'd with Cyder before the Spirits be drawn off; but the Spirits of Beer and Ale will do well to mix with the same kinds, and add very much to their strength, being a mixture much used of late, with Derby and Nottingham Ale, and with strong Beer.

Only note, first, that the stale and sour Cyder which is scarce fit to drink, will make the greatest quantity of Spirits, and the best tasted, and that the longer the Spirits are kept, the less taste they will have of the Fire; which is the greatest inconveniency that attends this way of making of Cyder; and therefore I should propose, when you design to be any thing curious, to take only the first running of your Spirits to mix with your Cyder, and to let the small part only be distill'd again, to which it will be best to allow as much age as you can to take off the burnt taste: one Gallon of strong Cyder will yield a Pint of Spirits.

As to the time of putting of your Spirits into your Cyder, observe, that the staler your Cyder is before the Spirits are added to it, the more time it will take to incorporate, and the sooner they are put in, the sooner it will be fit for use, only be sure that your Cyder has done working before you put it in.

The best way to order your Sugar before you put it into your Cyder is, to make it into a kind of Syrup or Sweets, by dissolving of it in Water; one hundred weight will make sixteen Gallons, and so proportionably. But before you put your Sugar into the Kettle, takes the Whites of thirty or forty Eggs, the more the better; which being well beaten with a thing like a Rod or a Whisk, in eight or ten Gallons of Water; put four Gallons of this Egg-Water so prepar'd, into your Kettle, where your Sugar is to be dissolved, then hang it over a gentle Fire, and stir it about till it is dissolved: But be sure when it boils, put in more Egg-water, to keep it from boiling too high, and so continue putting it in, one Quart after another, until all your Egg-water be spent. But to prepare your Egg-water in parcels, *viz.* a Quart or two at a time, as you use it, is the better way. Now the use of these Eggs is only to raise such a Scum as will carry away not only all the foulness and grossness of the Sugar, but all the Egg also. And when the Scum hath done rising, and is clear taken off, then fill up your Kettle with as much Water as will make up your Quantity, and let it boil to the size of a Syrup, and being cold put it into your Cyder. But if you put in a little Coriander Seed bruised and tied up in a fine Linen Bag whilst it's boiling, it will give it a fine grateful Scent.

Of these Sweets you may put in two or three Gallons, more or less, into an Hogshead as your Palate invites you, or as the Tartness of your Cyder requires. But put them not in till you have racked your Cyder the last time, and that it is past the Fermentation. And before you put your Sweets into the Cask, mix your Sweets and the Spirits you intend to put in, together with a like quantity of Cyder, and stir them well together; then put all into your Cask of Cyder, and stir them with all your strength with a strong staff in the

Bong-hole for one half quarter of an Hour; after that stop it close, and draw none off till two, three or four Months, by which time it will be answerable to what hath been propos'd, only remember, that if you would have it resemble Canary, you must add the greater proportion of Spirits and Sweets; but if French Wine, the less Sweets, or none at all.

As to the sort of Sugar, if the Sweets be made with white, the Cyder will remain pale; if of brown Sugar, it will raise it to an higher colour: And in my Opinion, the latter is as good as well as the cheapest, since the coursest, by the aforesaid Preparation becomes as pure as the finest; and Sweets being thus made, will cost but five Pence *per* Quart.

And thus every Man may merrily make his Varieties of Drink with that which he knows to be good, cheap and wholesome, which is more than he is sure to have at every Tavern, altho' he pay three times as much for it: Nor hath he so much reason to suspect these Liquors in these Houses to be so much adulterated as the others, because none of like goodness to the Eye, Scent, and Palate can be afforded so cheap to the Pocket.

The husky part of the Apples, after Cyder is pressed out, being steep'd two or three Days in as much Water as will cover it, and then press'd clean out, and kept in a Vessel until it hath well fermented, as also the Lees of all your Cyder will afford Spirit or Brandy, so much, that being added to the Cyder of the same Apples, will make it as strong as French Wine, which is a thing of great Advantage.

Spirits being put into Bottles amongst Cyder, or of the aforesaid Liquors will not drink well. I was a long time troubled to find how to make this Drink as palatable and pleasing as it was become strong and chearing, until I put both Cyder and Spirits into a wooden Cask. The first I compleated was in a Vessel of six Gallons, into which I put two Quarts of the Sweets, and three Quarts of the Spirits of Cyder, which after it had lain two or three Months I found to be as strong and pleasing as Canary.

By adding Wormwood to Cyder-Royal as you do to Wine, you may make it as good and grateful to the Stomach, both for procuring Appetite, and causing Digestion, as the best Purl-Royal, or Wormwood Wine. Thus you may have of your own growth Cyder-Royal, Goosberry, Currant, Cherry, &c. from the size of the smallest Wines to the strength and goodness of the best Canary, suitable to all Seasons of the Year, and to the Constitutions of all Persons, and Humours of all Palates, and agreeable to all Ages, from Children of twelve Months old, to the heighth of Old Age.

This Cyder-Royal, or New-Wine thus prepar'd, may be kept in the Cask two or three Years, and be better'd thereby, provided you keep the Cask full; which to do, you must observe, that in two Months time the Liquor will waste a Quart more or less, as the Vessel is bigger or lesser, which you ought to fill up again with Liquor of the same strength, or if stronger the better: And by this means it may be kept; and grow better and better some Years without putting into it (as some are said to do into their Liquors) Stum, or other unwholsome Ingredients. And,

Suppose by keeping Cyder-Royal too long it should become unpleasant, and as unfit to Bottle as *Old Hockamore*, take but one Hogshead of that, and one of tart new Cyder, and before the latter be quite clear or fine, mix them together in two other Hogsheads well perfum'd, and add of Spirits and Sweets a due proportion to the quantity of your new Cyder: Suppose it be in the Month of *October* or *November*, you may be sure to have it full as good, if not better, than ever it was, and a most excellent Cyder-Royal to drink, or to bottle, by or before *Christmas*; and your new Cyder cannot be made half so good by that time of the Year.

As to the Objections made against this sort of Cyder, and the other particulars relating to it, I shall refer

you

you to a small Treatise of Sir *Jonas Moore* on this Subject.

Some commend very much the boiling of Cyder, as what gives a mighty strength to it; but it is much better for some sort of Fruits than others. The best sort of Cyder for boiling being what is made of Pippins, Harvey-Apple, the Bitter Sweet (a *Dorsetshire* Apple) whose Juice is much mended by boiling, especially when kept to two Years old: The way of doing which is, to boil it as soon as it is pressed; for if it ferments, the boiling will cause the Spirits to fly away instead of strengthning it, strain the Juice as it comes from the Press, and in boiling of it let it continually be scum'd, and observe the colour of it as it boils; so as not to boil it longer than till it comes to the colour of Small-beer: And as soon as it is cold, tunn it, leaving only a small Vent in the Cask, the rest being close stopped; and when it begins to bubble out of the Vent, bottle it, only make it not of Fruit that hath been gather'd long.

But as Cyder is apt to contract an ill flavour from the Vessel it is boiled in, it is best to boil it in Tin or an Earthen-pot that is wide and open at the top, for the more expeditious wasting of the aqueous and phlegmatick part of the Liquor.

Of Mixtures with Cyder.

Tho' Cyder needs not any, it is yet a very proper Vehicle to transfer the vertue of any Aromatick or Medicinal thing, such as Ginger, Juniper, &c. The Berries dried, six or eight put in each Bottle, or proportionably in the Cask, is very good: But this is not so palatable as wholesome.

Ginger renders it brisk, and corrects its Windiness; dried Rosemary, Wormwood, Juice of *Corinths*, &c. whereof a few drops tinge and add a pleasant quickness; Juice of Mulberries, Blackberries, and (preferable to all) Elder-berries pressed among the Apples; or if to the Juice you add Clove-Gilliflowers dried and macerated, both for Tincture and Flavour, 'tis an excellent Cordial. Thus may the Vertues of any other

things

things be extracted: Some Pump *Malaga* Raisins; putting Milk to them, and letting them percolate thro' an *Hippocrates's* Sleeve; a small quantity of this, with a spoonful or two of Syrup of Clove-gilliflowers to each Bottle, makes an incomparable Drink.

Honey or Sugar mix'd with some Spices, and added to Cyder that is flat, revives it much, let the proportion be more or less, according to the quantity of your Cyder.

Mixture of Fruits is of great Advantage to Cyder, the meanest Apples mix'd being esteemed to make as good Cyder as the best alone, always observing, that they be of equal Ripeness. But the best mixture, Mr. *Worlidge* says, is Red-streak and Golden-Rennets together. The Bartlet Queening mixed with Golden Pippins makes an excellent Cyder.

If you intend a mixture of Water to your Cyder, let it be done in the grinding, and it will better incorporate with the Cyder, than if put in afterwards.

Some Cyder will bear a mixture of Water without injury to its Preservation, others will not; therefore be not over-hasty with too much at once, till you understand the Nature of the Fruit.

How to make Water-Cyder.

Boil'd Water suffer'd to stand (till cool'd) is best for this use, as being more defecated. This small Beveridge, or Cyderkin and Puree (as it is called) is made for the common drinking of Servants, &c. supplying the place of Small-beer, and to many more agreeable. It is made by putting the More into a Fat, adding what quantity of Water you please, namely, about half the quantity of press'd Cyder, or more; as you desire it stronger or smaller. Note, that the Water should stand 48 hours on it before you press it; when 'tis press'd, tun it up immediately, and it will be fit to drink in a few days, by clarifying of it self. It is fortified by adding to it the Lees or Settling of better Cyder, putting it to the Pulp before Pressure, or by some superfluous Cyder which your Vessels could not contain, or by grinding some fallen and refuse Apples.

Cyderkin will be made to keep long by being boiled after Pressure with such a proportion of Hops as is usually added to Beer; in which case, you need not to boil the Water before.

Some put in Ginger, *Jamaica* Pepper, and Bay Leaves, instead of Hops; which doth very well.

Some Observations relating to Cyder.

'Tis not good to grind or beat Apples in Stone Troughs, because it bruises the Kernel and Stalks, which give an ill savour to the Cyder.

Let not your Apples be ground too small, so as that too much of the Pulp may pass with the Liquor, it being good to strain it from the gross Particles of the Apples before you put it into the Fat.

Fining of Cyder.

Upon which account 'tis that the Juice of ripe pulpy Apples, as Pippins, Rennetings, &c. that are of a syrupy tenacious nature, do detain in them more of the dispers'd Particles of the Fruit that by the Pressure comes out with the Liquor; which Particles, or flying Lee being part of the flesh or body of the Apple, is (equally with the Apple it self when bruised) subject to Putrefaction, by which means by degrees the Cyder becomes hard or acid; whereas the Red-streaks, Gennet-moil, &c. that more easily part from their Liquor without the adhesion of so much of the Pulp, are not so subject to reiterated Fermentation, nor to Acidity, as the other sorts.

For Wine, Ale, Beer, and other Liquors, according as they tend more or less to Acidity, become clearer by the Precipitation of the gross Lees, which being subject (as I said before) to Putrefaction, according as the corrupt Particles are more or less in it, the Liquor becomes so much the sooner or later Vinegar.

As for instance, in Beer, which when 'tis design'd for Vinegar is never fermented, nor the Faeces precipitated, as 'tis when preserv'd for drinking.

And

And therefore if you intend your Cyder shall retain its full strength, abstract it from the gross Parts, as I said before.

Also Cyder made of green immature Fruit will not fine kindly; and when it doth, it abides not long good; but suddenly becomes eager.

Generally the Cyder that is longest in Fining is the strongest and most lasting, especially if the Fruit hath been kept some time.

But Cyder, or any other Liquor, will be much longer in clearing in mild moist Weather, than cold dry Weather or Frost. And therefore, the best time to make Cyder is in cold Weather; Frost being apt something to check the freting or overworking of it.

If your Cyder or other Liquor doth not fine, you *Ising-glass;* may take of Water-Glue, or Ising-glass, as 'tis commonly called, about the Proportion of three or four Ounces to a Hogshead, beat it thin on some Anvil, or Iron Wedge, and cut it in small pieces, laying of it in steep in White-wine (which will more easily dissolve it self than any other Liquor except Spirits) let it lie therein all Night, the next day heat it some time over a gentle Fire till you find it well dissolved, then take a part of your Cyder, as about 1 Gallon to 20 Gallons, in which boil your dissolved Glue, and put it into the whole Mass of your Liquor, stirring of it well, and stopping of it close, so let it stand to ferment eight or ten hours as you please; during which time the Glue being dispers'd through the whole Mass of the Liquor, it will precipitate the Lee. When you observe it hath done working, you may draw it out gently at a Tap below the Scum, or you may first gently take off the Scum, as you please: Or you may do it thus, Steep your Ising-glass in White-wine, enough to cover it; after 24 hours beat the Ising-glass to pieces, and add more Wine to it, and four times a day squeeze it to a Jelly, and as it thickens add more Wine to it: When 'tis reduc'd to a perfect Jelly, take about a Pint or Quart to a Hogshead, and add it to three or four Gal-

lons

lent of the Cyder you intend to fine, and mix well with the Jelly; putting of it into your Vessel of Cyder, stir it well with a Staff. This cold way is much better than the other; for boiling part of the Cyder makes it apt to decay the sooner.

This Liquor, thus gently purified, you may in a full Vessel preserve a long time, or draw it and bottle it in a few days, there being no more Lee in it than is necessary for its Preservation.

If Cyder be fine, the sooner you draw it off the Lee the better, lest any change of Weather should alter it.

When your Cyder begins to look white on the top, draw it off into another Vessel, but not hastily; let your Tap so that it may drop out by degrees.

I am told that Figs put into Cyder improves it very much.

A friend of mine had a hogshead of Cyder that proved sowre, and added some water to it, and brewed it as you brew other Liquor, and it made (as he told me) excellent Drink.

Casks for Cyder. A great occasion of spoiling of much Cyder, is the not having of good Casks for it, it being a Liquor very apt to attract any ill favour from the Vessel; and therefore new Casks very much affect the Cyder, with an ill favour and deep colour: Wherefore if you cannot obtain Wine casks, which are the best, scald your Casks with Water wherein a good quantity of Apple-pomice hath been boil'd, before you put your Cyder into them.

Put not Cyder into a Vessel wherein Strong-beer, or Ale hath lately been, especially Strong-beer; for it gives a very rank unpleasant Taste to Cyder, so doth a Cyder-Vessel to Beer; therefore a Small-beer Vessel is to be preferr'd.

If your Vessel be tainted with any ill favour, boil an Ounce of Pepper in Water, enough to fill the Vessel; put it in scalding hot, and let it stand therein two or three Days.

Or take some quick Lime, and put into the Cask, which sleck with Water, keeping of it close stopp'd,

tumble

tumble it up and down, till the Commotion cease and be sure your Cask be dry before you put your Cyder into it. But the most effectual Cure is to take them to pieces, and pare away the Film that is on the inside, and when air'd set them together again.

If your Vessel, before the Cyder is turn'd up into it, be fum'd with Sulphur, it much conduces to the *Fuming a Cask* Preservation of this or any other sort of Liquor; which may be done by dipping of a Rag in melted Brimstone, and by a wire letting of it down into the Cask, and fir'd so as to fill it full of Smoak: Upon which pour in your Liquor, which will give it no ill Taste, and is an excellent preserver of Health, as well as of the Liquor, and will much help to fine it. Or you may give your Cask a fine scent by taking of Brimstone four ounces, of burnt Alum one ounce, of Aqua Vitæ two ounces, melt these together in an Earthen Pan, on hot Coals, and dip therein pieces of new Canvas, and instantly sprinkle thereon Powder of Nutmegs, Cloves, Coriander and Annifeeds: Set this Canvas on fire, and let it burn to fume the Vessel.

But the better way for this Operation is to have a little Earthen-pot to burn the Brimstone in, to the Cover of which have one Pipe to go into the Cask, and another to come into your Mouth, with which you may blow the Fume into the Cask.

After you have closed up your Bung, you ought to leave open a small Vent-hole, or but loosly put in the Peg, lest the Cyder break your Cask: In case the Liquor be unquiet, you may sometimes try the state of your Cyder by often opening of the Vent.

Cyder pressed from pulpy, or thorough-ripe, or mellow Fruit, having lain long in hoard, is not so apt to emit its Spirits as the other, and so is more easily preferr'd.

The upright Cask is most commended for Cyder, because 'tis apt to contract a Skin or Cream on the top, which helps much to its Preservation, and is in other Forms broken by the sinking of the Liquor; but in

in this 'tis kept whole, which occasions the briskness
of the Drink to the last.

If Cyder do not work well, put a small quantity of
Lime to it, and it will cause it to ferment, not only
by reason of its warmth, but of the quick Salt that is
in it: The Powder of calcined Flints, Alabaster, white
Marble, Roch-alom, &c. is also good ; but then the
Cyder must be drank or bottled quickly.

The Shavings or Chips of Fir, Oak, or Beech, are
great Promoters of Purification or Fermentation ; and
therefore a new Cask many times occasions Cyder to
ferment too much.

Ginger accelerateth the Maturation of Cyder, and
gives it more brisk Spirits, helpeth Fermentation,
promoteth its Duration, and corrects its Windiness.

If Cyder hath any ill savour or taste from the Vessel,
or any other cause, a little Mustard-seed (ground with
some of the Cyder, and put to it) will help it.

Deadness or Flatness in Cyder, which is often occasion'd by the too free admission of Air into the Vessel,
for want of right stopping, is remedied by grinding a
small parcel of Apples, and putting of them into it,
stopping of it up close, only you must sometimes open
the Vent that it force not the Vessel ; but then you
must draw it off in a few Days, either into Bottles or
another Vessel, lest the Murc corrupt the whole Mass ;
which may also be prevented in case you press your
Apples, and only put in the Juice. The same may be
done in Bottles, by adding about a Spoonful or two of
new Must to each Bottle of dead Cyder, and stopping
of it again. Cyder that is dead or flat will oftentimes
revive again of it self, if close stopp'd, upon the Re-
volution of the Year, and approaching Summer.

But Cyder that hath acquir'd a Deadness or Flatness
by being kept in a Beer or Ale Vessel, is not to be re-
viv'd again.

Wheat unground, about a Gallon to a Hogshead, or
Leaven or Mustard, ground with some part of the Cyder, or rather with Sack, and put into the Cask, is
us'd

us'd either to preserve Cyder, or to recover it when acid; but the best Addition to preserve it, is a Decoction of Raisins of the Sun, or the new Lees of *Spanish* Wine.

Wheat boiled till it begin to break, and when cold put into the Cyder, but not in too great quantity, and stirr'd well, will help it much; the like doth Cinnamon: The Vessel must be kept close stopp'd.

But there is a difference between sharp or acid Cyder, and a Cyder that is eager or turn'd: The first hath its Spirits free and volatile, and may easily be retriev'd by a small addition of new Spirits, or some edulcorating matter; but the latter hath some of its Spirits wasted and decay'd, so that all Additions are but vain Attempts to recover it.

Thick Cyder may, by a second Fermentation, be made good and clear; but acid Cyder is rarely recover'd, except it be in cold Weather; and then, tho' it be a litile pricked, it will recover when warm Weather comes in.

Mustard beat with Sack, and put to boil'd Cyder, preserves it, and gives it good Spirits.

Two or three Eggs put into a Hogshead of Cyder that is sharp, sometimes lenifies it; and two or three rotten Apples will sometimes clarifie thick Cyder.

Wheaten-bran cast into a Cask after Fermentation thickens the Coat or Cream, and much conduces to its Preservation.

Bottling of Cyder is the only way to preserve it long; *Bottling* and it may be bottled two or three Days after 'tis well *of Cyder.* settled, and before it hath throughly fermented, if it be for present drinking; or you may bottle it in *March* following, which is the best time.

Bottles may be kept all Summer in cold Fountains, provided you pitch the Corks to prevent their rotting; or in Cellars, in Sand, if they are well cork'd: The longer they are kept the better, if the Cyder be good, and have a body,

After

After Cyder hath been bottled a Week (if 'tis new Cyder, else at the time of bottling) you may put into each Bottle a piece of white Sugar as big as a Nutmeg, this will make it brisk ; but if the Cyder be to keep long, it will be apt to make it turn sour.

If your Bottles are in danger of the Frost, cover them with Straw ; and about *April* put them into the coldest Repositories.

If your Bottles are musty, boil them in a Vessel of Water, putting of them in whilst the Water is cold to prevent their cracking ; and then set them on Straw, and not on a cold Floor, when you take them out.

When your Cyder is thus bottled, if it were new at the bottling, and not absolutely fine, it is good to let the Bottles stand a while before your stop them close, or else open the Corks two or three Days after to give the Cyder Air, which will prevent the breaking of the Bottles against the next turning of the Wind into the South.

The meaner Cyder is more apt to break the Bottles than the richer, being of a more eager Nature, and the Spirits more apt to fly, having not so solid a body to detain them as the rich Cyders ; and observe, that when any of the Bottles break thro' the Fermentation of the Cyder, to open your Corks, and give them vent, and stop them up again a while after, lest you lose many for want of this Caution.

Great care is to be had in choosing good Corks, much good Liquor being absolutely spoil'd through the only defect of the Cork ; therefore some much commend Glass-stopples.

If the Corks are steep'd in scalding Water a-while before you use them, they will comply better with the Mouth of the Bottle than if forc'd in dry, also the moisture of the Cork doth much help it to keep in the Spirits.

Therefore the laying of your Bottles side-ways where your Liquor is very fine, so as that the raising of them may not disturb the Settling, nor the Lee be-

gat

get any new Fermentation in them, is a great advantage to any Liquor.

Chap. VI. *Of Perry.*

THE next Liquor in esteem after Cyder, is Perry, the ordering of which being much the same with that of Cyder, I need not say much of it; only you must observe, not to let your Pears be over-ripe before you grind them, because of their Pulpiness, which makes them not easily to part with their Juice: And with some sorts of Pears, the mixing of a few Crabs in the grinding of them, in proportion to the sweetness of the Pear, is of great advantage to it; making some sorts of Perry equal to that of Red-streak Cyder.

Chap. VII. *Of making other sorts of Wines or Drinks of Fruits.*

BEsides Cyder and Perry, there are many other Drinks prepar'd out of our *British* Fruits, as of Cherries, &c. which are a Fruit as easily propagated as any, nor is there any Fruit that commonly bears better, nor that yields more Juice, which mix'd with the richest *Spanish* Wine, makes a very fine Drink, by the addition of some Sugar to it. *Cherry Wine.*

Or the Juice it self, press'd out and mix'd with a due quantity of Sugar and Water, makes a very rich Wine, that is very comfortable to the Stomach and Nerves.

The Plumb is also easily propagated; and no doubt but some of the more juicy sort of them, especially the Damascen, would yield an excellent Liquor, but scarcely durable unless boiled with Sugar, and well purified, or else the Sugar boil'd before-hand in Water, and then added. The Juice of the Plumb being of a thick Substance, will easily bear Dilution. This is easily experimented where Plumbs are in great plenty. *Plumb Wine.*

The Red *Dutch* Currant, or *Corinth*, yields a very rich and well colour'd Juice, and a vinous Liquor, which

which is to be diluted with an equal quantity of Water boil'd with refin'd Sugar, about the proportion of one pound to a Gallon of your Wine (when mixed with the Water); and after the Water and Sugar so boil'd together is cold, then mix it with the Juice of the Currants, and purifie it with Ising-glass dissolved in part of the same Liquor, or in White-wine, as is before directed for the purifying of Cyder, after the rate of an ounce to eight or ten Gallons, but boil it not in a Brass Vessel for the Reasons beforemention'd: This will raise a great Scum on it of a great thickness, and leave your Wine indifferent clear, which you may draw out either at a Tap, or by your Siphon into a Barrel, where it will finish its Fermentation, and in three Weeks or a Month become so pure and limpid, that you may bottle it with a piece of Loaf-sugar in each Bottle in bigness according to your Discretion, which will not only abate its quick Acidity, that it may as yet retain, but make it brisk and lively.

At the time you bottle it, and for some time after, it will taste a little Sweet-sowre, from the Sugar and from the Currant; but after it hath stood in the Bottles six or eight Weeks, it will be so well united, that it will be a delicate, palatable, rich Wine, transparent as the Ruby, of a full body, and in a Refrigeratory very durable; and the longer you keep it, the more vinous will your Liquor be.

Let your Currants hang on the Trees until they are thorough ripe, which is long after they are become red, to digest and mature their Juice, that it need not that large addition of Sugar that otherwise it would do in case the Fruit had been gather'd when they first seem'd to be ripe, as is vulgarly us'd, and the common Receipts direct: Also it makes the Liquor more Spirituous and Vinous, and more capable of Duration than otherwise it would be if the Fruit had not receiv'd so great a share of the Sun.

The Gooseberry-Tree, being one of the greatest Fruit-bearing Shrubs, yields a pleasant Fruit, which
although

although somewhat luscious, yet by reason of its gross Lee, whereof it is full, it is apt to become acid, unless a proportion of Water sweetned with Sugar (but not with so much as the other acid Liquors) be added unto it. This Liquor, of any other, will not bear a Decoction, because it will debase its Colour, and make it brown.

There is no Shrub yields a more pleasant Fruit than the Rasberry Tree, which is rather a Weed than a Tree, never living two Years together above-ground. Nor is there any Fruit yields a sweeter and more pleasant Juice than this, which being extracted, serves not only to add a Flavour to most other Wines or Liquors, by a small addition of Water and Sugar boiled together, and when cold added to this Juice, and purified, makes one of the most pleasant Drinks in the World. The same way Apricot, and the Wines of other Fruits may be made.

Having given you a taste of most Wines made by pressure of the Juices out of the Fruits, you may also divert your self with the Blood of Grapes, or any other of the before-mentioned limpid Liquors tinged with the spirituous Flavour of other Fruits, that cannot so easily and liberally afford you their Juices; as, of the Apricot, which steeped in Wine gives the very taste of the Fruit; also Clove-gilliflowers, or sweet-scented Flowers do the like. You may also make Experiment of some sort of Peaches, Nectarines, &c. what Effect they will have upon those sorts of Drinks.

Chap. VIII. *Of the making of some other Drinks or Wines, usually drank in this Island.*

THere are several other pleasant, wholesome, and necessary Drinks made of Trees, Leaves, Grains, and other things, besides such Drinks or Liquors as are commonly made of the Fruits of Trees or Shrubs.

As Mead, or Hydromel, that is prepared out of Honey, being one of the most pleasant and universal Drinks the Northern part of *Europe* affords; and one of the most ancient Drinks of the Northern parts. Honey being to be had from the Southerly parts of *Spain* and *Italy*, &c. to the Arctick Circle or Frozen Zone.

Those that lived formerly in the more Southern parts (as *Pliny* reports) made a drink compounded of Honey and tart Wine, which they termed Melitites, by the addition of a Gallon of Honey to five Gallons of their Wine: 'Tis also an excellent Ingredient mixed with Cyder.

In *Swedeland, Mufcovia*, and as far as the *Caspian* Sea, they make great Account of this Drink; to which Liquor they give a great advantage by the addition of the Juice of Rasberries, Strawberries, Mulberries, and Cherries.

They also keep Rasberries in *Aqua-vitæ*, twenty-four Hours, and add it to their Hydromel, which is a great Amendment of it.

There are very great variety of Receipts for the making of Metheglin or Hydromel; but the best that I have met with, is to

Take twelve Gallons of Water, and put in the Whites of six Eggs, mix them well with the Water, and twenty Pounds of Honey; boil it an Hour, and when boiled add Cinamon, Ginger, Mace, Cloves, and a little Rosemary; and when 'tis cold put a spoonful of Yeast to it, and tun it up, keeping of it filled up as it works; when it hath done working, stop it up close, and when fine, bottle it.

But the finest Mead is that made of what they call Live Honey, which is what naturally runs from the Combs (but that from swarms of the same Year is the best.) and add so much Honey to clear Spring-water, as that when the Honey is diffolved throughly, an Egg will not sink to the bottom, but eafily swim up and down in it. Boil this Liquor in a Copper Veffel for

about

about an Hour or more, and by that time the Egg will swim above the Liquor, about the Breadth of a Groat, then let it cool; the next Morning you may barrel it up, adding to the Proportion of fifteen Gallons an Ounce of Ginger, half an Ounce of Cinamon, Cloves, and Mace, of each an Ounce, all grosly beaten; for if you beat them fine they will always float in your Mead, and make it foul; and if you put them in whilst it is hot, the Species will lose their Spirits. You may also add a Spoonful of Yeast at the Bung-hole to increase its Fermentation; but let it not stand too cold at first, that being a principal Impediment to its Fermentation; as soon as it hath done working, stop it up close, and after a Month bottle it; and the longer 'tis kept, the better it will be.

By the floating of the Egg you may judge of its Strength, and you may make it more or less strong as you please, by adding of more Honey, or more Water; and by long boiling of it, it is made more pleasant and durable.

The Sycamore and Wallnut-tree are said to yield an excellent Juice; but that which we have the most Experience of, is the Birch-tree.

The Juice of which may be extracted in very great Quantities; where those Trees are plenty, many Gallons in a Day may be gathered from the Boughs of the Trees by cutting them off, leaving their Ends fit to go into the Mouths of a Bottle, and so by hanging many Bottles on several Boughs, the Liquor will distill into them very plentifully.

The Season for this Work is from the end of *February* to the end of *March*, whilst the Sap rises, and before the Leaves shoot out from the Tree; for when the Sap is forward, and the Leaves begin to appear, the Juice by a long Digestion in the Branch grows thick and coloured, which before was thin and limpid: The Sap also distils not in cold Weather whilst the North and East Winds blow, nor in the Night time,

but

but very well and freely, when the South or West Winds blow, or the Sun shines warm.

That Liquor is best that proceeds from the Branches, having had a longer time in the Tree, and thereby better digested, and acquiring more of its Flavour than if it had been extracted from the Trunk.

When you draw out the Sap of Trees for any use, and desire to have a quantity, what you gather first put into Glasses or other fit Vessel, and set it in the Sun 'till the rest be ready, and put into it a hard Toast of Rye-bread cut thin, and it will cause it to ferment when it works, take out the Bread, and bottle the Liquor.

In Birch Trees the Sap rises out of the least Twigs or Fibres of the Roots, but from Branches and Roots that bend downwards will Issue more Sap than from those that are erect, and a Branch cut quite off will yield Sap.

Thus many Hogsheads may soon be obtained. Poor People (where Trees are plenty) will draw it for two or three Pence the Gallon : to every Gallon whereof add a pound of refined Sugar, and boil it about a quarter, or half an Hour, then set it to cool, and add a very little Yeast to it, and it will ferment, and thereby purge it self from that little Dross that is in the Sugar and Liquor. Put it into a Barrel, and add thereto a small proportion of Cinamon and Mace bruised, about half an ounce of both to ten Gallons, stop it very close, and about a Month after bottle it. Its Spirits are so Volatile, that they are apt to break the Bottles unless placed in a cool place, without which Conveniency it will not keep long.

Instead of every pound of Sugar, if you add a quart of Honey, and boil it as before, and adding Spice to it, and fermenting of it as you do Mead, it makes an admirable Drink, both pleasant and medicinal.

Ale also brewed of this Juice or Sap is esteemed very wholesome.

Mum

Mum being become a common Drink, and being very wholesome, and what may be made of our own Product, I should hope it might be made a home Commodity instead of a Foreign : And therefore, for the Encouragement of it, I shall give you the Receipt, as recorded in the Town house in *Brunswick*; which is thus :

Take sixty-three Gallons of Water that hath been boiled to the Consumption of a third part; brew it according to Art, with seven Bushels of Wheat Malt, one Bushel of Oatmeal, one Bushel of ground Beans; and when 'tis tunned, let not the Hogshead be too full at first; and when it begins to work, put into it, of the inner Rind of Fir three Pounds, tops of Fir and Birch one pound, *Carduus Benedictus* three handfuls, Flowers of Rosa solis a handful or two, Burnet, Betony, Marjoram, Avens, Penny-royal, Wild-thyme, of each a handful and a half, of Elder Flowers two handfuls or more, Seeds of Cardamom bruised three ounces, Barberries bruised one ounce. Put the Herbs and Seeds into the Vessel when the Liquor hath wrought a while; and after they are added, let the Liquor work over the Vessel as little as may be. Fill it up at last, and when 'tis stopped, put into the Hogshead ten new-laid Eggs, unbroken or cracked; stop it up close, and drink it at two Years end.

But our *English* Brewers use Cardamom, Ginger, and Sassafras, which serves instead of the inner Rind of Fir, also Wallnut-Rinds, Madder, Red Sanders, and Enula Campana; and some make it of Strong-Beer, and Spruce-Beer; and where 'tis designed mostly for its Physical Virtues, some add Water-cresses, Brook-lime, and Wild-parsley, with six handfuls of Horse-radish rasped to every Hogshead, according to what their Inclinations and Fancy most lead them.

IF this Month prove cold, 'tis feaſonable to kill the Ver-min and Weeds, and to mellow the Ground : And is the chief time to plow up Lays, to fallow the Ground you intend for Peas, to water Meadows and Paſtures, to drain Arable Grounds where you intend to ſow Peas, Beans, Oats, or Bar-ley, to lay Dung on heaps to carry on the Land in froſty Weather, to make Hedges Ditches, to cut Ant bills, and to fill up the Holes in Meadows and Paſture-ground, to ga-ther Stones, &c.

Rear Calves, Pigs, &c. have eſpecial Care of Ewes and Lambs ; Houſe Calves ; geld young Cattle ſoon after they are fallen ; feed Doves ; repair Dove-coats.

Plant Timber Trees, Coppice-wood, or Hedge-wood, and alſo Quick-ſets ; cut Coppices and Hedge-rows, lop and prune greater Trees.

Work to be done in the Orchard and Kitchen Garden.

Prune Vines and forward Fruit Trees : If the Weather be open and mild, dig and trench Gardens or other Ground for Peas, Beans, &c. Againſt the Spring, dig Borders ; pre-pare your Soil or Manure ; and ſuffer no Weeds to grow on them : Uncover Roots of Trees where need is, and add ſuch Manure to them as they require, not laying of it too near the Roots : You may alſo, if the Weather prove mild, ſet Beans and Peas : As yet Roſes may be cut and removed.

Gather Pears, Cherries, and Plumb Cions for Grafts about the latter end of this Month, before the Bud ſprouts, which ſtick in the Ground for ſome time, becauſe they will take the better for being kept a ſmall time from the Tree ; and graft them the beginning of the next Month. Cleanſe Trees of Moſs, the Weather being moiſt.

Make hot Beds, and ſow therein your choice Sallets, as Chervil, Lettice, Radiſh, &c. Sow early Colliflowers : Se-cure your choice Plants and Flowers from the injuries of the Weather by Covers, Straw, or Dung. Earth up the Roots of ſuch Plants as the Froſt hath uncovered.

Set Traps to deſtroy Vermin where you ſow or have ſuch Plants or Seeds as they will injure. Take Fowls ; deſtroy ſparrows in Barns, and near them kill Bull finches, &c.

and

and in wet or hard Weather clean, mend, sharpen, and prepare your Garden-tools.

Dig up weedy Hop Gardens.   Hops.

Turn up your Bee-hives, and sprinkle them with warm Apiary.
and sweet Wort. Also you may remove Bees.

Fruits in prime, and yet lasting.

Kentish, Russet, Golden, and French Pippins, Kirton Apples.
Pippins, and Holland Pippins, John Apples, Winter Queen-
ings, Marigold, Harvy Apple, Pome-water, Golden Doucet,
Renneting, Love's Pearmain, Winter Pearmain, &c.

Winter Musk (bakes well,) Winter Norwich (excellent Pears.
baked,) Winter Burgamot, Winter Bon Chrestien, both
Mural, the great Surrein, &c.

In the Flower Garden.

If the Weather in this Month prove cold, great
Care must be taken of your Flowers ; especially such
as least endure the Cold, or that are in danger of being washed out of the Ground, or overchilled with
extream Frosts. Likewise Earth up your Flowers
with fresh Mould. Plant Anemony Roots and Ranunculus's, which will be secure without covering : Except such as you sowed in October or November for
earlier Flowers, which should be secured both from
Frosts and Rains, and about the end of the Month put
Mould about the Roots of your Auricula's that have
been uncovered by Frost.

If the Weather be extream you must be careful of
your tenderest Plants, and mind to tend your Fire in
your Conservatory or Green-house, so as to keep
them in a moderate Heat (too suddain, or too much
Heat or Cold being apt to spoil most of your Plants)
and to keep the Windows and Doors well closed and
lined with Matts, &c. to keep the Air out of the
Cracks, especially where the Orange Trees are.

Flowers in prime, or yet lasting.

The Præcoce Tulip, Winter Aconite, some sorts of Anemonies,
Black Hellebore, Winter Cyclamen, Oriental Jacinths, Brumal Hyacinth, Levantine Narcissus, Laurustinus, Mezereon, Primroses, &c.

Note, That these Flowers are more forward or backward according to the Soil, and the Situation of the Place they grow in.   This

THIS is a principal Seed Month, for ſuch as they commonly call Lenten Grain, and is uſually ſubject to much Rain or Snow, which is not unſeaſonable.

Now ſow all ſorts of Grey Peas, Fitches, Beans, and black Oats; carry out Dung, and ſpread it before the Plough, and alſo on Paſture Ground; this being the principal Month for that purpoſe.

This is the beſt time to plant Trees and Quick, as alſo to plaſh it; to ſet Willows, Plants, or Pitchers, and alſo Poplars, Oſiers, and other Aquaticks; and to ſhroud or lop Trees, or cut Coppices.

Sow Muſtard Seed, and Hemp Seed, if the Spring prove mild: Feed your Swans, and make their Neſts where the Floods cannot reach them.

Soil Meadows that you cannot overflow or water, catch Moles, and cut Mole-hills, and take great care of Ewes and Lambs where they are forward.

Work to be done in the Orchard and Kitchen Garden.

Prune, trim, and nail up Fruit Trees, and cleanſe them from Moſs and Cankers: Now is a good time to graft the more forward ſort of Fruit, if the Weather be temperate.

Do not prune your tender Wall Fruit 'till you think the hard Froſts over, though it ought to be done before the Buds and Bearers grow turgid, and mind to ſpread your Wall Trees well at the bottom.

Plant Vines, or any ſort of Fruit Trees in open Weather; trim up your Palliſado Hedges and Eſpaliers; ſet Kernels, Nuts or Stones of Fruit, and ſow other hard Seeds.

Lay Branches to take Root, or place Baskets, &c. of Earth for the Branches to paſs through. Graft in the Cleft, and ſo continue to the latter end of the next Month.

Sow Anniſe, Beans, Peas, Radiſh, Parſnips, Carrots, Potatoes, Onions, Parſly, Spinage, Corn Salleting, and other hardy Herbs or Seeds, and plant Cabbage Plants and Colliflowers in warm places; alſo Liquorice, and ſow Aſparagus if the Spring be mild. Now the Bull-finches do the moſt hurt to Fruit Trees. This is the beſt time to raiſe any thing that will grow of Slips.

Make

Make hot Beds for Melons, Cucumbers, &c. continue *Vermin Traps*, and pick up all the Snails you can find, deſtroying the Frogs and their Spawn.

This is a good time to few Fiſh ponds, and to take Fiſh, moſt Fiſh being now in Seaſon.

You may, if the Weather prove mild, plant *Hops*, and Hops. dreſs them that are out of Heart. And alſo dig up your Hop Ground, if 'tis waedy.

Half open the paſſage for Bees, and now you may remove them, but continue to feed weak Stocks.

Fruits in prime, or yet lafting.

Kentiſh, *Kirton, Ruſſet,* Holland *Pippins, Deuxans,* Apples *Winter Queening, Harvey, ſometimes Pome-water, Pomeroy, Golden Doucet, Renneting, Love's Pearmain, Winter Pearmain,* &c.

Bon Chreſtien of Winter, Winter Poppering, Little Pears. *Dogobert,*

In the Flower Garden.

If the Weather is feaſonable, Air your houſed Carnations in warm Days, and mild Showers, but ſet them in again towards Night; and ſo you may do by other Flowers that are not too tender : Sow Alaternus Seeds, Lark Spurs, &c.

Flowers in prime, or yet lafting.

Some Anemonies, Winter Aconite, Hyacinthus Stellatus, Præcoce Tulip, Perſian Iris, Leucoium bulboſum, Dens caninus, Black Hellebore, Vernal Crocus, Single Hepatica, Vernal Cyclamen Red and White, early Daffadils, the great white Arnithogals, Mezereon, large leaved Yellow Violets, &c.

H

IF this Month prove cold 'tis seasonable to check the pregnant Buds till a more safe Season; and if it prove dry, the Country-man esteems it to presage a happy born Year. You may yet prune or plant Trees, tho' 'tis of the latest of all sorts, except Winter-Greens.

Let Cattle feed no longer on Meadows that you intend to mow; have especial regard to the Fences both of Meadow, Corn and Woods and take care of Ewes and Lambs.

About the end of this Month you may begin to sow Barley, earlier in Clay than Sand; you may roll Wheat if the Weather prove dry; make an end of sowing of all sorts of Pulse. You may now shroud or lop old Trees, and fell Coppice-wood before the warm part of the Month come in.

This is the only time to raise the best Poultry.

It is now a good time to set Osier's, Willows and other Aquaticks; sow the Rye, called March-Rye and Oats, and plant Saffron, Woad, Weld, Madder and Liquorice.

In this Month and the next sow all sorts of French Grasses or new Hays, as Clover, S. Foyn, &c. also now sow Hemp, and Flax, if the Weather be temperate.

This is the principal Month in the Year for the destruction of Moles.

Work to be done in the Orchard and Kitchen Garden.

This is the chief Month for grafting, beginning with Pears, and ending with Apples; only if the Spring proves forward, be the earlier. Prune last Year's Grafts, and cut off the Heads of your budded Stocks. Now cover the Roots of all such Trees as you laid bare the preceding Winter, and remove such young Trees as you omitted before, if the Bud is not too forward.

Plant Peaches and Nectarines, but cut not off the Tap-root as you do of other Trees, because it will prejudice them.

Carry Dung into your Orchards, Gardens, &c. turn your Fruit in the Room where it lies, but open not yet your Windows. Smoak your Orchards.

Top your Rose-trees near a Leaf-bud, and prune off the dead and wither'd Branches, keeping of them to a single
Stem.

Stem. Slip and set *Sage, Rosemary, Thyme, Lavender,* &c.

You may now transplant most sorts of *Garden-herbs, Sweet-herbs,* and *Summer-Flowers.* Now is the best time to make *Hot-beds* for *Cucumbers, Melons,* &c.

Now sow Alisander, Basil, Beets, Borrage, Bugloss, Cabbage, Carrots, Chervil, Cresses, Endive, Fennil, Garlick, Leeks, Lettice, Marigolds, Marjoram, Onions, Orach, Parsnips, Parsley, Peas, Purslain, Radish, Sellery, Smallage, Spinage, Skirrets, Sorrel, Succory, Turneps, Tobacco, &c. and *Samphire* to replant in May, which will grow well of *French* Seed.

About the middle of this Month dress and string *Strawberries,* uncover *Asparagus* Beds, and dig about them. You may also now transplant *Asparagus* Roots to make new Beds. Slip and plant *Artichoaks* and *Liquorice.*

Stake and bind up weak Plants against the Wind: Sow *Pinks, Carnations,* &c. In this Month sow *Pine-kernels,* and the Seeds of all *Winter-greens.*

Plant all *Garden-herbs* and *Flowers* that have fibrous Roots. Sow choice *Flowers,* that are not natural for our Clime in hot Beds this Month.

You may now plant *Hops:* This is a very seasonable time to dress them. Now the Bees sit, keep them close Night and Morning: If the Weather prove ill, you may yet remove *Bees.*

Fruits in prime, or yet lasting.

Golden Ducket (*Doucet*), *Pippins, Rennetings, Lasie Pearmains, Winter Pearmains,* John *Apples,* &c.
Later Bon Chrestien, Double-blossom Pear, &c.

MARCH

MARCH.

In the Flower-Garden.

STake and Bind up your Weak Plants and Flowers, Plant Box, Sow Pinks, Sweet-Williams, Carnations, Pine Kernels, Firz Seed, Bayes, Alaternus, Phillyrea, and most Perennial Greens, &c. or you may stay somewhat later. Sow Auricula-Seeds in Pots or Cases in good Earth somewhat Loamy.

Plant some Anemony Roots to bear late and successively, and if the Season be dry, Water them once in two or three Days. Transplant Ranunculus's and Fibrous rooted Flowers, as Primroses, Hepatica's, Auricula's, Camomil, Hyacinth, Tuberose, Matricaria, Gentianella, Hellebore, and other Summer Flowers; set Leucoium ; slip Wall-Flowers, Lupines, Convolvulus's, Spanish or ordinary Jessamine, and prune Pine and Fir Trees, &c.

Towards the Middle or latter end of *March*, sow on your Hot-Beds such Plants as are late bearing in our Climate ; as Balsamine, and Balsamum mas, Pomum Amoris, Datura, Æthiopic Apples, some choice Amaranthus's, Dactyls, Geraniums, Hedysarum Clypeatum, humble and sensitive Plants, Lentiscus, Myrtle Berries, Capsicum Indicum, Canna Indica, Flos Africanus, Mirabile Peruian, Nasturtium Ind. Indian Phaseoli volubilis, Myrrh, Carobs, Matococ, &c.

About the end of this Month carry into the Shade such Auricula Seedlings or Plants, as are for their choiceness preserv'd in Pots. Transplant Carnation Seedlings, Earth up your Layers, cut off the infected Leaves.

Cover your choice Tulips, and take the like care of your best Anemonies, Chamæ Iris, Auricula's, early Cyclamen, Brumal Jacinth, &c. Cover with Straw or Pease-Haum your Seedlings of Fir, Pine, Phillyrea, Bayes,

Bayes, Cyprus, and all other Winter Greens, till they have paffed 2 or 3 Years in the Nurfery, and are fit to be tranfplanted; which Rules fhould be obferved in all extream Weathers during the Winter, minding to uncover them in good Weather, and in fharp Winds neither fow nor tranfplant. Sow Stock-Gilliflower Seeds in the full of the Moon.

You may now towards the end of the Month, fet Oranges, Limons, Myrtles, Oleanders, Dates, Lentifcs, Aloes, Amomums, and other tender Trees and Plants in the Portico, and open the Windows and Doors of your Confervatory or Green-houfe, to acquaint 'em gradually with the Air, but truft not too confidently to the Night. Now alfo is the Seafon to raife Stocks, to bud the Orange and Limons on, and to tranfplant fome of the hardieft Ever-Greens, efpecially if the Weather be moift and temperate.

Flowers in prime and yet lafting.

Anemonies, Winter Aconite, Auricula's, Arbor Judæ, Brumal Crocus of all Colours, Chelidonium, Crown Imperial, Spring Cyclamen, Dens Caninus, Fritillaria, Grape-Flower, Black and White Hellebore, fingle and double Hepatica, Hermodactyls, Hyacinth, Perfian Iris, Chamæ Iris, Tuberous Iris, Junquils, Spanifh Junquils, Leucoion, Dutch Mezereon, Narciffus, Ornithogalum max. Primrofes, Rubus odoratus, Præcoce Tulips, Violets, Dutch Yellow Violets, Zeboin, &c.

A Dry

A Dry Season in this Month is best to sow Barley and
White Oats in, to prevent Weeds, and likewise to
fallow in.

Fell the Timber you intend to hew, if the Spring be forward: Cleanse and rid your Coppices, and preserve them from
Cattle: Keep Geese and Swine out of Commons and Pastures, and Water new-planted Trees, if the Weather prove dry.

Pick up Stones in the new-sown Land: Sow Hemp and
Flax.

Cleanse Ditches, and get in your Manure that lies in
Streets or Lanes, or lay it on heaps.

Set Osiers, Willows, and other Aquaticks, before they are
too forward.

You may throughout this Month sow Clover-grass, St.
Foin, and all French and other Grasses or Hays; and plant
Madder, and be selling of your Winter-fed Cattle.

Work to be done in the Orchard and Kitchen Garden.

You may yet graft some sorts of Fruit in the Stock the beginning of this Month.

Now sow all sorts of Garden-seeds in dry Weather, and
plant all sorts of Garden-herbs in wet Weather.

Plant Cucumbers, Melons, Artichoaks, and Madder; and
sow such tender Seeds as could not abide the harder Frost.
Set French Beans, gather up Worms and Snails after Even-
ing-showers, and early in the Morning.

Sow Turneps, to have them early, and your annual Flowers
that come of Seeds, that you may have Flowers all the Summer; and transplant such Flowers with fibrous Roots as you
left unremoved in March. Sow also the Seeds of Winter-
greens.

Now bring forth your tender Plants you preserved in your
Conservatory, except the Orange-tree, which may remain
till May.

Smoak your Orchard with Straw towards the Evening.

Transplant and remove your tender Shrubs, as Jessamines,
Myrtles, Oleanders, &c. Towards the end of this Month,
also in mild Weather, clip Phillyrea and other tonsil Shrubs,
and transplant any sort of Winter-greens.

Hop gar
den

Plant Hops, and Pole them the beginning of April, and
bind them to the Poles.

Open

Open the Doors of the Bee-bives, for now they hatch, that they Apiary. *may reap the benefit of the flowry Spring, and be careful of them.*

Fruits in prime, or yet lasting.

Pippins, Deuxans, Westberry Apple, Russeting, Gilliflower, Apples. *Flat Rennet,* &c.

Lazar Ben chreftien, Oak-Pear, Double Bloffom, &c. Pears.

About the beginning of this Month fow fweet Marjoram, Hyfop, Thyme, Scurvy-grafs, Bafil, Winter Savory and tender Plants ; fow alfo Purflain, Colliflowers, Lettuce, Radifh, Carrots ; plant Artichoak Slips ; fow Turneps ; fet French Beans ; flip Lavender, Sage, Pennyroyal, Rofemary, Lavender, &c.

In the Flower Garden.

Sow Divers Annuals to have Flowers all Summer, as Belvidet, Digitalis, Delphinium, Cyanus of all forts, Columbines which renew every four or five Years, Candy Tufts, Marigolds, Mufcipula Medica, Hollyhocs, Garden Panfey, Scabious, Scorpoides, Tufts, &c.

Tranfplant fuch Fibrous Roots as were not done laft Month, as Violets, Hepatica, Primrofes, Hellebore, Matricaria, &c. Sow Pinks, Carnations, and trim off the dead rotten Leaves and old Roots. Sow Sweet-Williams, &c. After Rain place Auricula Seedlings in the Shade. Sow Leucoium, remove it often, and replant it in moift Weather the Spring following. Set Lupines, &c.

Sow alfo Pine Kernels, Fir Seed, Phillyrea, Alaternus, and moft perennial Greens. Vide Sept.

Take out your Tuberofe, parting the off Setts, and plant them in natural Earth, under which is a Layer of rich Mould. Set your Pots in a hot Bed temperately Warm, and give them no Water till they Spring, and then fet them under a South Wall in dry Weather ; Water them freely. You may treat the Narciffus of Japan or Guernfey Lily, and the Luca after the fame manner, only mix a little Sea-Sand with the Earth you plant them in near the Surface.

Set out your Flos Cardinalis ; flip and fet Matrums, Water Anemonies, Ranunculus's, &c.

Towards the end of this Month tranfplant your tender Shrubs, as Spanifh Jafmines, Myrtles, Oleanders, Young Oranges, Cyclamen, Pomegranate, &c. but firft let them begin to Sprout.

At this time (if the cold Winds are paft) after Showers clip Phillyrea, Alaternus, Cyprus, Box, Myrtle, Barba Jovis, &c.

Now is the Seafon to bring out in a fair Day, your choice and tender Shrubs, fuch as you durft not adventure out in March, only Orange Trees you may keep in till next Month to prevent Danger. You may now graft by approach your tender Shrubs, as Oranges, Limons, Pomegranates, Jafmines, &c.

Flowers in prime, or yet lafting.

Anemonies, Auricula urfi, Almonds, Arbor Judæ, Acanthus, Bell Flowers, Chamæ Iris, Crown Imperial, Caprifolium, Cyclamen, Cowflips, Caltha Paluftris, Cochlearia, Dens Caninus, Double Dazies, Fritilleria, Floreted Iris, Gentianella, Geranium, Musk-grape-Flower, Hypericum Frutex, Hepatica's, Starry Jacinth, Perfian Jafmine, Double-Junquils, Perfian Lilies, Ladies-Smock, Leucoium, Lilacks, Narciffus, Mufcaria reverfed, Primrofe, Pulfatilla, Parietaris, Peonies, Perfian Jafmine, Parietaria Lutea, Rofemary, Ranunculus's, R. dix Cava, Tulips, &c.

IF this Month prove dry, it gives great hopes of a full Barn; and if cold, 'tis an Omen of good Health. The pleasure of Angling is now in its splendor, especially for the Trout and Salmon.

Now wean those Lambs you intend to have the Milk of their Ewes: Forbear cutting or cropping of Trees you intend shall thrive till October: Kill Ivy.

If your Barley be too rank, now you may mow it, or feed it with Sheep or Hogs, before it be too forward: Weed Corn. In some places Barley may be sown in this Month. Begin to fold Sheep, and put your Mares to the Horses.

Now sow Buck-wheat or Brank: Sow Latter Peas: Also Hemp and Flax may yet be sown: Put fatting Cattle and milch Cows into fresh Pasture, and let nothing be wanting in the Dairy.

Weed Quicksets, drain Fenns and wet Grounds, twifallow your Land, carry out Soil or Compost, gather Stones from the Fallows, turn out the Calves to grass, over-charge not your Pastures, lest the Summer prove dry, get home your Fewel, begin to burn beat your Land, and stub or root up Goss, Furze, Broom, Bushes, &c. that you intend shall not grow again.

Sell off your Winter-fed Cattle: About the end of this Month mow Clover-grass, St. Foyn, and other French Grasses. Now leave off watering your Meadows, lest you gravel or rot your Grass.

Look well after your Sheep if this Month prove rainy lest the Rot surprize them.

Work to be done in the Orchard and Kitchen-Garden.

Plant all sorts of Winter-greens, and sow the more tender Seeds, as, Sweet-marjoram, Basil Thyme, and hot aromatick Herbs and Plants: Set Sage and Rosemary.

Smoak your Orchard as before, thin your Sailetting and other Herbs, that what you leave may thrive the better.

Cover no longer your Cucumbers, Melons, &c. excepting with Glasses: Sow Purslain, Lettuce, &c. and distil Plants for Waters, Spirits, &c.

At the end of this Month take up such Tulips as are dried in the Stalk.

Bind Hops to their Poles, and make up the Hills after Rain.

Now

Now set your Bees at full liberty, and expect a Swarm.
Fruits in prime, and yet lasting.

Pippins, Deuxans, or John Apple, Westberry Apples, Russet- Apples,
sing, Gilliflower Apples, the Maligar, &c. Codling.

Great Kairoille, Winter Bon-Chrestien, Black Pear of Wor- Pears
cester, Surrein, Double-blossom Pear, &c.

The May Cherry, Strawberries, &c.   Chetries, &c.

In the Flower Garden:

Be careful to keep under the Weeds; spread and bind down the
Branches of Arbors, and clip your Hedges and Trees. Shade your
Carnations and Gilliflowers, and plant Stock Gilliflowers in Beds;
continue watering Ranunculus's, and plant forth your Amaranthus's.

You may now bring the Orange Tree out of the Conservatory
when you see the Mulberry Tree put forth, and also transplant and
remove them. Let your Cases be filled with natural Earth taken up
half a Spit deep, under the Tuft of the best Pasture Ground that
has been Fother'd on; mix it with rotten Cow-dung, or very mellow
Soil screen'd and prepar'd sometime before. If this be too stiff sift a
little Lime amongst it, and put a few rotten Willow Sticks; cut
the two biggest Roots a little, especially at the bottom; and set your
Plants not too deep. At the bottom of the Case lay some Brush Wood
to give a free Passage to the Water, that it may not rot the Fibres. Set
them in the Shade a Fortnight; and then bring them into the Sun by
Degrees; Water them with Water in which Sheeps Dung is dissolv'd by
stirring of it, and setting it in the Sun some days before you use it; but
do not drench them too much at first, and do not let it touch the Stem.
If you cut off any Branch make a Searcloth of Rosin, Turpentine,
Bees Wax and Tallow, and put it on the Wound till it is heal'd.

Now give fresh Earth that is rich to all your housed Plants, laying of it on the Surface about four Inches deep, loosening the Earth
with a Fork. Brush and cleanse them from Dust.

Flowers in Prime or yet lasting.

Late set Anemonies, Anpodophyton, Augustifol. Asphodel, An-
tirrhinum; Blattaria, Bulbous Iris, Bellis double white and red, Bee
Flowers, Buglofs, Barba Jovis; Chamæ Iris, Cyanus, Cytifus Maranthe,
Cyclamen, Columbines, Caltha Palustris, Double Cotyledon, Cinna-
mon, Centifol. Cherrybay, Cowslips, Chalcedons, Crowfoot, Campa-
nula's; Digitalis Deptford Pinks; Fraxinella; Gladiolus, Geranium,
Guelder Rose: Helleborine, Horminum Creticum, Hesperis, Honey
Suckles; Jacea; Lychnis, Yellow Lilies, Lilium Convallium, Ladies
Slipper, Persian Lily, Leucoium, Laurus, Millefolium Luteum, Red
Martagon, Homer's Moly; Sea Narcissus; Oleaster, Oxyacanthus, Or-
chis; Phalangium Pink, Pansies, Prunella, Peonies; Ranunculus, Roses,
Rosemary; Syringas, Sedum, Stock Gilliflowers, Starflowers, Sisym-
brium, Sambucus. Stæchas, Satyrion; Tulips, Trachelium, Tha-
lictrum, Tamariscus; Valerian, Veronica, Musk Violets, &c.

VOL. III.   B b   A

A Shower at this time of the Year is welcome, and if the Weather be calm it makes the Farmer smile on his hopeful Crops.

This Month is the prime Season for the Washing and Shearing of Sheep; in forward Meadows mow Grass for Hay.

Cast Mud out of Ditches, Pools, Rivers: This is the best time to raise Swine for Breeders.

Twy-fallow your Land in hot dry Weather, it kills the Weeds, and sweetens the Land, one plowing then being worth three or four in rainy Weather.

Carry Marl, Lime and Manure of what kind soever to your Land, bring home your Coals, and other necessary Fuel fetched far off, before the Teams are busie at Harvest.

You may continue to weed Corn, the beginning of this Month, but not longer; sow Rape and Coal-seed, and also Turnep-seed. Now Mill-dews or Honey-dews begin to fall.

Mind your Sheep, as was advised before in May*, and make the first return of fat Cattle.*

Work to be done in the *Orchard* and *Kitchen-Garden.*

Now begin to inoculate, and beware of cutting of Trees other than the young Shoots of this Year, pluck off Buds where you are not willing they should Branch forth.

Water your latter planted Trees, and lay moist Weeds, &c. at the Roots of them, having first cleared them of Weeds, and a little stirred the Earth, and hoe up all such Weeds as grow in your Nursery.

This is a seasonable time to distil Aromatick and Medicinal Herbs, Flowers, &c. and to dry them in the shade that you design to keep dry for the Winter, gathering of them in the Full of the Moon, also to make Syrups, &c.

Gather Snails, Worms, &c. and destroy Ants; kill Insects and other Vermin; set Saffron, plant Rosemary and Gilliflowers, sow Lettuce, Chervil, Radish, and other Sallets for latter Salleting.

Gather

Gather Seeds that are ripe, and preserve them cool and dry, water dry Beds, and take up the Bulbous Roots of Tulips, Anemonies, &c.

Inoculate Jessamines, Roses, &c. *also transplant any sort of bulbous Roots that keep not well out of the Ground. Now plant Slips of Myrtle, and sow latter Peas.*

You may now also (or in May before) *cleanse Vines of exuberant Branches and Tendrils, cropping* (not cutting) *and stopping the second Joint immediately above the Fruit, especially in young Vineyards, when they first begin to bear, and thence forward binding up the rest to Props.*

Dig your Ground where you intend to Hop Garden, and bind such Hops to the Poles as the Wind hath shaken off.

Bees now swarm plentifully, therefore be very vigilant over them; they will requite your care.

Fruits in prime, or yet lasting.

Jennetting (first ripe,) *Pippins*, John *Apples, Robillard,* Apples. Red *Fennouil,* &c.

The Maudlin (first ripe,) *Madera, Green-Royal,* St. Pears. Laurence *Pear,* &c.

Duke, Flanders, *Heart Black, Red, White.*   Cherries.

Luke, *Ward, Early* Flanders, *the Common Cherry,* Spanish *Black,* Naples *Cherries,* &c.

Rasberries, Corinths, *Strawberries, Melons,* &c.

J U N E.

TRanfplant autumnal Cyclamens if you Defign their Removal, and gather the Ripe Seeds of fuch Flowers as you intend to fave, and preferve them dry : Shade your Carnations from the Sun ; take up the beft Anemonies and Ranunculus's after Rain, when the Stalk is withered, and dry the Root well. You may now begin to lay your Gifliflow-ers, and to inoculate Jafmine, Rofes, &c. Take up your Tulip Roots ; fet flips of Myrtle in a moift place, alfo flips of Cytifus Lunatus that are of that Years Shoot : You may alfo take up fuch Plants and Roots as endure not well, out of the Ground, and replant them again immediately ; fuch as the early Cyclamen, Jacinth, Oriental, and other Bul-bous Jacinths, Iris, Fritillaria, Crown Imperial, Mar-tagon, Mufcaris, Dens caninus, &c. Water fuch Things as require it. Trim your Knots, and get your Garden into Order.

Flowers in prime, and yet lafting.

Amaranthus, Antirrhinum, After Atticus, Ame-rius, Allobrogium, Afphodils, Amaranthus ; Blat-taria ; Campanula, Cyclamen, Cyanus, Campions, Corn Flag, Citratum, Clamantis, Creticum, Car-nations ; Digitalis ; Fraxinella, Ficus Indicus ; Gla-diolus, Gentiana, Genifta of *Spain* ; Horminum Cre-ticum, Hieracium, Hefperis, Hellebore, Honey-Suckle, Hollyhoc, Hedifarum ; Jafmine, Bulbous Iris, Len-tifcum,

tifcum, Lychnis, Larks Heels, Lime Tree, Lilies ; Martagon, Millefoliums, Mufcaria ; Nafturtium Indicum, Nigella, Night Shade ; Oranges ; Panfy, Phalangium, Pilofella, Palma Chrifti, Pomegranate, Poppies ; Rofes, Rofemary ; Stock-Gilliflowers, Serpillum citratum ; Trachelium, Thlafpi Creticum ; Veronica, Viola pentaphyl.

B b 3   *T H E*

THE Earth now would be glad of refreshing showers to moisten the scorched Vegetables. Tempests now much injure the laden Fruit-trees and standing Corn, to the great detriment of the Husbandman.

Now is the universal time for Hay-making: Lose not a good opportunity, especially if Fair-weather be scarce.

Mow your Head-lands and try-Fallow where the Land requires it. Gather the Fimble or earliest Hemp and Flax.

At the latter end of this, Month Corn-Harvest begins in most places in a forward Year. Still carry forth Marl, Lime, and other Manure: Bring home Timber, Fewel, and other heavy Materials.

Wheat and Hops are now subject to much damage by Mildews.

Sow Turnep-seed in this Month, and sell such Lambs as you have fed for the Butcher.

Work to be done in the *Orchard* and *Kitchen-Garden*.

This is the chief time to inoculate choice Fruit, Roses, &c. and for the Summer pruning of Wall-trees for the making of Cherry-wine, Rasberry-wine, &c.

Re-prune Apricocks and Peaches, saving as many of the young likeliest Shoots as are well plac'd, for the new Bearers commonly perish; the new ones succeeding cut close, and even purging your Wall fruit of superfluous Leaves, which keep the Fruit from the Sun; but do it discreetly.

Graft by Approach, and inoculate Jessamines, Oranges, &c.

Cut off the Stocks of such Flowers as have done blossoming, and cover their Roots with new fat Earth.

Sow Sallet Herbs for latter Salleting, and also Peas.

Take away Snails from Mural-trees, Slip stocks, and other lignous Plants and Flowers. Lay Gilliflowers and Carnations for increase, watering of them, and shadowing of them from the fervent heat of the Sun-beams. Lay also Myrtles and other curious Greens. Clip Box and other tonsil Plants.

Let your Olitory Herbs run to Seeds that you design to save. Transplant or remove Tulips or other bulbous Roots:
some

fome may be kept out of the Ground, and others immediately planted.

Towards the latter end of this Month, vifit your Vines a-gain, &c. and ftop the exuberant Shoots at the fecond Joint above the Fruit (if not done before) but not fo as to expofe it too much to the Sun without fome Umbrage.

Keep down Weeds that they grow not to Seed, and begin your Work of Hoeing fo foon as they begin to peep; by this means you will difpatch more in a few hours than afterwards in a whole day, in that the ftirring of the Earth will but help the increafe of the Seed.

If the Seafon be dry, the watering of Hops will very much advantage them, and make them the more fruitful; if it prove moift, renew and cover the Hills ftill with frefh Mold.

Now Bees caft their latter Swarms, which are of little ad-vantage; therefore 'tis beft to prevent them: Streighten the Entrance of the Hives, kill the Drones, Wafps, Flies, &c,

Fruits in prime, or yet lafting.

Deux-ans, Pippins, Winter Ruffetting, Andrew Ap- Apples. *ples, Cinamon Apples, red and white Jenneting, the Mar-garet Apples, &c.*

The primate Ruffet Pear, Summer Pears, Green Chiffel Pears. *Pear, Pearl Pear, &c.*

Carnation, Morella, Great Bearer, Morocco Cherry, Cherries. *the Egriot, Bigarreaux, &c.*

Nutmeg, Ifabella, Perfian, Newington, Violet, Muf- Peaches. *cat, Rambouillet.*

Primordial, Myrobalan, the red, blue, and amber Violet, Plumbs, *Damask, Denny Damask, Pear-plumb, Violet or Chefon-* &c. *plumb, Apricock-plumb, Cinamon-plumb, the King's-plumb, Spanifh Morocco plumb, Lady Elizabeth's plumb, Tawny, Damafcen, &c.*

Rasberries, Goosberries, Corinths, Strawberries, Melons, &c.

JULY.

FROM this time to *Michaelmas* you may lay Gil-liflowers and Carnations, and flip Stocks and o-ther Lignous Plants and Flowers. The Layers will (in a Month or Six Weeks time) Strike Root. Plant Six or Eight in a Pot to save room in Winter; if it prove Wet lay your Pots side-long, but shade those that blow from the Afternoon Sun.

You may still lay Myrtles, Laurels, and other Greens.

Water young planted Shrubs, and Layers, &c. as Orange Trees, Myrtles, Granades, especially Amo-mum, which Shrub you can hardly water too often; It requires also abundant compost; as does also the Myrtle and Granade Trees. Clip Box, &c. after Rain, and graft by approach *Inarch;* inoculate Jaf-mine, Oranges, &c.

Take up your early autumnal Cyclamen, Tulips, and bulbous Roots, gather Tulip Seeds, but let them lie in the Pods, and also early Cyclamen Seed, and sow it presently in pots.

Remove seedling Crocus, and place them at wider Intervals: Likewise you may take up some Anemo-nies, Ranunculus's, Crocus, Crown Imperial, Persian Iris, Fritillaria and Colchicum, but plant the three last as soon as taken up, or you may stay till *August* or *September* before you do it. Remove Dens Caninus, &c. and take up your Gladiolus the Blades being dry, and about the latter end of this Month sift your Beds for off-sets of Tulips and other Bulbous Roots; also for Anemonies, Ranunculus's &c. which will pre-pare it for replanting. You may sow some Anemo-nies, keeping of them temperately moist.

Continue to cut off the wither'd Stalks of your Lower Flowers, and cover with Earth the bared Roots.

You

You may now water your Gravel Walks with Brine, Pot Afhes, or a Decoction of Tobaco Stalks to deftroy the Worms and Weeds.

Flowers in prime, or yet lafting.

Amaranthus, Afphodil, Antirrhinum, Flos Africanus, Alkekengi, Agnus caftus, Arbutus, Amomum Plinij; Balfome Apple; Campanula, Clematis, Cyanus, Convolvulus, Corn Flower, Caryophyllata, Flos Cardinalis; Digitalis; Eryngium Planum; Fraxinella; Geranium Trifte, Gladiolus, Gentiana; Hefperis, Hedyfarum, Hollyhoc; Jacea, Jafmine, Jacinth; Lupines, Limonium, Linaria Cretica, Liguftrum; Millefolium, Musk Rofe, Myrrh, Peru, Monthly Rofe; Nafturtium Ind. Nigella; Oleander; Orange, Olive; Ind. Phafeolus, Prunella, Phalangium, Periploca, Flos Paffionis; Scorpion Grafs, Stock-Gilliflower, Scabious, Spartum Hifpanicum; Tilia Ind. Tuberous; Veronica Volubilis, &c.

This

THis Month returns the Country man's Expences into
his Pocket, and encourages him to another Years Adventure. If it proves dry, warm, and free from high Winds,
it saves a great deal of the Husbandman's Expence.

You may yet trifallow, also lay on your Compost or Soil,
as well on your Barley Land as Wheat Land.

Carry Wood, or other Fuel home, before Winter.

Provide good Seed, and picked well, against Seed time.

Put your Ewes and Cows you like not to fatting.

This is the principal Month for Harvest for most sorts of
Grain, therefore make use of good Weather whilst you have it.

About the end of this Month you may mow your After
grass, and also Clover, St. Foyn, and other French Grass.
Geld Lambs, and make the second return of your fat Sheep
and Cattle.

Work to be done in the Orchard and Kitchen Garden.

The former part of this Month is the best time to inoculate.
You may now make Cyder of Summer-fruits. Prune away
superfluous Branches from your Wall fruit-Trees; but leave
not your Fruit bare, except the Red Nectarine, which is
much meliorated by lying open to the Sun, nailing up what
you design to spare to cover the Defects of your Walls.

Pull up Suckers from Roots of Trees, unbind the Buds you
inoculated a Month before, if taken.

Plant Saffron; set Slips of Gilliflowers; sow Annise.
Now is beginning a second Season for the increasing and
transplanting of most Flowers, and Garden Plants and
Herbs, Strawberries, &c.

The Seeds of Flowers and Herbs are now to be gather'd:
Also gather Onions, Garlick, &c.

Sow Cabbage, Colliflowers, Turneps, and other Plants,
Roots and Herbs, for the Winter, and against Spring; and
also Endive, Angelica, Scurvy-grass, &c.

Now sow Larks-heels, Candy-tufts, Columbines, &c. and
such Plants as will endure the Winter.

You may yet slip Gilliflowers, and transplant bulbous Roots:
About Bartholomew-tide, some esteem the only secure season

son for removing of Perennial, or Winter-green, as Philly-reas, Myrtles, &c. It is also the best time to plant Straw-berries ; and 'tis not amiss to dress Roses that have done bearing, and plant them about this time.

Prop up those Poles the Wind blows down : Also neat the end of the Month gather Hops.

Towards the end of the Month take Bees, unless the good-ness of the Weather provoke you to stay till the middle of the next Month. Destroy Wasps and other Insects ; and straiten the Passages to secure them from Robbers.

Fruits in prime, and yet lasting.

The Ladies-longing, Kirkham Apple, John Apple, Seam-ing Apple, Cushion Apple, Spicing May-flower, Sheeps-snout. *Apples.*

Windsor, Sovereign, Orange, Bergamot, Slipper Pear, Red Catharine, Ring Catharine, Penny Pear, Prussia Pear, Summer Peppering, Sugar Pear, Lording Pear, &c. *Pears.*

Roman Peach, Man Peach, Quince Peach, Rambouillet, Musk Peach, Grand Carnation, Portugal Peach, Crown Peach, Bourdeaux Peach, Lavar Peach, Peach Despot Savoy, Malacota. *Peaches.*

The Muroy Nectarine, Tawny, Red Roman, Little green Nectarine, Cluster Nectarine, Yellow Nectarine. *Necta-rines, &c.*

Imperial Blue, Dates, Yellow Late Pear-Plumb, Black Pear-plumb, White Nutmeg, Late Pear-plumb, Great An-thony Turkey-plumb, the Jane-plumb. *Plumbs.*

Cluster Grape, Muscadine, Corinths, Cornelians, Mul-berries, Figs, Filberts, Melons, &c. *Other Fruits.*

AUGUST.

AT the beginning of this Month is the time to inoculate Orange Trees, to take up Bulbous Iris's, or you may sow their Seed, and also of the Larks Heels, Candy Tuft, Columbines, Fox-Gloves, Hollyhoc, Narcissus, Oriental Jacinth, and such Plants as endure the Winter, and replant such as will not do well out of the Earth, as Fritillaria, Lilies, Hyacinths, &c. Plant some Anemony Roots to have Flowers all Winter, and take up your Seedlings, of last Year, which now transplant for bearing; also plant Dens Caninus, Autumnal Crocus, and colchicum. Gilly-Flowers may be yet slipt. Continue to take up Bulbs.

Gather Alaternus Seed as it grows black and ripe, and the Seeds of most other Shrubs. Water well Balsame Fœm.

About the middle of this Month transplant Auricula's. Pick out your Seedlings: You may likewise sow the Seeds.

Towards the latter end sow Anemony Seeds, Ranunculus's, &c. also Cyclamen, Jacinths, Iris, Hepatica, Primroses, Fritillaria, Martagon, Fraxinella, &c.

About Bartholomew-tide is the best time to remove and lay your Perennial Greens, as Oranges, Limons, Myrtles, Phillyrea's, Oleanders, Jasmines, Arbutus, Pomegranates, Monthly Roses, and whatsoever is most obnoxious to Frost. Water them during Summer, &c.

Flowers in Prime and yet lasting.

Amaranthus, Anagallis Lusitanica, After Atticus, Amethystinum, Africanus Flos, Asphodils, Agnus Castus; Blattaria, Spanish Bells, Belvedere, Balsamine Fœm.; Carnations, Campanula, Clematis, Cyclamen Vernum,

Vernum, Convolvulus's, Colchicum, Catch Fly; Da-
tura Turcica, Daſies, Delphinium; Eliochryſon, Eryn-
gium Planum; French Mary-golds; Geranium Creti-
cum, Geranium noƈe olens; Hieracion minus Alpe-
ſtre, Tuberoſe Hyacinth, Hearts Eaſe, Heliotrope,
Holly-hoc, Helleborus; Spaniſh and Indian Jaſmine;
Limonium, Linaria, Lychnis, Lupines, Leucoion;
Malva Arboreſcens, Martagon, Myrtles, Maracoc;
Nigella, Naſturtium Indicum; Oranges; Mirabile.
Peruvian Panſies, Pomegranates; Musk and Monthly
Roſes, Roſemary; Thlaſpi Creticum, &c.

Gentle

GEntle Showers now do well, and make the Earth mellow, preparing of it for Wheat, which delights in a moiſt Receptacle; but ſtill Weather and dry is moſt ſeaſonable for the Fruits yet on the Trees. The Salmon and Trout in moſt Rivers go now out of Seaſon 'till Chriſtmas.

This Month is the moſt univerſal time for the Farmer to take Poſſeſſion of his new Farm. Get good Seed, and ſow Wheat in the Dirt, and Rye in the Duſt.

Mend the Fences about the new-ſown Corn; ſcare away Crows, Pidgeons, &c.

Geld Rams, Bulls, &c. Sew Ponds: Put Boars up in the Sty.

Beat out Hemp-ſeed, and water Hemp; gather Maſt, and put Swine into the Woods.

Carry home Brakes, ſaw Timber and Boards, and manure your Wheat Lands before the Plough. Thatch your Stacks and Ricks, and make an end of Carting.

Work to be done in the Orchard and Kitchen-Garden.

You may now make Cyder and Perry of ſuch Fruits as are not laſting, and gather your forwardeſt Fruit, but not your laſting Winter Fruit 'till after Michaelmas. Alſo gather your Windfalls every day that is dry.

Releaſe inoculated Buds, if not done before, eſpecially if they pinch.

Sow Cabbages, Colliflowers, Turneps, Onions, &c. Tranſplant Artichoaks and Aſparagus Roots, and Strawberries out of the Woods. Plant forth your Cabbages and Colliflowers that were ſown in Auguſt, and make thin the Turneps where they grow too thick.

Now plant your Tulips, and other bulbous Roots you formerly took up; or you may now remove them: You may alſo tranſplant all fibrous Roots.

Now withdraw your choice Plants into the Conſervatory, and ſhelter ſuch Plants as are tender, and ſtand abroad.

Towards

Towards the end of this Month you may gather Saffron, and earth up your Winter Plants and Sallet Herbs; and prepare Compost to trench your Earth and Borders with.

Now finish the gathering and drying of your Hops, cleanse the Poles of the Hawn, and lay up the Poles for next Spring.

Take your Bees, in time; straighten the Entrance into the Hives, and destroy Wasps, &c. Also you may now remove Bees.

Fruits in prime, and yet lasting.

The Belle-boone, the William *Pearmain, Lording-Apple, Pear-Apple, Quince-Apple, Red-greening, Ribbed bloody Pippin, Harvey Violet Apple,* &c. **Apples.**

Hampden's Burgamot (first ripe,) Summer Bon Chrestien Norwich, Black Worcester (baking,) Green-field, Orange Burgamot, the Queen-hedge Pear, Lewis Pear, (excellent to dry,) Frith Pear, Arundel Pear (also to bake,) Brunswick Pear, Winter Poppering, Bing's Pear (baking,) Diego, Emperor's Pear, Bluster Pear, Messire Jean, Rowling Pear, Balsam Pear, Bezy D' Hery, &c. **Pears.**

Malacoton, and some others, if the Year prove backward, Almonds, Quinces, &c. **Peaches.**

Little blue Grape, Muscadine Grape, Frontiniac, Parsly, great blue Grape, the Verjuice Grape (excellent for Sauce, &c.) Barberries, &c.

SEPTEM-

SEPTEMBER.

YOU may now plant some sorts of Anemonies for early Flowers; but if you stay 'till next Month they will be more certain of growing. This is the best time of sowing Auricula Seeds, setting of the Cases in the Sun 'till *April*.

Plant Daffodils and Colchicum, and transplant Hepatica, Camomile, Capillaries, Matricaria, Violets, Primroses, Iris, Chalcedon, Cyclamen, &c.

You may continue to sow Alaternus, Phillyrea, Iris, Primroses, Crown Imperial, Martagon, Tulips, Delphinium, Nigella, Candy Tufts, Poppy, and such Annuals as are not prejudiced by Frosts.

Remove Seedling Digitalis, and plant the Slips of Lychnis; remove your Tuberoses into the Conservatory, and keep them dry. 'Tis best to take them out of the Pots, and to preserve them dry in Sand, or wrap them up in Papers, and keep them in a Box near the Chimney.

Bind up your Autumnal Flowers and Plants to Stakes for fear of suddain Gusts. Take off your Gilliflower Layers, Earth and all, and plant them in Pots or Borders shaded.

Crocus may now be raised of Seed.

In fair Weather when your Greens and Plants are dry, as Oranges, Limons, *Indian* and *Spanish* Jasmine, Oleanders, Barba Jovis, Amomum Plinii, Cityfus Lunatus, Chamelæa Tricoecos, Cistus Leden. Clusii, Dates, Aloes, &c. remove them into the Conservatory, and take away some of the Top exhausted Earth, and stirring up the rest, fill the Cases with what is rich, and well prepared, to nourish the Roots in Winter; but as yet leave the Doors and Windows of the Conservatory open to give them Air, provided the Winds be not too sharp and high, nor the Weather
sher

ther foggy. Myrtles will endure abroad near a Month longer.

When the Cold advances, fet fuch Plants as will not endure the Houfe into the Earth two or three Inches lower than the Surface, in a warm Place, and cover them with Glaffes, having firft cloathed them with dry Mofs; but upon all warm and benign Emiffions of the Sun, and fweet Showers, give them. Air by taking off their Covers. Thus you may preferve Marum Syriacum, Ciftus's, Geranium noĉte olens, Flos Cardinalis, Marococs, choiceft Ranunculus's, Anemonies, Acacia Ægypt. &c.

Flowers in prime, or yet lafting.

Amaranthus, Anagallis, Antirrhinum, Africanus Flos, Amomum Plinii, After Atticus, Afphodils; Belvedere, Bellis, Balauftia, Campanula's, Colchicum, Autumnal Cyclamen, Clematis, Chryfanthemum Auguftifol. Convolvulus diverfi Generis, Candy Tufts, Crocus, Capficum Ind. Dature; Eupatorium of Canada, Ethiopic Apples; Geranium Creticum, Gentianella annual, Gilliflowers; Hieracion Indicum; Jacinth, Jafmine; Linaria Cretica, Lychnis, Conft. Limonium, Ind. Lily; Marvel of Peru, Millefolium, Moly, Malva Arborefcens; Nafturtium, Narciffus of feveral forts; Oranges; Phalangium, Ind. Phafeolus, Poppy, Ind. Pink, Paffion Flower; Portugal Ranunculus, Rhododendron; Veronica, &c.

C c   OCTOBER

OCTOBER *often gives an earnest of what we are to expect from the succeeding Winter.*

If it prove Windy, as it usually doth, it finishes the fall of the Leaf, and also shakes down the Mast and other Fruits, leaving neither Leaf nor Fruit.

Lay or Plow up your Barley Land as dry as you can. Seed time yet continues, especially for Wheat.

Well Water-furrow, and Drain new sown Corn Land. Now it is a good time to sow Acorns, Nuts, or other sorts of Mast or Berries for Timber, Coppice-wood or Hedges. You may still gather Saffron.

Sow Peas in a warm fat Land, you may plant Quick-sets, and all sorts of Trees for Ornament or for Use, and also plash quick Hedges.

Wean the Foals that were foaled of your Draught Mares; at Spring put off such Sheep as you have not Wintering for. Follow Malting, this being a good time for that Work.

Spare your private Pastures, and eat up your Corn Fields and Commons, give over folding of Sheep, and separate the Lambs from the Ewes that you design to keep for your own use.

Work to be done in the Orchard and Kitchen-Garden.

Make Cyder and Perry of Winter Fruit throughout this Month; now is a very good time to plant all sorts of Fruit-trees, or any other Trees that shed their Leaf.

Trench stiff Grounds for Orcharding and Gardening to lie for a Winter fallowing. Now is the time for Ablaqueation, or laying open of the Roots of old or unthriving Trees, or such as spend themselves too much or too soon in Blossom.

Gather the residue of your Winter-Fruit, also gather Saffron.

Sow all sorts of Fruit-Stones, Nuts, Kernels and Seeds, either for Trees or Stocks; some also sow Peas in a rich warm Soil to be early in the Spring; and you may yet sow Genoa Lettuce, which will last all the Winter, and Parsnips. Choose no Trees for a Wall that are not above two Years grafting.

Many of the September Works may be yet done, if the Winter be not too forward.

Now plant your bulbous Roots of all sorts, and continue planting and removing several Herbs and Flowers, with fibrous Roots, if a former and better season be omitted.

This

This Month is the best time to plant Hops, and you may Hops.
bag or pack those you dried the last Month.
Now you may safely remove Bees.   Bees.

Fruits in prime, or yet lasting.

Belle and Bonne William *Costard*, Lording, *Parsley Ap-* Apples.
ples, Pearmain, *Peat-apple*, Honey-meal, &c.
  The Law pear (baking,) Green Butter-pear, Thorn-pear, Pears.
Clove-pear, Rouset-pear, *Lombert pear, Russet-pear, Saf-
fron-pear, Violet-pear*, Petworth-pear, *or Winter*·Windsor.
  Bullace, and divers of the September *Plumbs and Grapes,*
Pines, Arbutus, &c.

Continue sowing what you did in *September*; likewise
Cyprus may be sown, but not during Frosts. Plant
Ranunculus's, vernal Crocus's, &c. Move Seedling
Holly-hocs, or others.
  Plant now your choice Tulips, *&c.* This is the se-
curest time; and take care your Carnations do not get
too much Wet. All sorts of Bulbous Roots may now
be bury'd, likewise Iris's, *&c.*
  You may yet sow Alaternus and Phillyrea Seeds.
Beat, Roll and Mow your Walks and Camomile. Fi-
nish your last Weeding, and cleanse your Garden of
Trumpery.

Flowers in Prime or yet lasting.

Amaranthus Tricolor, *&c.* After Atticus, Amo-
mum, Antirrhinum, Arbutus; Balaust; Colchicum,
Cyclamen, Clematis; Gilly-flowers, Geranium triste;
Heliotropes; Tuberose Jacinth, Jasmine; Limonium,
Lychnis; Marvel of Peru, Millefolium luteum, Myr-
tles; Nasturtium Persicum; Oranges; Phalangium,
Pilosella; Stock Gilly-Flowers; Violets, Veronica,
&c.

THIS Month generally proves dry, and the Earth and. Trees are wholly uncloathed. Sowing of Wheat on a concluſion is yet allowable on very warm rich Land, eſpecially ſuch as are Burn-baited.

The Countryman now generally forſakes the Fields, and ſpends his time in the Barn and Market.

Fat Swine are now fit for ſlaughter, leſſen your ſtocks of Swine and Poultry.

Thraſh not Wheat to keep till March, leſt it prove foiſty.

Lay Straw or other waſt ſtuff in moiſt places to rot for Dung ; alſo lay Dung on heaps.

Fell Coppices, Wood and Trees for Mechanick Uſes, as Plough-boot, Cart-boot, &c. and plant all ſorts of Timber or other Trees; break Hemp and Flax.

Now you may begin to overflow your Meadows that are fed low, and to deſtroy Ant-Hills.

Work to be done in the Orchard and Kitchen-Garden.

Peas and Beans may now be ſet to be early in the Spring. Trench or Dig Garden Ground.

Remove and plant Fruit trees, and furniſh your Nurſeries againſt Spring.

Lay up Carrots, Parſnips, Cabbages, Colliflowers, &c. either for your Uſe, or to tranſplant for Seed at Spring ; cover Aſparagus Beds, Artichoaks, Strawberries, and other tender Plants with long Dung, Horſe litter, Straw or ſuch like, to preſerve them from Froſt. Dig up Liquoriſh.

Now is the beſt Seaſon to plant the faireſt Tulips, if the Weather prove not very bad.

Cover with Mattreſſes, Boxes, Straw, &c. your tender Seedlings. Plant Roſes, Lilac, and ſeveral other Plants and Flowers, the Weather being open.

Take up your Potatoes for Winter ſpending, trench and fit your Ground for Artichoaks, &c. as yet you may ſow Nuts, Stones, &c.

Hops.   Now carry Dung into the Hop-Garden, and mix it with Earth that it may rot againſt Spring.

Apiary.   You may this Month cloſe ſtop up your Bees, ſo that you leave breathing Vents, or you may Houſe them till warm Weather.

Froits

Fruits in prime, or yet lasting.

The *Belle Bonne*, the William, *Summer Pearmain*, Apples.
*Lording Apple, Pear Apple, Cardinal, Winter Chesnut;
Short Start*, &c. and *some of the former Months*.

Messire Jean, *Lord Pear, Long Burgamot, Warden* (to Pears.
Bake,) *Burnt Cat, Sugar Pear, Lady Pear, Ice Pear, Done
Pear, Deadman's Pear, Winter Burgamot, Belle Pear*, &c.
Arbutus, Bullaces, Medlars, Servises.

Sow Auricula Seeds. Cover your peeping Ranunculus's, *&c.* and your over Green Seedlings, especially
if the Snows be long and the Winds Sharp.

Now is the best time to plant your fairest Tulips
in places of Shelter. Transplant ordinary Jasmine, *&c.*

About the middle of this Month uncover your tender Plants in your Conservatory, secluding all entrance
of Cold. If the Weather prove very Cold, so as to
freeze Water in your Conservatory, kindle some Charcoal: At other times, while the Sun shines on the
House and no longer, shew them the light. You must
never water Aloes or Sedums during the whole Winter. If they grow too dry expose them a while to the
Air when 'tis clear, and you can hardly be too sparing
of Water to most of your housed Plants, which should
be done only when the Leaves shrivel and fold up. If
pale and whitish the fault is in the Roots.

House your best Carnations, or rather set them in a
Pent-house against a South Wall, to keep them only
from the extremity of the Weather.

Prepare also Matrasses, Boxes, Pots, *&c.* for shelter for your new Sown under Plants and Seedlings.
Plant Roses, Althæa frutex, Lilac, Syringas, Cityfus,
Peonies, *&c.* and all fibrous Roots, set stony Seeds,
&c. and cleanse your Walks and other Places of Autumnal Leaves, *&c.*

Flowers in Prime or yet lasting.

Anemonies, Antirrhinum; Bellis; Clematis, Carnation; Jasmine; Myrtles, Musk Rose; Pansies; Meadow Saffron; Stock Gilliflowers; Violets, Veronica, *&c.*

THE Earth is now commonly locked up under its frozen Coat, that the Husbandman hath leisure to sit and spend what Store he hath before-hand provided.

Now is the time to house old Cattle, and to cut all sorts of Timber, and other Trees for Building, or other Utensils ; to fell Coppices, &c.

Let Horses blood, fatten Swine and kill them. Destroy Ant-hills.

Plow up your Land that you design for Beans ; drain Corn-fields where Water offends, and water or overflow Meadows.

Put your Sheep and Swine to the Peas-rick, and fat them for a Market. Cut Hedges and Trees.

Work to be done in the Orchard and Kitchen-Garden.

You may now set or transplant such Fruit or other Trees as are not very tender, nor subject to the Injuries of Frost in open Weather.

Also you may plant Vines, or other Slips or Cions, and Stocks for Grafting ; and also prune Vines, if the Weather be open.

Cover the Beds of Asparagus, Articboaks, and Strawberries with Horse-litter, &c. if not covered before.

Sow Beans and Peas, if the Weather be moderate. Trench your Ground, and dress it against Spring.

Set Traps for Vermin, and pick up Snails out of the holes of Walls, &c.

Sow or set Bay-berries, Laurel-berries, &c. dropping ripe. This Month you may dig up Liquorice.

Hop-garden.   Dig a weedy Hop-garden, and carry Dung into it, which mix with Earth.

Apiary.   Feed weak Stocks.

Fruits in prime, or yet lasting.

Apples.   Russeting Pippin, Leather-coat, Winter red Chesnut Apple, Great-belly, the Go-no-farther, or Cats-head, with some of the precedent Month.

Pears.   The Squib Pear, Spindle Pear, Doyoniere, Virgin, Gascoigne-bergamot, Scarlet Pear, Stopple Pear, White, Red, and French Wardens (to bake or roast), &c. Deadman Pear, excellent, &c.

DECEM-

DECEMBER.

COntinue to deſtroy Vermin, and preſerve from too much Rain and Froſt your beſt Anemonies, Ranunculus's, Carnations, &c.

You may ſow for early Beans, Peas, &c.

Be careful to keep the Doors and Windows of your Conſervatory well matted and guarded from the piercing Air; for if your Orange Trees and o-ther tender Plants take cold, it will be difficult to recover them; therefore temper the Cold with a few Charcoal, but never accuſtom your Plants to it but in the extreameſt Colds, for if the place be very cloſe, they will even then hardly require it.

Set Bay-berries, &c. prepare Litter to lay over your choice Plants that are to continue abroad, and to cover your Pipes with in caſe of Fróſts.

Flowers in prime, or yet laſting.

Anemonies, Antirrhinum; Winter Cyclamen; black Hellebore; Iris Cluſii; Lauruſtinus; Prim-roſes; Snow-drops, Stock-Gilliflowers; Yacca, &c.

A
TABLE·

A TABLE

A TABLE

A TABLE

Ising-

A TABLE

F I N I S.

Milton Keynes UK
Ingram Content Group UK Ltd.
UKHW020821050923
428087UK00008B/768